J. B. (John Butler) Johnson

Engineering Contracts and Specifications

J. B. (John Butler) Johnson

Engineering Contracts and Specifications

ISBN/EAN: 9783743686977

Printed in Europe, USA, Canada, Australia, Japan

Cover: Foto ©berggeist007 / pixelio.de

More available books at **www.hansebooks.com**

Engineering Contracts and Specifications

INCLUDING

A Brief Synopsis of the Law of Contrac

AND

Illustrative Examples

OF THE

General and Technical Clauses

OF VARIOUS KINDS OF

Engineering Specifications

DESIGNED FOR THE USE OF STUDENTS, ENGINEERS, AND CONTRACTORS

BY

J. B. JOHNSON, C. E.

PROFESSOR OF CIVIL ENGINEERING, WASHINGTON UNIVERSITY, ST. LOUIS, M

MEMBER OF THE INSTITUTION OF CIVIL ENGINEERS.

MEMBER OF THE AMERICAN SOCIETY OF CIVIL ENGINEERS.

MEMBER OF THE AMERICAN SOCIETY OF MECHANICAL ENGINEERS, ETC.

FIRST EDITION.

ENGINEERING NEWS PUBLISHING CO., NEW YORK.

1895.

INDEX.

1

TABLE OF CONTENTS.

I

PART II.

ENGINEERING SPECIFICATIONS AND ACCOMPANYING DOCUMENTS.

PART IV.

ILLUSTRATIVE EXAMPLES OF COMPLETE CONTRACTS AND SPECIFICATIONS.

PREFACE.

The leading American Engineering Schools have long needed a text-book on the subject of the Law of Contracts and Engineering Specifications. In the absence of any such text, this department of engineering practice has received scant and meagre treatment at the hands of these schools. This work has been written primarily to serve this purpose. After it was completed, however, it seemed to the author it might prove of value to the profession at large and also to contractors, especially those portions of it treating of the Law of Contracts and of the General Clauses in Specifications.

While the author makes no pretension to a knowledge of the law, yet he has had to impart instruction on this subject for many years, and has attended lectures in this field in the St. Louis Law School (a department of Washington University). He has tried to follow strictly the recognized authorities in all he has said in this work, and while he thinks his synopsis may serve as a good general guide to the fundamental principles of the subject, he recommends that the reader refer all important particular cases to his attorney, or else consult the standard works themselves. If a single volume is desired containing a general review of the Law of Contracts, the layman can not do better than obtain that of John D. Lawson, of the Law Department of the Missouri State University. Another similar, and perhaps better work for the young lawyer, is that of J. P. Bishop; while Parson's three-volume work is the recognized standard authority for the lawyer.

Since this work is designed only for laymen, however, the author may well quote the maxim that "the man who is his own lawyer has a fool for a client." The brief synopsis of the law herein given, therefore, is not intended to remove the necessity of consulting a lawyer on àll important matters, but only to enable one to steer clear of some of the legal pitfalls which lie in the way of every business man and especially of engineers.

Since custom has laid on engineers and architects the duty of writing specifications and contracts, it is well for them to know something of the legal ground they are forced to traverse. The first part of this work is intended, therefore, to serve as a cautionary warning against legal entanglements, rather than as a counselor or guide through such difficulties. The synopsis of the Law of Contracts as here given has been revised by a very competent legal authority, and the author is indebted. to him for many valuable suggestions and corrections. It probably will not mislead one into trouble, though it may not always point the way out.

The author also wishes to acknowledge his indebtedness to the many prominent engineers who have kindly sent him copies of their latest specifications for use in this work, and he has acknowledged this debt in the body of the book by appending to each quotation the initials of the person quoted. A key to these initials is given on page 6.

The illustrative examples of engineering specifications given in Part III are selected so as to cover a wide field with as little repetition as possible. They are not given to be blindly copied, but rather as illustrating a good method of treating the subject, and to serve as patterns as to manner as well as to matter. As the best engineers seldom copy their own specifications or use them unchanged a second time, much less can one safely copy unchanged the specifications of another. In fact the writing of engineering specifications is wisely left for engineers of large experience, but as the younger men have to enforce them and serve as inspectors under them, they should in all cases understand fully why they have been drawn in a particular way.

That this somewhat crude effort may serve to help engineers and architects to a more efficient and satisfactory performance of their professional duties, is the hope and aim of

THE AUTHOR.

KEY TO SUBSCRIPT INITIALS.

The following gentlemen have kindly furnished the author copies of their specifications from which he has freely quoted in parts II and III. In every case he has appended the initials of the writer of the specifications used, the key to which is here given:

Onward Bates, Engineer Bridges and Buildings, C., M. & St. P.
    R'y, Chicago, Ill.... ....   O. B.
A. P. Boller, Consulting Engineer, New York City............   A. P. B.
G. Bouscaren, Consulting Engineer, Cincinnati, Ohio.........   G. B.
Wm. H. Bryan, Consulting Engineer, St. Louis... .... W. H. B.
Col. Wm. P. Craighill, Corps of Engineers U. S. Army, and
    Past President Am. Soc. C. E....... ....   W. P. C.
J. T. Fanning, Consulting Engineer, Minneapolis, Minn.......   J. T. F.
Alphonse Fteley, Chief Engineer Aqueduct Commission, New
    York City, N. Y..... ....   A. F.
E. A. Fuertes, Director School Civil Engineering, Cornell Uni-
    versity, Ithaca, N. Y.......... .. ..   E. A. F.
John W. Hill, Consulting Engineer, Cincinnati, Ohio........   J. W. H.
M. L. Holman, Water Commissioner, St. Louis, Mo........... M. L. H.
Johnson & Flad, Engineers, St. Louis, Mo ....   J. & F.
Emil Kuichling, Chief Engineer Water Works, Rochester, N. Y.   E. K.
Milwaukee City Specifications.... ....   M.
George S. Morison, Consulting Engineer, New York City, N. Y.,
    President (1895) Am. Soc. C. E....... ....   G. S. M.
W. D. Pence, Instructor in Civil Engineering, Champaign, Ill.. W. D. P.
Pennsylvania Railroad Co., Wm. H. Brown, Chief Engineer,
    Philadelphia, Pa...... ....   P. R'y.
Col. O. M. Poe, Corps of Engineers, U. S. Army.............   O. M. P.
St. Louis City Specifications ....   St. L.
Union Pacific Railway, George H. Pegram, Chief Engineer,
    Omaha, Neb.... ...:... .. ..,. .... U. P. R'y.
J. A. L. Waddell, Consulting Engineer, Kansas City, Mo......J. A. L. W.

ENGINEERING CONTRACTS & SPECIFICATIONS

PART I.

BRIEF SYNOPSIS OF SUCH PORTIONS OF THE

Law of Contracts

AS BEAR ON THE CARRYING OUT OF

ENGINEERING OR ARCHITECTURAL CONSTRUCTION.

---

1. **Introductory.** The Law of Contracts is said to be as simple and as readily. comprehended by the layman as any department of the law. Two standard single volume works on the law of contracts are those of Bishop and of Lawson,* to which the reader is referred for a more complete treatment of the subject, and from which the following synopsis has been principally derived. In this synopsis only such rules and principles are incorporated as may be profitably presented to undergraduate students in our leading engineering schools. The practicing engineer or architect may also find them valuable, however, as furnishing to him certain guiding principles, the recognition of which will frequently enable him to avoid legal complications and inherent weaknesses in the drawing of specifications and other documents pertaining to contracts. This work is intended to emphasize the necessity of consulting competent legal authority in all important matters rather than to enable one to dispense with such reliance.

2. **Essential Elements of a Legal Contract.** A contract is a promise to do or to refrain from doing some act

---

*The layman will probably find the work of Judge Lawson better suited to his wants.

which the law will enforce. The law will not enforce an agreement unless the following essentials are fulfilled.

First. The parties must be *competent* to make the agreement.

Second. The subject-matter must be *lawful.*

Third. The parties must have mutually *assented* or *agreed* to the conditions named, or they must have been of the same mind and intention concerning the subject-matter.

Fourth. Except in the case of sealed contracts there must be a *valuable consideration.*

The four essentials of a legal contract, therefore, may be grouped under the four words, Competency, Legality, Agreement, and Consideration.

3. Two General Classes of Contracts. There are in general two kinds of contracts, namely: contracts made under seal, called *sealed* contracts or *specialties* (see Art. 28), and simple written or oral agreements unaccompanied with the formality of a seal, called *parole* contracts.

A sealed contract is a written agreement signed by the parties, the signatures, having appended to them what is commonly known as a seal. Formerly a seal consisted of "An impression on wax, or paper, or some other tenacious substance capable of being impressed." Now, however, an impression of a seal on the paper itself is commonly construed as a proper seal, and in many states by statute a mere scroll enclosing the word "seal" made opposite the name of the signer is sufficient.

Engineering contracts are often executed under seal, though preferably not, while the bond which holds the sureties for the faithful performance of the work by the contractor must be under seal. This is necessary because the agreement of the bondsmen to become responsible for the faithful performance of the contract by the contractor is not usually supported by a valuable consideration.

The principal difference between a sealed contract and one not under seal is that in the former case a valuable consideration is not required to support the agreement, while in the latter case the contract is invalid unless such a consideration can be shown to exist.*

The affixing of a seal to a signature implies a special care and deliberation on the part of the signer, more than can be assumed in the case of a simple signature. It is for this reason that a consideration is not required to support a sealed contract.

The mere existence on the document of a printed scroll or word "seal" on the lines provided for signatures does not constitute a sealed document unless these words or scrolls were so intended by the signers.

COMPETENCY.

4. Competency of Individuals. A sane person who has attained his majority is competent to make any legal agreement or contract. The disabilities of married women in the matter of contracts are numerous, but will not here be entered upon. Neither will any reference be made to those disabilities pertaining to aliens, convicts, infants, insane persons, and drunkards.

5. Competency in Governmental Relations. The national or any state government may become a party to a contract, and such government may sue on its contracts and enforce them, but the converse of this is not true. *Neither the United States nor any state can be sued without its consent.*†
The only remedy for a person who seeks the enforcement of a contract with such a government is an appeal to congress or to the state legislature. Many of the states of the south have repudiated

---

*See subject of Consideration, Art. 24.

† The state may consent to be a party to a suit in order to have the rights of the parties passed upon by the courts.

their contracts in the matter of state bonds, issued during the periods of reconstruction, and the bondholders have no remedy. Neither are public officers who negotiate contracts on the part of the state personally liable on contracts made in their own names, when these are signed in their official capacities. This freedom from all legal necessity to carry out its contracts is an essential element of sovereignty, and applies to kings and other more or less absolute rulers in their official relations.

All public corporate governments, subordinate to that of the state, as of the county, or township, or village, or city, can be sued upon their contracts, and such contracts enforced whenever these lie within their legal corporate powers. Thus a county, or town, or city can not repudiate its legal obligations, as the state has the privilege of doing, but these obligations can be enforced through the agency of the courts. For instance, if a county organization should wish to repudiate a particular issue of bonds, which have been issued and sold, because of some real or fancied grievance connected therewith, and if the county commissioners who represent the county in its corporate capacity should refuse to levy taxes for the payment of the interest or principal, the courts could order them to do so, and if they should refuse they could be fined and imprisoned for contempt. In some cases city charters have been repealed by the state legislature and the city changed into a "taxing district" in order to more readily enforce orders of the courts, in requiring them to fulfill the terms of some legal contract or obligation.

6. Competency of Semi-Public and Private Corporations. A corporation has no powers for entering into or performing contracts beyond those given it by the state in its charter. Its capacity for transacting business, however, is not limited to the specific privileges granted in its charter, but is of necessity extended by implication to include such other powers as may be necessary for the complete consummation of its specific purposes. For instance, if a corporation requires the use of certain real estate for the transaction of its business, it can evidently buy and sell such property when this is intended for its own uses. It may also borrow money and issue therefor various kinds of obligations, and, in fact, it may make any contract which it is lawful for an individual to make, provided such contract relates to a subject which is within the sphere of its operations.

When a contract or agreement on the part of a corporation does not fall within its express or implied powers, it is termed *ultra vires,* and such contracts can not be enforced. The official acts of the officers or agents of a corporation bind it much the same as such acts would bind an individual when made in a private capacity, and this applies both to oral and to written agreements, unless the corporation charter specifically requires certain kinds of agreements to be in writing.

7. Contracts by Agents.* A contract by an agent is not valid unless the principal is himself competent to enter into a contract. On the other hand, a contract by an agent is valid, provided the principal is competent, even though the agent be incompetent to enter into a contract as a principal. Thus a minor may be a competent agent, but not a competent principal. The agent, however, must have no adverse interest from that of his principal under the contract negotiated. For instance, he must not be interested on both sides of the agreement, if these interests are supposed to be adverse.

The legality of the acts of an agent is similar to the legality of the acts of a corporation. As a corporation receives its authority for the transaction of a particular kind of business from the state, and its capacity in the formation of contracts is limited thereby to the express and implied powers under its charter, so an agent receives his authority from his principal,

---

* An engineer or architect is the agent of the owner (person or corporation), and as such has the express powers given him in the contract itself or in his agreement with his employer, and also many customary implied powers.

his legal acts are limited to the scope of the authority conferred upon him by his principal, and, as in the case of a corporation, he will be justified in the law in the making of any contract, as agent, which may prove to be necessary or essential to the carrying out of his more specific instructions, or for the transaction of the business for which he has received special authority.

Unlike a corporation, however, an agent may exceed both his express and implied authority in the making of a contract, and yet this contract will become binding on the ratification of it by his principal. This ratification may also be either express or implied, an implied ratification consisting of a failure to object or protest or to annul the contract on learning of its existence, or of acting under it as though consent had been given.

A ratification, whether express or implied, of the acts of an agent operates always so as to include the whole of the agent's acts pertaining to the particular transaction in question, and can not operate for the acceptance of a part, and the rejection of other parts. By adopting a part, the principal is bound by the whole. If it appear, however, that the express or implied ratification was due to a mistake of fact, the principal may repudiate the action of the agent on learning of the facts.

If the agent wishes to avoid personal responsibility in the entering into a contract, it must be understood by the other party that he is acting as an agent, and not in his own behalf. He may, however, enter into contract in his own name, not as an agent, when in fact he is the agent of another party. In this case, however, the other party to the contract on learning of the principal, has his option to enforce the contract against the agent or against the principal as he may choose. In all cases of contracts with agents the other party to the contract must know of the agent's authority aside from the agent's own testimony in the case, as this latter is not received as evidence of the fact. Whatever the agent's pretended authority may be, if it should prove that he has exceeded both his express and implied authorization, the principal is at liberty to repudiate his acts, and the other party to the contract has no remedy except against the agent himself. The agent's authority is evidenced, however, by the usual and customary transactions of such agent which have been accepted by his principal, and which have become known to the other party in a proposed new contract. Therefore as to third persons the authority of the agent may be implied from previous performances of similar acts which have come to the knowledge and received the consent of the principal.

In the case of sub-agency, or of the appointment of an agent by an agent such authorization must proceed originally from the principal, or be afterwards ratified by him before the principal can be bound by the acts of the sub-agent.

In order that an agent may relieve himself from responsibility in the signing of a contract, the document must reveal, either in its body or in the signature, who the principal is; a mere signing of a contract by a person as "agent" will not relieve the party so signing from personal responsibility unless the document does reveal the principal.

If an agent enters into contract in a matter beyond his express and implied authorization, he becomes personally liable to the third party, unless he reveal to such party, at the time of the signing of the contract, the exact relation between himself and his principal in such a way that this third party becomes aware of the dubiousness of the agent's authority. In this case the principal may repudiate the act of his agent and the third party will not be able to hold either principal or agent to the contract. If, however, the agent does not disclose his exact relations with his principal, and assumes authority beyond his authorization, he does become personally liable for such damage as may result from failure of performance on the part of his principal.

The principal is also liable for all the frauds, deceits, and negligent acts of his agent so long as these pertain to the business he is authorized to perform. In this case, of course, the agent himself is liable both to his principal and to any third party. While if such fraud or deceit or negligent act pertains to matters outside the scope of his authority, the agent alone is liable.

Acts of an agent continue to be binding upon the principal as to third persons, even if the agent's authority has been revoked and the agency ended, until such termination of the agency comes to the knowledge of such third person. This applies to all kinds of continuous agencies, but does not apply, of course, to an agency for the performance of a particular act.

The death of the principal always acts to terminate the agency, which termination occurs at the instant of the death of the principal. This nullifies even such acts of the agent after the death of his principal as may have occurred before such death came to the knowledge of the agent; but when the agent enters into contracts for his principal after the decease of the latter, with or without the knowledge of such decease, the contract is void as against the estate of the principal, and, generally speaking, as against the agent himself, and the third party is without remedy. In this case no notice of the termination of the agency is required. In a few states, however, the rule has been adopted that the *bona fide* acts of the agent after the death of his principal and before he becomes aware of the fact, and which do not require the principal's signature are valid in favor of third parties.

LEGALITY OF THE AGREEMENT.

8. Kinds of Illegal Subject-Matter. No contract can be enforced in the courts which involves an agreement to perform an act which is (*a*) forbidden by statutory law, or (*b*) is contrary to the rules of common law, or (c) which is opposed to public policy.

9. Contracts in Breach of Statute Law.

This subject will not here be entered upon at length. It may be said, in short, that all acts which are expressly prohibited by statute law, or all acts for which specific penalties are attached in national, state, or municipal laws, if made the subject of a contract, such contract can not be enforced. Without here mentioning the acts which would be criminal or immoral, it may be well to call attention to a certain class of contracts which can not be enforced at law because the plaintiff in the suit has no legal standing in court. Thus where the state statute requires a diploma or license for the practice of medicine or surgery, or a license to act as attorney at law, or as a surveyor, or as an engineer, a person not having such legal authorization can not collect in the courts the price of his professional fees.

Under this head also fall agreements to pay usurious interest, which in some states involves the forfeiture of the entire interest, and in a few states the entire contract becomes void even to the sacrifice of the principal.

In most states all kinds of wagers are declared unlawful by statute and can not be collected.

While all contracts for fire or life insurance are in a certain sense wagers, they are valid and lawful when the person for whose benefit the insurance is made can be shown to have a suitable interest in the property or person insured.

In all states where Sunday labor, with the exception of "works of necessity and charity," is prohibited, contracts made on Sunday are illegal and can not be enforced.

Where contracts in breach of statute law have been fully executed, in other words, where the act has been done and the compensation received, the law will not recognize such transactions for the purpose of annulling them. Thus, in the case of a wager which has been paid, the law will not enforce the return of the money.

10. Immoral Acts. The courts will not enforce an agreement, the object of which is forbidden either by statute or by common law, or which in law may be regarded as immoral or wrong. Such agreements might relate to such subjects as the commission of crime; all kinds of frauds upon creditors, either by way of fraudulent assignments, or by way af agreements with certain creditors to the disadvantage of others; all kinds of transactions under false pretenses, as the selling of articles under false labels; fraudulent conveyance of real estate to defraud creditors; changes in contracts after they have been signed, either by one party without the consent of the other, or by the two principals without the consent of the sureties; all acts of officers of corporations in their official capacity, in furtherance of their private ends; fictitious bidding at auctions for the purpose of raising the bids of *bona fide* purchasers; collusion between the auctioneer and private individuals to defraud owner, and the like.

The particular class of illegal acts in this category which has especial interest to engineers, is that referring to changes in contracts agreed to by the principals without the consent of the sureties or bondsmen. In all such cases if the changes are material, that is to say, if they are such as may be said to have a money value, then if these changes be made without the consent of the surety, such surety can no longer be held for any damage resulting from failure of his principal to fulfill his agreement. Since such changes are almost always made in all contracts after they are signed and before the work is fully executed, and since it is very common to neglect to obtain the consent of the sureties when making all such changes, these sureties or bondsmen are nearly always relieved from liability in the manner here indicated. Furthermore, if such sureties are consulted in regard to the proposed changes and they should not choose to give their consent, then if they are still to be held for the fulfillment of the contract their consent to such changes must be purchased, the same as must be done with the principal himself as provided for in the specifications or contract. Because of this common oversight and the resulting relief of these sureties, or of their opposition to allowing changes to be made in case they are consulted, it is becoming customary to provide some other kind of guarantee of fulfillment, other than that of personal sureties.

11. Contracts Opposed to Public Policy. These contracts may relate to all such acts as may be shown to be detrimental to the public welfare. Such as acts which would tend to injure the public service, or to obstruct the course of justice, or to encourage litigation, or as have an immoral tendency, or as will restrain the freedom of trade, or as will diminish the security of property and life. As such contracts in general have no immediate bearing upon the work of engineers they will not be further enlarged upon here. There is, however, a class of agreements commonly entered into by the principals to an engineering contract which are often construed in the courts as against the public policy, which will be discussed in the following article.

12. Agreements Which Refer to Arbitration. The following discussion of this subject is taken bodily from Lawson on Contracts, being article 318 of that work.

"An agreement that matters which have arisen or may arise between the parties shall be referred to an arbitrator or arbitrators is not binding and either party may have recourse to the courts notwithstanding it. The reason of the rule is by some traced to the jealousy of the courts and a desire to repress any attempt to encroach on the exclusiveness of their jurisdiction, and by others to an aversion on the part of the courts from reason of public policy to sanction contracts by which the protection which the law affords the citizen is renounced.

"But when a contract contains a condition which provides that disputes arising out of it shall be referred to arbitration,

2

the validity of such a condition depends upon rather a fine dis-
tinction. Where the *amount of damage sustained by a
breach* of the contract is to be ascertained by specified arbitration before any right of action arises, the condition is good;
but where all matters in dispute, of whatever sort, are to be
referred to arbitrators and to them alone, the condition is illegal. The one imposes a *condition precedent* to a right of
action accruing, the other endeavors to *prevent* any right of
action accruing at all. As well put by an English judge:
'If a tenant covenant that he will cultivate the demised land in
a husband-like manner and also covenants that if any dispute
shall arise in respect thereof it shall be referred to arbitration,
an action may nevertheless be maintained; but where the covenant is to pay such damages as shall be ascertained by an arbitrator, no action will lie until he has ascertained them.'

"The principle is frequently applied in the United States
to contracts for the construction of buildings, railroads, canals
and other works involving numerous details. These contracts
give rise to many questions which a court of law might reasonably send to a referee, and the parties may agree that such questions
shall be determined by an architect or engineer or by arbitrators, and that such determination, or a *bona fide* effort to
obtain it, shall be a *condition precedent* to the right to bring
an action on the contract. Contracts of insurance usually contain similar clauses. Thus an insurance policy provided that,
in case of differences arising touching any loss or damage, the
matter might at the request of either party be submitted to
impartial arbitrators whose award in writing should be binding
on the parties to the amount of such loss or damage, 'but
shall not decide the liability of the company under this policy;'
also, 'it is furthermore mutually agreed that no suit or action
against this company for the recovery of any claim by virtue of
this policy shall be sustainable in any court of law or chancery
until an award shall have been obtained fixing the amount of
such claim in the manner hereinabove provided.' It was held
that no suit could be sustained against the objection of the
company until an award had been made, although neither party
previous to the suit had requested arbitration.

But it must be expressly stipulated in all cases that the award
or determination is a *condition precedent* to the right of action
on the contract, or the agreement to arbitrate will be of no
effect.

Agreements of a similar nature have been held illegal, as
aiming to oust the jurisdiction of the courts; as, for example,
a provision in the by-laws of a benefit association that the
decision of the officers on the claim of a member shall be final
and conclusive. And parties are not allowed by contract to
vary the procedure in the courts prescribed by statute. In Illi-

nois a lease contained a provision that the landlord should have the right to take immediate judgment against the tenant in case of a default on his part without giving the notice and demand for possession and filing the complaint required by the statute. · It was held that such a provision was illegal."

13. The Engineer as Arbitrator. In the carrying out of engineering and building contracts, the specifications are usually so written as to make the engineer or architect an arbitrator on almost all questions which can possibly arise under the contract, and it is usual to specify that his decisions thereon shall be final and conclusive. In view of what is given in the previous article it is evident that such a clause can not operate to prevent a case being brought before the courts under such a contract, but when it has been so brought it evidently will operate to sustain the decisions of the engineer on all points which may be construed by the court as "*conditions precedent*" to final settlement. On all questions of fact, however, which the court can pass upon as well as the engineer, and on all questions of law the court would retain its jurisdiction notwithstanding the agreement of both parties to submit *all* questions to the arbitration of the engineer, whose decisions were to be "final and conclusive." On all questions which the engineer is, from the circumstances of the case, especially competent to determine, as to quantities and classification, as well as all questions which are more or less matters of opinion as classification of materials and perfection of work done, the decision of the engineer will be sustained under such a clause, provided it be not shown that he has acted fraudulently in the matter. See Articles 85 and 108.

AGREEMENT.

14. Mutual Assent. In order that a contract shall be binding on both parties to an agreement it must have been understood and assented to by both in the very same sense. However clear the agreement would appear to be on its face, if it can be shown that the proposition was not mutually understood in the same sense it can not, in general, be enforced. It must not be understood, however, that all pleas of having misunderstood the plain and express provisions of a *written* contract will relieve the party making such claim from liability under it. In other words the mental agreement is evidenced by the language used in expressing such agreement, and the law will presume that such words were understood, provided their meaning is plain and evident. Furthermore whatever a man's real intention may be, if he so acts as to lead another person to reasonably suppose that he was assenting to a given proposition, and this person proceeds on this assumption, the other party so acting becomes bound by the proposition.

The agreement is not consummated until each party has communicated to the other, either orally, by letter, or by overt acts, his intention in the matter. The secret or mental acceptance of a proposition by one or both of the parties to it does not complete a legal agreement, until this mental act has been communicated to the other party.

A person making an offer, whether orally, by messenger, by mail, or by telegraph, or by public advertisement, must allow a reasonable time for its acceptance, provided no time limit is stated in the proposition. If the acceptance is returned by the same agency used in sending the offer, the contract is completed at the time such acceptance is delivered to such agency, whether the party sending the offer ever receives such acceptance or not. A person is bound by the acts of the agent of his own selection, and the failure of this agent to deliver to him the acceptance does not operate to prevent the completion of the contract. For instance, a proposition sent by mail is accepted at the time the letter of acceptance is deposited in the postoffice or letter box, and a proposition sent by telegraph is accepted and the contract completed at the time of the delivery of a telegraphic reply at the telegraph office or to a telegraphic messenger.

If the person receiving the offer wishes it to remain open for a definite length of time, longer than might be construed as reasonable, if no time is specified, he must pay to the other party something which may be construed as a consideration for the privilege of acceptance for such specified time. On the other hand, the party accepting can withdraw his acceptance if he can succeed in having his withdrawal presented to the first party before his acceptance has been received. Thus an acceptance by mail may be withdrawn by telegraph, provided the telegram is received before the letter.

A mere offer may be withdrawn at any time before it is accepted, unless a consideration has been paid for the privilege of acceptance for a definite time as above described. A formal notice of withdrawal is not always necessary, as when the party receiving the offer becomes aware of the sale of the property in the mean time to another.

When an offer is made by mail or telegraph the means used for communicating the offer become the recognized agent of the party making such offer, and the party receiving it is at liberty to accept it as received, even though a mistake may have been made in the transmission of the same. .Thus if an offer is made by telegraph, and an error has been made in transmitting the same, the erroneous proposition may be accepted either by mail or .telegraph, and the party making such offer is bound. His only remedy is to sue the telegraph company for damages. This is because the party making the proposition assumed all responsibility for the correctness of the transmission by the agency selected by him.

When an offer has been made and no consideration paid to keep it open for a given time, it is supposed to stand for what the law will consider a reasonable time, the actual length of time depending altogether on the nature of the transaction.

15. Qualified assent. Whenever a proposition made by one party is accepted by another with any kind of qualification or change of the conditions or wording of the original proposition, such an acceptance is simply the making of a counter proposition to the first party, and does not constitute an agreement until such party has in turn assented fully to the entire proposition as last stated, and if he again assents to the proposition with further changes or amendments, it becomes again a new proposition, which must be agreed to by the second party, before it becomes binding on the party to whom it is sent. The assent which finally makes of the offer or proposition a binding contract, is the full, absolute, and unconditional acceptance of its terms.

16. Qualified offers. The party making the offer has the right to prescribe in it the time, place, form, or other condition of acceptance, in which case such offer can be accepted only in the manner prescribed. This privilege on the part of the proposer does not enable him to impose the condition, however, that a failure to receive an acceptance by a certain time will be construed as an acceptance. In other words, he may not impose the conditions of refusal.

17. Implied acceptance. An offer may be accepted by merely acting upon it, the act becoming an acceptance from the time it was performed. Thus an offer to purchase goods may be accepted by simply shipping the goods, or in the case of a published offer of a reward for the apprehension of a criminal, the act of apprehending is construed as both an acceptance and fulfillment of the contract.

18. Failure of agreement by mistake. The parties to an agreement are bound to the fulfillment of the same in accordance with the plain intent and meaning of the language used, whether oral or written, provided the meaning of this language be clear, and neither party is allowed to plead either carelessness in the reading of the terms thereof or ignorance of the meaning of the language used. It goes almost without saying, however, that apparent or evident mistakes in the use of language will be corrected by the court. However, the following kinds of mistakes will lead to a decision that no contract was really entered into because of utter failure of the parties to agree to the same thing.

(a) **Mistake concerning nature of transaction,** as where a person, by mistake, signs a document of an entirely different character from that which he intended to sign, as where he signs a bond instead of a petition, or a deed instead of a lease, the two documents being similar in form and appearance. In such cases it is held that the "mind of the signer did not accompany the signature" and therefore he never agreed to such a proposition. If it can be shown, however, that the mistake resulted from negligence to read the terms of the proposition, and that while the signer understood in a general way the character of the document, but did not read it over carefully, or perhaps did not read it at all, he will he held to the contract because of his culpable negligence. This only holds where the plea of fraud on the part of the other party to the contract is not maintained.

(b) **Mistake concerning person with whom contract is made,** as "where A contracts with B, thinking that he is contracting with C, there can obviously be no contract, for B not being present to A's mind, A can not be a consenting party to a contract with B." This, of course, does not affect contracts made through agents, when the agency is declared.

(c) **Mistake concerning Subject-matter of Contract.** If the parties contracting engage themselves concerning a thing which does not exist, such a mistake avoids the contract, because of the nonexistence of the subject-matter. This applies to property which may have ceased to exist before the contract is signed, although both of the parties were ignorant of the fact. In all cases, however, where the existence of the subject-matter is in the mind of the proposer more or less doubtful and yet the offer which he makes is unconditional, he can be held for damages resulting from failure, even though the subject-matter be nonexistent.

A mistake prevents the consummation of a contract also, where each of the parties has in mind a different subject-matter from the other. This applies to mistakes which may be considered legitimate, as where the same words apply with equal force to different things, or in the case of an oral contract where the language was clearly misunderstood, from a failure to hear the words which were really spoken.

The remedy for a contract entered into under a mistake is the privilege of repudiating it on the part of the person who made the mistake, or the privilege of recovering, by a suit at law, part payment which may have been made, it being understood that the contract has not been fully executed by a complete payment. Or the case may be brought into a court of equity, and justice obtained by having the court correct the terms of the agreement, it being assumed in all these cases that no fraud has been committed.

19. **Misrepresentation in the Contract.** In order that a misrepresentation of facts may make a contract invalid, it must have been made with a fraudulent intent. The mere falsity of a statement of a material fact in the contract, however much it may have misled the other party, will not invalidate the document, unless a fraudulent motive accompanied the misrepresentation. It is necessary, therefore, to prove the motive of the misrepresentation before the validity of the contract can be passed upon.

If, however, a particular term in the contract or some integral part of it is based upon a misstatement of fact, which term or integral part can be passed upon separately from the body of the contract, such a misrepresentation is equivalent to a promise by the party making it, and if not fulfilled the other party can recover damages. This, however, does not invalidate the contract as a whole.

A nondisclosure of fact. is equivalent to a misrepresentation of fact, provided the disclosure properly accompanied the transaction.

Fraudulent misrepresentation will be treated in the following article.

Contracts for insurance, whether marine, fire, or life, and contracts for the purchase of stock in corporations, also contracts between parties occupying intimate and confidential relations with each other may be invalidated by misrepresentation of fact, which would not invalidate ordinary business or engineering contracts. (See a general treatise on the Law of Contracts for these cases.)

20. Invalidity of Contract through Fraud. As a general rule fraud vitiates all contracts. That is to say, fraudulent misrepresentation by one party enables the other or injured party to declare the contract void from fraud, or he may enforce the contract against the defrauding party, at his own option. The defrauding party, however, has no option or privilege in the way of declaring the contract void. In other words, should he find that the contract was adverse to his interests, he could not plead his own fraudulent act to his own benefit.

Fraud in the sense here used consists in a "false representation of fact made by the party who is charged with it, with a knowledge of its falsehood, or in reckless disregard whether it be true or false, with the intention that it shall be acted upon by the complaining party, and actually inducing him to act upon it to his damage."

"From the above definition the following essential elements of fraud may with profit be stated separately: (*a*) A false *representation*. (*b*) A misrepresentation of *fact*. (*c*) A representation made by the *party charged*. (*d*) *Knowledge* of its falsity or a *reckless indifference* in the matter. (*e*) An intention that it *shall be acted upon* by the other party. (*f*) A *reliance* upon it by the other party. (*g*) *Damage* to the party deceived." Unless each and every one of the above essential elements of fraudulent misrepresentation be proved, the contract can not be avoided on the score of fraud.

The false representation also must refer to some material fact. Furthermore the concealment or nondisclosure of a material fact in an active manner, that is to say, an active prevention of the disclosure of material facts, may constitute fraudulent misrepresentation.

Where the one party knows that he is being trusted by the other party and relied upon for the disclosure of material facts, as is the case when a contractor relies upon the engineer or architect to disclose to him the material facts pertaining to the work to be done, this rule will be more rigidly applied than when no such confidence is imposed.

A concealment of the real value of goods shipped by express or freight, or by other agency, in order to obtain such shipment at a lower rate, is a fraudulent misrepresentation.

While the vendor or seller of an article is not obliged to make known to the purchaser the defects of the article. when such defects can be discovered by the buyer, yet a deliberate hiding of such defects on his part will be considered a fraudulent misrepresentation. As to defects which can not be discovered by the buyer, the vendor is bound to make known to him such defects as he himself may be aware of.

While known false representation of the quality or defects of an article made by either seller or buyer, for the purpose of gaining the advantage in a transaction is fraudulent, the purchaser, however, is not obliged to disclose his knowledge of the real value of an article, which is offered to him below its actual value.

On questions which may be considered matters of opinion rather than questions of fact, misrepresentations do not constitute fraud; only misrepresentations of known facts fall in this category. What is not really known may be misrepresented without invalidating the contract. Neither do false representations of future intention, or of questions of law, constitute fraud.

The fraudulent misrepresentation must have been made by the party charged or by his agent, or with his connivance and knowledge. The fraud of a third person does not invalidate the contract between two others.

The misrepresentation must be known to be false. Sometimes a contract may be set aside because of violent injustice resulting from a false representation, which was, however, believed to be true. Frequently, however, a party may make extravagant statements in a reckless manner for the purpose of influencing the other party to a transaction, not knowing whether his statements be true or not. In such a case the willful negligence or recklessness as to the truth of his positive statements will act to invalidate the contract, provided such statements prove to be false, the same as though he had known them to be false.

The false statements must also be made with the expectation that they are to be believed and acted upon. Extravagant affirmations made in a jocular manner, and not expected to be believed, would not constitute fraud.

The misrepresentation must be accepted as true, and be acted upon before the fraud is perfected. The burden of proof here lies upon the party charging fraud, to show that he did really act upon the statements made. It is not necessary that he shall show that the fraudulent statements formed the sole basis of his action. He need only show that they contributed materially to that end, and that the action would not have been consummated without them.

A false representation as to one of several material matters in a contract operates to vitiate the entire agreement.

The party claiming fraud must also show that some actual damage has been suffered. It would not be sufficient cause for the annulling of a contract to show that one was fraudulently led to the payment of a just debt, since no damage has in this case been sustained.

21. Remedy of party defrauded. Immediately on discovering the fraud the party defrauded should take action, and he has his option of the following courses, the last two being remedies.

(*a*) He may enforce the contract against the defrauding party, or take no action whatever, and allow the contract to be enforced against himself. If he does not take action on discovery of the fraud, he will be supposed to have consented to the enforcement of the contract, notwithstanding such fraud, and he will lose his option of resisting such enforcement.

(*b*) He may at once give notice of the rescission or rescinding of the contract on his part, because of fraud claimed to have been perpetrated by the other party, and he may bring suit to recover damages, or he may either by word or act give evidence to the other party of his intention to treat the contract as null and void.

(*c*) If it be practicable to reinstate the parties in their original relative relations, he may sue for such restoration. That is to say, if goods have been delivered, they may be recovered if practicable.

Any action under the contract in the way of acknowledging its force by the party defrauded, after he has discovered the fraud, will operate to make the contract binding upon him, as he will be assumed to have deliberately forfeited his right of rescission. It must be understood, also, that he can not consent to the operation of a portion of the contract with the privilege of rejecting another portion of it to which the fraud may more directly relate. Since fraud vitiates the entire contract, the defrauded party must forfeit his privilege of rejection entirely by agreeing to its terms notwithstanding the fraud, or he must reject it entirely and in all its parts. He can not obtain the benefit of a part and reject another part.

Outside of the rights described above, arising under the contract itself, the defrauded party has the privilege at common law of bringing action for deceit to cover such damage as he may have sustained as a result of such fraudulent misrepresentation. This action is in addition to his privilege of avoiding or rescinding the contract itself.

Furthermore a party defrauded of his property may recapture it, if he is able to do so without unnecessary violence and without a breach of the peace, without recourse to the law and its agencies.

22. Invalidity of Contract through Duress. "A person is said to have acted under duress when he does or promises to do any act not of his own free will, but in consequence of unlawful physical restraint imposed by another, or in consequence of threats made by another, either to do him some great bodily harm, or to unlawfully destroy his property, or deprive him of the same. Promises made under duress will not be enforced, and money paid, or property transferred under duress may be recovered."

Contracts entered into under duress as above defined are voidable at the option of the constrained party, the same as though fraud had been perpetrated upon him. The contract is not voidable, however, at the option of the other party.

23. Invalidity of Contract through Undue Influence. Where the parties occupy a confidential relation to each other, or from long association and other peculiar circumstances affording the proper and sufficient opportunities, courts of equity take cognizance of what may be called undue influence, which may act the same as fraud in persuading the person to enter into an unfair and unreasonable contract. Such are the relations of the members of one family, or those of guardian and ward, attorney and client, priest and parishioner, physician and patient, as well as those where mental weakness from old age or sickness and the like, furnish suitable opportunities.

The remedies in the case of undue influence are the same as those in the case of fraud, except that the influenced party does not lose his rights of choice of remedies by delay in action, since it is unfair to assume that such a party can suddenly recover his normal independence.

CONSIDERATION.

24. Consideration Defined. All business contracts such as an engineer will be called upon to enforce must always be supported by a *valuable consideration;* otherwise they are not enforceable. As such a consideration is always named and specifically determined in all engineering contracts, it is not necessary to go into that subject here very fully.

A "valuable consideration" in the eyes of the law is *"some right, interest, profit, or benefit, accruing to one party, or some forbearance, detriment, loss, or responsibility given, suffered, or undertaken by the other."*

Such a consideration is necessary to enforce a written agreement the same as would be necessary with an oral agreement.

A contract under seal, however, does not require a consideration to enforce it. This is the principal and characteristic difference between contracts under seal, and ordinary written or oral contracts, both of which latter class constitute what is known as simple or parole agreements.

It is not necessary that the consideration be named in the agreement, or that the fact of consideration should appear in the agreement; it is only necessary that there shall be a consideration in fact.

In cases of promissory notes and other negotiable paper, the presumption is that there was in fact a consideration, whether named in the document or not, and the burden of proof rests upon the maker of the note to show that there was in fact no consideration.

In the states of California, Iowa, Indiana, Kansas, Kentucky and Missouri, an agreement made in writing is presumed by statute law to be founded on a consideration, and is therefore placed on the same basis as holds generally for negotiable paper. Here again the burden of proof rests upon the defendant to show that there was in fact no consideration. In both the above cases if it can be shown that the promise was not supported by what the law will construe as a valuable consideration, the agreement or contract fails.

A promise made without a valuable consideration is construed by the law to be gratuitous, and not enforceable, even though the party to whom it was made has acted upon it, and has sustained serious loss or damage thereby.

25. Adequacy of Consideration. It is not necessary that the consideration named, or implied, or shown to exist by any acceptable evidence shall be adequate to support the promise. So long as it is valuable at all, in the sense defined in the previous article it will support the full promise. The question of adequacy of consideration will not be allowed to be put in evidence except for some ulterior purpose, as when it is attempted to prove fraud.

Here the law seems to conflict with the principles of right and justice; but for the court to inquire into the adequacy of a consideration would make nearly all contracts subject to litigation and the freedom and rights of the individual would be greatly curtailed. This rule as to adequacy, however, does not apply to the exchange of sums of money, for instance, whose absolute values are fixed and known. In this case the consideration must be adequate and equal.

When the consideration is grossly inadequate, suit may be brought in equity and the courts will in that case sometimes vary the terms of the agreement in the interest of justice.

Neither is it necessary that the consideration should have any assignable money value, as is plainly implied in the definition of a consideration given in the previous article. Thus mutual promises are each a consideration for the enforcement of the other, but when the promise of one of the parties includes no more than it was already his legal duty to perform, such a promise will not support a promise made by the opposite party. For example, if A owes B a sum of money, and interest thereon which *is due*, and A promises to pay B the interest if he will extend the loan, which B promises to do, here B's promise to extend the loan is not supported by a valuable consideration and is therefore not enforceable. Again a promise to pay to a public officer more than his lawful fee for the performance of a public duty, is not enforceable.

26. Agreement to Take Less than is Due. A very common case in the execution of contracts is that of an agreement by one of the parties to receive or accept less than the contract calls for. It is important here to distinguish between sums of money or matters which are in dispute, and sums of money or matters which are not in dispute.

If one of the parties agrees to accept a sum of money less than that which is avowedly due him, such an agreement is not enforceable, because of failure of consideration, unless some condition of performance accompanied the offer which may be construed as a consideration. If, however, the sum of money claimed by A is disputed by the other party B, and never has been acknowledged by B as being the amount owed, then and in that case an agreement on the part of A to accept less than his claim, when accepted by B, is enforceable. This is because no agreement had been made previous to this compromise arrangement.

Similarly an agreement on the part of the owner to accept a less amount of work or a cheaper construction on the part of the contractor than that contained in the written specifications is not enforceable, unless it is supported by some further act on the part of the contractor, or by a corresponding change in the price of the work, which may be construed as a consideration. If, however, the original contract provided for such changes as these by agreement without further consideration, such further agreements simply modify the terms of the original contract and become a part thereof without a new consideration being required.

Where several creditors enter into a mutual agreement among themselves and with the debtor to take less than is acknowledged to be due them, and to discharge their several debts, such an agreement is held to rest on a sufficient consideration, since these mutual promises are evidently for their mutual benefit, and therefore all do receive a valuable consideration in support of such promises.

If it is desired or intended that an agreement shall hold without a corresponding consideration, such as have been referred to above, it is only necessary to execute the new agree-

3

ment under seal, in which case a consideration is not required.

When a person brings suit against another or threatens to do so, for a sum of money claimed in good faith to be due, not evidenced by a note or promise to pay, the dismissal of such suit, or a promise not to bring it, is a sufficient consideration for a promise by the party sued, or threatened to be sued, to pay to the claimant a sum of money, or for a promise to do any other lawful act.

In the case of an engineering contract, an agreement by the parties to vary the terms of the original contract, which variation may not have been provided for in such contract, an agreement to vary the conditions in one particular must be supported by a consideration in the way of an agreement to vary the terms of the original contract in some other particular which may be accepted as a consideration, or some other and new consideration must be provided for in order to support such agreement. Thus, if A has agreed to build a house for B in accordance with certain plans and specifications with no provision for changes of plan, if B consents to a change in the plans by which the cost is reduced, without any consideration being agreed upon or mentioned for such change of plans, B has the privilege of changing his mind, and of enforcing the original agreement, since the second agreement was not supported by a consideration. In like manner, should A consent to a change of plans without naming a consideration he can not be held to such an agreement even though it be made in writing, but may continue to carry out the original agreement, which alone is binding. In other words, all subsequent or auxiliary agreements or changes in the original contract not provided for in the original document are in fact new contracts and must each and all be supported by a consideration.

Promises or contracts which have been fully executed can not be inquired into by law, as to whether there may have been no consideration. Thus money which has been paid without consideration can not be recovered, and for work executed before a promise to pay has been given or implied, no recovery or compensation can be obtained.

Of this general character are gifts which have been made, the article having been delivered to the donee. They can not be recovered, neither can payment for them be enforced. An exception to this is where property has been given away to defraud creditors. In this case the person receiving the gift may be compelled to restore it to the creditor, or so much thereof as will discharge the creditor's claim against the donor.

27. As to Waiver of Legal Rights. An apparent exception to the general rule that a promise must be supported by a consideration is an agreement to waive a statutory right of defense. Thus "a promise to pay a debt barred by the statute of limitations, or by a discharge in bankruptcy, though made without consideration is enforceable, and a promise by an endorser of a bill or note to pay it, although the endorser knows that he has been released from all liability, from the note not having been protested when due, is likewise binding." In these cases the new promise is equivalent simply to waiving the legal rights of the party, after which the old promise is again restored to its legal status, which former promise was supported by a consideration.

"But when a debt has been canceled by the act of the parties, as by a release under seal, which would require no consideration, a subsequent promise to pay the debt, notwithstanding the release, is not valid unless supported by a consideration." In this case the former promise or agreement had been obliterated by a subsequent release under seal, and hence a new contract would have to be made.

CONTRACTS UNDER SEAL.

28. Classes of Sealed Contracts. While any contract may be executed under seal, and so become a scaled contract, *under the common law*, the following must be executed under seal to become binding, namely: (*a*) Gratuitous promises. (*b*) Contracts with corporations. (*c*) Conveyances of real estate. (*d*) Bonds.

(*a*) If it is the purpose to make a gratuitous promise legally binding on the parties, it must be executed under seal, and when so executed the absence of a consideration will not invalidate it.

(*b*) The common law rule that contracts with corporations must be executed under seal no longer obtains in the United States. Here a contract entered into with the proper officers of a corporation is valid without being sealed, the same as though made with an individual, unless the charter of the corporation specifically requires all contracts to be made under seal.

(*c*) Deeds and mortgages do still in this country require the presence of a seal, except where a special statute provides otherwise.*

(*d*) A bond is an instrument under seal whereby one acknowledges himself indebted to another in a specified sum, generally but not necessarily conditioned on the performance of some act. Thus bondsmen or sureties in the case of an engineering contract are those who sign an obligation or acknowledgment of indebtedness in favor of the party letting the work, in a specified sum, conditioned on the faithful execution of the work which the contractor has undertaken to perform. Such an instrument, called a bond, should be executed under seal.

*This is true in Ohio, Indiana, Iowa, Kansas, Nebraska, Tennessee, Texas, Dakota, Kentucky, and Mississippi.

The significance of a seal is losing its force in America. In some states a seal no longer has any significance, whatever, so that even when present in due form, it does not import a consideration, but such consideration must be proved the same as for a simple or parole contract. The plain intent of the parties is the controlling factor. The rules given above are the common law rules and still have more or less force in some states.

PAROLE CONTRACTS.

29. Oral and Written Contracts. All contracts, either oral or written, not executed under seal, are called simple or parole contracts.

An oral contract has all the force of a simple written contract, but it is subject to difficulties in the way of establishing or proving its terms, which a written contract is comparatively free from. A large proportion of the litigation arising from the nonfulfillment of contracts is caused by a failure to reduce the terms of the contract to writing.

An oral or written contract can be modified by subsequent agreements, and such subsequent agreements become a part of the original contract. A written contract, however, has this advantage over an oral contract: It is presumed in law to embody all understandings and agreements made at the time of, or previous to, the signing of the contract. No oral evidence can be admitted therefore as to agreements or understandings made at the time of the written agreement or antecedent thereto which would modify its terms. Evidence will be received, however, as to oral or written agreements made subsequent to the signing of the written contract which may modify its terms.

It is allowable, however, to admit testimony as to oral agreements or understandings made prior to, or contemporaneous with, the signing of the contract, for the purpose of proving fraud and deception.

Such evidence may also be introduced for the purpose of proving duress or mistake in the drafting of the contract. It can not be introduced for the purpose of modifying its terms, since it must be assumed that all the essential or material matters in the agreement were embodied in the written contract.

Subsequent oral or written agreement modifying the terms of the original contract requires a separate and distinct consideration to support it, unless the original contract contained special provision for such changes, in which case they must be made in accordance therewith, and may or may not require a new consideration.

ASSIGNMENTS OF CONTRACTS.

30. When Assignments Can Be Made. All contracts and agreements can in general be assigned by either party, and the contract enforced by the assignee, except such contracts or agreements as involve a personal trust or confidence in one or both of the parties. Evidently trust and confidence in the skill or professional ability of another can not be assigned, and when such trust is a material element in the contract there can be no assignment. Of such character are nearly all kinds of personal services, except, perhaps, the most common labor.

All building and engineering contracts are assignable, unless the writings themselves contain conditions denying such privilege. It is, however, common to insert such a clause in all engineering contracts by which they then become non-assignable.

31. Notice of Assignment Necessary. While an assignment is effectual as between the assignor and assignee, from the moment it is made, it does not bind the other party to the original contract until he has received notice of the assignment. Without such notice any performance on his part in favor of the original party or assignor releases him to that extent with the assignee. It is necessary, therefore, to give prompt notice of all assignments to all the parties concerned. After such notice has been given, all parties become bound to the assignee, the same as they had formerly been to the assignor.

An exception to the rule of the necessity of giving notice obtains in the case of what is called negotiable paper. The transfer of such contracts is not called assignment, the document itself carrying with it its own evidence of ownership. Such documents are bills of exchange, bank checks, promissory notes, bills of lading, certificates of deposit, certain kinds of bonds and coupons, warehouse receipts, and bank bills.

CONSTRUCTION OF THE CONTRACT.

32. The Original Contract. An original written contract is presumed to embody all the agreements made at, or previous to, the time of its signing. No oral evidence will be admitted to explain or supplement the terms expressed in the written contract, provided these are clear and plain. It is permissible, however, to modify the terms of any written contract by subsequent oral or written agreements. It is also permissible to submit evidence as to contemporaneous oral agreements which supplement or explain the terms imposed in the contract, provided these be not inconsistent with the terms of the written document.

Oral evidence is also admissible to explain the identity of the parties, or the existence of an agency, the identity of the subject-matter, and the sense in which certain unusual or technical words have been used.

Oral evidence is also admissible to explain any latent ambiguity in the instrument, as where more than one meaning may be given to a word or phrase; but in the case of a *patent* ambiguity, that is to say, an ambiguity apparent on the face of the instrument itself, and which is meaningless without oral explanation, such an ambiguity will make the contract void.

33. The Explanation of Technical Terms in Contracts. In all cases where either common or uncommon words are used in a technical sense, or in a sense peculiar to a given trade or business, in which custom has given to such expressions particular and definite meanings, oral testimony can be received for the purpose of explaining the real meaning of such terms. Furthermore, the meaning which the law will enforce is that which such a term has in that neighborhood, or with the parties to the contract. In such cases the common usage or custom will fix the meaning of the technical words used.

In other cases oral evidence may be introduced to explain the real meaning of a contract, where custom or usage caused the meaning to be clear to the parties themselves when the contract was signed, but which would not be understood by strangers to such usages.

In order that a contract may be interpreted in the light of custom or usage, such custom or usage must be certain, definite, and uniform in that district, or between the parties to the contract. Unless it is a universal custom or usage as between the parties, it can not be received as positive evidence of meaning. Furthermore such custom and usage must have been continuous and uninterrupted up to and including the time of the transaction in question. Thus one or more acts do not establish a custom as between the parties, and a few illustrative examples will not serve to establish a usage.

When the explanation rests upon usage in the neighborhood, such usage must be general and a knowledge of it must be common, so that it may have been presumed to have been known to the parties to the contract.

Such custom or usage must be reasonable, and must have been generally assented to, and complied with without protest, in order to become binding in explaining the terms of a contract.

Such custom or usage, also, must not be repugnant to any of the express terms of the contract itself, neither must it contravene a state statute, city ordinance, or conflict with the law of public policy.

34. Rules of Construction.

1. The first and principal rule to be followed in the construction of contracts is to ascertain the real intention of the parties at the time the contract was signed. In fact all rules are merged in this one, and have for their object the determination of the original real meaning of the document.

2. In arriving at this real meaning, the words used must be understood in their ordinary and popular meaning, when these do not have a technical significance, as indicated in the previous article. In all other cases, the language is supposed to mean what it would ordinarily be understood to mean under the given circumstances of time and place, and as between the given parties.

3. Furthermore the whole instrument must be looked to, and all the terms thereof made effective if possible. The whole instrument will be construed, also, in construing any latent ambiguity which may pertain to any given part. Where more than one document enters into a general agreement they shall all be taken into account in the construction of the entire

contract. Words may be wholly rejected which are inconsistent with the manifest intention of the parties.

If a portion of the contract is printed and other portions written, the latter will take precedence over the former, when they are found to conflict.

Where both general and specific terms have been used in describing the same thing, the agreement will be limited to the scope of the more specific terms, and may not be applied to the more general.

Doubtful words will be construed more strongly against the party who used them. This is based on the principle that a man is responsible for ambiguities in his own expressions. Thus a deed is construed most strongly against the grantor, and a clause in a promissory note will be construed most strongly against the maker. Such a prejudice, however, is never exercised against either party, if possible to avoid it.

CONTRACTS REQUIRED TO BE IN WRITING.

35. The Statute of Frauds. In the year 1676 the English Parliament passed "An act for the prevention of frauds and perjuries" which has become common law for this country where it is not replaced by statutes in the various states covering the same ground. In either case where reference is had to the original English enactment or to the corresponding statutes in the various states, this law is commonly referred to as the "Statute of Frauds." The object of such a law is to prevent litigation and fraud by requiring certain kinds of contracts to be in writing. These in general relate to the official acts of executors or administrators, marriage contracts, to the sale and transfer of real estate, to agreements which can not be performed inside of one year, and other contracts for the sale of goods, wares, and merchandise of a value greater than $50.

As only the last two of these pertain to the character of the present work they alone will be discussed here.

36. Agreements Which Can Not Be Performed within One Year.

The English statute provides that "no action shall be brought whereby to charge any person upon any agreement that is not to be performed within the space of one year from the making thereof, *unless* the agreement upon which such action shall be brought, or some memorandum or note thereof, shall be in writing and signed by the party to be charged therewith, or some other person thereunto by him lawfully authorized." That is to say, contracts which can not be performed within one year from the date thereof must be in writing. This is construed as meaning that the necessity for a written contract only holds when the performance within one year is demonstrably impossible. If by any possibility it may be fully performed within one year, an oral contract is valid.

Thus a contract for a year's service, to be entered upon at a future time, can not be performed within one year, and hence must be in writing. Or in the absence of a written contract to this effect an oral agreement can not be enforced, and either party is at liberty to annul the contract at pleasure. Where services have been rendered, however, under an oral contract which by this clause ought to have been in writing, the party benefited must pay for them.

Thus, also, a contract for the carrying out of any engineering construction, if it plainly can not be completed within one year, is not binding, unless it be in writing.

37. Contracts of Sale Where the Value Is More than $50.

Another clause in the same original English statute reads as follows: "No contract for the sale of any goods, wares or merchandise, for the price of ten pounds sterling or upwards, shall be allowed to be good; except the buyer shall accept part of the goods so sold, and actually receive the same, or give something in earnest to bind the bargain, **or** in part of payment. or that some note or memorandum in writing of said bargain be made, and signed by the parties to be charged by such contract or their agents thereunto lawfully authorized.''

Similar statutes have been enacted in this country in which the limit of the value of the goods sold is usually placed at $50, while in Maine and in New Jersey it is placed at $30. For all values over these amounts the contracts must be in writing or, as stated in the statutes, the buyer must give proof of his agreement by accepting and receiving a portion of the same or by part payment for the same.

It must be noted, however, that a contract for labor is not included in the statute. The limit of value here used applies not to individual articles but to the sum total of the articles named in the transaction.

SUBSEQUENT CHANGES AND AGREEMENTS.

38. The General Rule. In general any oral or written agreement may be altered at pleasure after it has been signed, when this is done by mutual consent. Alterations made at the time of, or previous to, the signing of the instrument become elements in the original contract.

Any change by mutual consent in the terms of an agreement after it has been signed makes a new contract out of the original agreement, and because of this a surety or a third party to the agreement not consenting to the change is released from all obligation. The new contract remains good as to those who consent to the change. In the case of engineering contracts where it is common to have sureties or bondsmen who guarantee faithful performance, such sureties must always be consulted and their consent obtained to any material change in the original contract which may be mutually agreed on by the principals. In default of such reference and consent on the part of the sureties, they become discharged from all liability.

Even though the written contract has a clause forbidding any oral alteration in it, and declaring that no change shall be made in it except in writing, such a provision is void, and the contract may be altered by oral agreement notwithstanding. This is because in law oral and written agreements are of the same class, both being simple or parole contracts, and hence are of equal force and effect. An agreement in writing, therefore, by the parties, to forfeit their legal rights, does not operate to change the law in this respect, and their rights can not be forfeited by such an agreement. One of these rights is the privilege of modifying the contract by oral agreement.

Where contracts are illegal except when they are in writing, as under the Statute of Frauds, it follows that such a written contract can not be modified by oral agreement, since this would circumvent the law as applied to such cases. This also applies to promissory notes and other commercial paper. Oral agreements in regard to them are invalid, as they would work injustice to innocent parties.

Written contracts executed under seal, not required by law to be so, may be modified or altered by either written or oral agreements, but when this is done the whole contract is reduced to the force and significance of a simple or parole agreement, and no longer remains a specialty.

Furthermore this can only be done in the case of a sealed contract, when the new agreement or alteration rests upon a new and separate consideration. Where the seal is required by law, alterations must also be made under seal. If, however, all the parties to the original agreement are together, and the instrument is changed by the principals who signed and sealed it, in the presence of all, and with the consent of all, the alterations are valid.

39. Results of Alterations of the Contract. This is one of the most important subjects connected with the execution of engineering work. Very seldom is an engineering or architectural project constructed strictly in accordance with the original plans and specifications. Usually the contract itself provides for changes in plans and specifications, and in general changes must be made in accordance with such provisions.*

While the court decisions are extremely various and frequently directly opposed to each other in their enforcements of contracts which have been changed more or less after they have been signed, it is thought the following is a fair interpretation of the intent of most of the authorities in passing on such cases.

(*a*) Changes in the contract will not operate to annul the original contract unless such was the plain intent of the parties, and so long as any portion of the original contract may fairly be construed as remaining in force.†

(*b*) In general, every change made in a contract after it has been signed, should be based on some kind of a legal consideration. Thus, if a change is made which involves an addition to the cost of the work, it should be accompanied and sustained by a corresponding increase in the compensation or price, or by a corresponding reduction in some other part of the work, or by a corresponding accommodation of some character in favor of the other party, which may be construed as a consideration for the change made. Otherwise the change agreed upon can not be enforced.

(*c*) Similarly, if the parties agree to a less performance than that required by the contract, unless there be also a corresponding reduction made in the price, or some other accommodation to the other party, which may be construed as a consid-

* One exception to this rule is given in the previous article, where the contract provides that changes shall be made only in writing.

† It is sometimes expressly stipulated in the original specifications that subsequent changes shall not operate to annul those portions of the contract with which these changes are not in conflict.

eration, the agreement can not be enforced against the party making the concession. He is at liberty to change his mind.

(*d*) In such changes as are mentioned in (*b*) and (*c*) the law will not inquire particularly as to the adequacy of the consideration, so long as a legal consideration may be shown to exist.

(*e*) An apparent exception to the above rules of construction is that in which the original contract provides for changes to be made in a specific manner, and without further consideration. Thus it is sometimes specified and agreed upon that the character of the materials or the methods described in the original contract may be changed at the pleasure of the engineer or architect, without further consideration. In this case any change made consistent with this provision would not be regarded as a new contract, but simply as a sort of construction of the old agreement. Under such a clause, however, the law would not allow a gross injustice to be worked against the contractor in the way of violent changes which would greatly increase the cost of the work, and which evidently were not anticipated by the parties to the contract at the time it was signed.

(*f*) In all cases where changes have been made in a contract, if such changes involve an increase in the time required for performance, the date of completion of the work fixed by the original contract will be extended by the courts for a period sufficient to cover the additional time required for the changes made. This the courts will do whether such extension of time be provided for, either in the original contract, or in the subsequent agreement.

(*g*) It is customary to include in the original specifications a clause describing the manner in which all changes in plans and specifications may be made, and the compensation for the same determined. In this case changes in the contract must be made in accordance with such provision, and such changes, when so made, are binding upon the contractor, whether he consent or not. They might operate, however, to release the bondsmen.

(*h*) All contracts, except those required by law to be in writing, whether sealed or unsealed, can be modified by oral as well as by written subsequent agreement, regardless of any provision to the contrary in the body of the original contract.

DISCHARGE OF CONTRACTS.

40. Methods of Discharge. Any contract entered into in any of the methods heretofore indicated may be discharged and the parties thereto freed from all obligations thereunder, in any one of the following ways: 1. By agreement. 2. By performance. 3. By impossibility of performance. 4. By operation of law. 5. By breach.

41. Discharge by Agreement. Any contract which has been entered into by mutual agreement may evidently by mutual agreement be dissolved. This may be done, (*a*) by a waiver or cancellation, (*b*) by a substituted agreement between the parties, or of the contract, (*c*) by a condition in the contract itself.

(*a*) An agreement to discharge the contract must be supported by a consideration the same as any other agreement. The usual consideration in this case is the mutual release from liabilities under the original contract.

(*b*) A contract may be discharged by the substitution therefor of a new agreement, the consideration in this case being as before the mutual discharge of obligation under the previous agreement. This new agreement may be either oral or written, and it will serve to replace or rescind the previous agreement if such were the intention of the parties. This is true whether the original agreement was a sealed contract or simply a parole agreement. If, however, the original contract was required by law to be in writing so must also the new contract which replaces it.

The rescission of the former contract may be implied, as where the terms of the latter agreement conflict with those of the old, the later agreement taking precedence and discharging the former. The intention to discharge the former, however, must be clearly implied from its being the only rational assumption in the premises.

The contract may be rescinded by the substitution of a new party to it in place of one of the original parties. This may be done only where all parties to the contract are agreed, this agreement being either express, or implied by subsequent acts.

(c) The contract may contain a provision for its own discharge on the happening of some event or contingency. This contingency may be the nonfulfillment of some specific clause in the contract itself, or on the occurrence of some particular event, or on the exercise by one of the parties of an option to determine it. When the event transpires which forms the condition of the discharge, the contract is thereupon rescinded.

Engineering contracts sometimes contain a clause to the effect that the work may be stopped at any time with a specified notice at the option of the party paying for the same.

42. Discharge by Performance. The usual method of discharging a contract is by each party fully performing the duties prescribed for him in the agreement. In this case the performance by each party must be strictly in accordance with the terms of the contract.

In engineering work it is seldom that the work is done in all details strictly in accordance with the plans and specifica-
4

tions. or with such plans as authoritatively modified by the engineer. While in law the contract requires a strict and full compliance with all the terms of the agreement, yet in equity a substantial compliance is accepted in place of a full and complete performance. Also in equity an imperfect compliance is often taken as a discharge of the contract subject to such damages as would equitably compensate for the degree of failure to fully and completely satisfy the agreement.

One of the essential requirements of the contract is the time specified for the completion of the work, when this is so named in the agreement. When no time limit is mentioned in the agreement, the element of time is not deemed to be of the essence of the contract, but performance will be required within a reasonable time. When a specific time or date is given for completion, a court of equity will examine as to whether the intent of the parties was to determine in a general way the time when performance was expected or whether such limit was intended to be a specific and essential part of the contract. If the former meaning is imposed no relief can be had in equity for nonperformance within the specified time.

43. Performance on Conditional Promises. In engineering contracts performance on the part of the owner is usually conditioned on a previous performance on the part of the contractor. On the other hand, the owner sometimes agrees to make payments, for instance, at specified stages of the work, in such a way that further performance on the part of the contractor may be conditioned on the making of such payments at the times specified. Performance may also be conditioned in various other ways, as after the lapse of a certain time, or upon the occurrence of a particular event or contingency which may be uncertain, or on the acts of a third party, or even on the will of the promisor. In this last case it is really no contract at all, so far as the promisor is concerned. That is to say, while he can enforce it against the other party, the other party can not enforce it against him. Such a case as this last is where one party agrees to do work to another's satisfaction. Here the party performing the work is wholly at the mercy of the party to be satisfied, and the plea of dissatisfaction relieves him from liability. Evidently no person should place himself thus at the mercy of another, unless he can rely implicitly upon the good faith of the other party.

A common instance of the operation of a "condition precedent" with reference to a third party is where a contractor binds himself to receive payments on a building or engineering work only on the certificate of the architect or engineer. Without such certificate which forms a "condition precedent" the owner is not obliged to make payment. Before the contractor can force the owner to pay him for his work, in the absence of such a certificate from the architect or engineer, he must be able to prove that the architect or engineer has acted fraudulently in withholding the certificate, or that he has acted under gross mistake, and in bad faith, or has negligently refused to honestly examine the work. As this is, of course, very difficult to establish, the refusal of the architect or engineer to give such certificate commonly acts as a bar to payment under the terms of the contract.

Also where the quality or quantity of the work to be done is, by the terms of the contract, to be left to the approval of a third person, such as the engineer, his decision in the premises is binding upon both the parties.

The agreement may be conditioned upon a notice being given to the promisor, as·where the engineer is required to give notice to the contractor to begin work at a certain time. In this case the proof of having given such notice is necessary to the enforcement of the contract.

44. Discharge by Payment. The discharge of a contract by full payment of money due upon it requires no further comment in this connection. This subject is further amplified in the works on the "Law of Contracts," but it is not necessary to elaborate it here.

45. Discharge by Tender. When the performance of a contract is frustrated or prevented by the act of the party to whom the performance is due the offering to perform is called a tender. As applied to engineering contracts, if the contractor is prevented from performance by the owner, the latter subjects himself to liability on a suit for damages sustained by the contractor by not being allowed to perform. In other words, the owner breaks the contract by his refusal and subjects himself to a suit for damages, the same as in any other case of breach of contract, while the contractor stands released from all further obligation under the contract, his tender being construed as performance so far as he is concerned.

46. Kinds of Impossibility Which Will Discharge a Contract. An agreement between parties to do what both know to be impossible is discharged when their knowledge of such impossibility is shown, but where the impossibility is known only to one of the parties, he is liable for damages to the party to whom it is unknown.

Where the subject-matter is nonexistent, or has ceased to exist, the impossibility of performance results from a mutual mistake of fact, and the contract is discharged.

Where performance is rendered impossible by what is called in law "an act of God or of the public enemy" the party so contracting is excused. By "an act of God" as used in law as discharging a contract is meant a manifestation of the powers of nature over which man has no control, such as fires caused by lightning (but not by accident or other cause), winds, floods, sickness and the like. In the performance of engineering contracts unusual difficulties will not be placed in this category, *so long as they are by any possibility under human control.*

While as stated above an "act of God or of the public enemy" making performance impossible, will discharge a contract, yet it must be clearly shown that such "act of God or of the public enemy" did in fact render the performance quite impossible, and not simply difficult or expensive. Thus if wind, flood, or lightning should destroy a partly completed engineering work, if it were possible to re-erect it within the time specified, the contractor would be held to full performance.

47. Kinds of So-called Impossibilities Which Will Not Discharge the Contract. "When a person contracts to do a given act he pledges himself as having the capacity to do it, and assumes the risk of being prevented from performing his contract by obstacles or accidents; against obstacles or accidents that may interfere with performance *he should protect himself by contract.* Having presumed generally to do a thing he can not allege that difficulties and obstacles prevented him from fulfilling his contract, although they did in fact render the doing of the thing *by him* impossible. He is bound to do whatever is within the scope of any human being to accomplish."

From the above which is quoted from Judge Amos Thayer, of the United States Court of Appeals, it is evident that if a contractor wishes to obtain release from full and complete performance for certain contingencies, as for instance, inability to obtain material, or. to place sub-contracts, or to get the subcontractors to comply with their agreement, or to provide against labor strikes, whether in the trades or on the railroads, or against the inclemencies of the weather which might make performance within the time difficult and very expensive, or against any other of the extraordinary contingencies which may arise to prevent performance except at great loss, he must evidently provide protection for himself in the body of the contract. ` In the absence of such a protection and under a simple agreement to perform certain work within a certain time, the law will hold him to a strict compliance, *so long as such compliance lies within the realm of human possibility, regardless of expense.*

48. Discharge of Contract by Operation of Law.

There are various methods by which a contract may be discharged through the operation of the law, as for instance, by merging one contract into another, by a fraudulent alteration of the written agreement, by the bankruptcy of one of the parties, or by death. In the case of the death of one of the parties, the contract is discharged only when this is made a condition in the contract, or when performance thus becomes impossible. It will become impossible when the performance is required to be of a personal character, as contracts for services, or such as require professional skill, marriage contracts, and the like.

49. Discharge of Contract by Breach.

While any material breach of the contract on the part of either of the parties furnishes a right of action to the injured party, it is only in exceptional cases that such a breach operates to discharge fully the other party from his obligations. The contract will be discharged as to the injured party by a breach by the other party:

(*a*) When one of the parties announces his positive renunciation of the contract, whether this be previous to a partial performance, or after a partial performance. In this case the injured party is entirely relieved from further obligation, or in other words, the contract is discharged. Suit may at once be entered for damages. When the renunciation is only partial, and does not affect a vital portion of the agreement, the contract remains in force, but a suit for damages will lie. The injured party is, however, not bound to treat a formal renunciation as a breach of contract, but may insist on performance until the specified time has elapsed.

(*b*)   By one of the parties making it impossible for him to perform his agreement.   When this impossibility of performance comes to the knowledge of the other party, he may at once consider the contract discharged, and may enter suit for damages.

(*c*)   By such a failure to perform in case of a "condition precedent" or failure which goes so to the root of the matter that a recovery of damages would not satisfy the agreement. When the performance of one of the parties is clearly made a "condition precedent" to performance on the part of the other, a failure to substantially perform on the part of the one operates to discharge the contract as to the other.

In engineering contracts a "condition precedent" to the final payment on the part of the owner is usually the certificate of performance to be given by the engineer or architect, he being a third party, and not one of the principals to the agreement.   In this case a failure to give such certificate does not operate to discharge the contract between the principals, but does excuse the owner from making a final payment unless it can be shown that the engineer or architect has failed to perform his duties in this respect.

In determining whether or not the failure to perform on the part of the contractor, for instance, is so vital as to operate to discharge the contract entirely as to the owner, and release him from all obligation to pay for the work done, we may distinguish between divisible and entire agreements.   A contract or agreement may be considered divisible, when a fulfillment in part is valuable to the other party so far as it goes, and when a failure as to a part does not operate to destroy the value of the partial performance.   Thus a contract to build two houses is a divisible contract, since the building of one would be a satisfactory performance as far as it goes, and a failure to build the second would not operate to destroy the value of the first; whereas a contract to build a house is an entire contract, since the building would not be serviceable until fully completed.

The degree of failure to perform, as in the case of agreements by contractors and builders, which will operate to discharge a contract on the part of the owner, must be determined by the court or by a jury. In general any substantial failure to perform an indivisible contract will operate to discharge the contract. If, however, the work done or goods delivered are accepted and used, the law will create a new and implied contract on the part of the recipient and beneficiary, by which the party supplying the service or goods can recover a fair price for the same. Such recovery, however, not being under the contract or in accordance with its terms.

A failure in minor details does not, as a rule, discharge the contract, but simply furnishes to the other party the right to obtain damages to the extent of the failure. It is very important to note, however, that such failures which give to the injured party only the right to recover damages must be in their nature insignificant, and of small relative importance, not in any sense going to the root of the matter or affecting the value of the parts which have been satisfactorily performed. The law is very severe in enforcing agreements literally and fully, especially where departures have been made intentionally and perhaps against the protest of the other party. In such cases even small failures to comply may be considered as a discharge of the contract. Where the contractor has evidently acted in good faith, much larger failures to perform may be remedied by a payment of damages instead of operating to discharge the entire contract.

When a failure to perform pertains to work which must either be accepted and used, or removed at great expense, as where a structure is built upon the owner's land, if such structure fails materially to comply with the terms of the contract, such failure to perform will operate to discharge the contract without compensation to the contractor, even though the owner does accept the work and use it. In this case the owner is not at liberty to refuse to accept, since this would involve him in great additional expense and delay. It is, of course, very different in the case of all kinds of personal or movable property. Here a refusal to accept does not involve the owner in any additional cost.

When a contractor has shown indifference, dishonesty, or incompetency in the execution of his contract, resulting in a material failure to perform, and this work is the building of a structure upon land which becomes part of the real estate, the owner may not only accept and use the structure without compensation to the builder, but, in extreme cases, he may even decline to allow such builder to reconstruct the work, even though he should offer to do so, since the owner thereby has no assurance that a second attempt will result any better than the first.

While the law gives to the owner such remedies as those stated above, he must be careful not to act in such a manner as to imply that he has waived his legal rights in the premises. Thus where a contract is to be performed within a given time, and the time elapses before complete performance, if the owner urges him, or requests him to go on and complete the work, he thereby waives his legal remedies for noncompletion within the time, so far as a *discharge* of the contract is concerned. He may, however, recover damages for the delay.

REMEDIES FOR BREACH OF CONTRACT.

50. Results of a Breach of Contract. When a contract has been broken, or not fully performed, the failure to perform may result either (*a*) in the discharge of the contract as described in the previous article, or (*b*) in a right of action by the injured party for damages sustained, or (*c*) a right of action to enforce specific performance.

The two remedies by which one either obtains damages or enforces specific performance will be discussed in the two following articles.

51. Damages for Nonperformance. *The foundation principle of damages is compensation.* Where there has been a partial or complete failure to perform, in accordance with the agreement, the law undertakes to require the party so failing to pay to the injured party such a sum as will cover the actual loss in money value which he has sustained on account of the breach. When the promise was the payment of a certain sum of money, nothing more than this sum with interest can be recovered. Where no decided loss in money value can be shown, the injured party can recover only a nominal sum. That is to say, "a sum of money such as may be spoken of but has no existence in point of quantity."

The leading case in determining the amount of damages which can be collected in the United States courts is that of Hadley v. Baxendale. In this case the court laid down the following rules, which have been followed in all the United States courts.

"Where two parties have made a contract which one of them has broken, the damages which the other party ought to receive in respect of such breach of contract should be:

(1) *Such as may fairly and reasonably be considered as arising naturally, i. e., according to the usual course of things, from such breach of contract itself.*

(2) *Such as may reasonably be supposed to have been in the contemplation of both parties at the time they made the contract, as the probable result of the breach of it.*

(3) *Such as arose out of the special circumstances under which the contract was made, where such circumstances were communicated by the plaintiff to the defendant.*

(4) *But, if these special circumstances were wholly unknown to the party breaking the contract, he, at the most, can only be supposed to have had in his contemplation the amount of injury which would arise generally, not affected by any special circumstances.*

It must be remembered that "damages in an action for breach of contract are always by way of compensation, and not a punishment, hence the plaintiff can never recover more than such pecuniary loss as he has sustained, nor can he recover for great disappointment, nor injury to the feelings, or vexation of mind, caused by the breach."*

The party who is injured by a breach of contract is required to make reasonable exertions to render the injury as light as possible, and if he carelessly or indifferently allows the damage to be unreasonably large, such increase falls upon himself.

52. Distinction between Liquidated Damages and Penalties.† "The parties to a contract not infrequently assess the damages at which they rate a breach of the contract by one or both of them, and introduce their estimate into the terms of the contract. This is perfectly legal, and on a breach the sum agreed upon becomes the measure of damages; as, for example, a stipulation in a building contract that if the building is not completed by a certain day the contractor will pay a certain fixed sum for each day or week or month he is in default, or an agreement in a contract of sale that a certain

---

*Breach of promise of marriage is an exception to this rule.
† This article is quoted from Lawson on Contracts.

sum shall be deducted from the purchase price if the quantity is not delivered as agreed. These are called '*liquidated damages.*'

"But the parties in affixing a fixed sum for the nonperformance of his promise by one, or each of them, may have intended not to assess the damages at which they rate the nonperformance of the promise, but to secure the performance by the imposition of a penalty in excess of the actual loss likely to be sustained. And in this case, the amount recoverable is limited to the loss actually sustained, regardless of the sum undertaken to be paid by the defaulter. These are called '*penalties.*'

"The courts will always construe the contract in harmony with the intention of the parties, and without regard to the terms used. If the general effect of the agreement shows that they intended to provide for a penalty they will restrict the recovery to the actual damages incurred although the words 'liquidated damages' are used in the instrument. So, where the parties have used the milder term 'penalty,' courts have sometimes held that the stipulated sum was, from the nature of the case, to be considered as liquidated damages and recoverable in full. Whether the sum mentioned in an agreement to be paid for a breach is to be treated as a penalty, or as liquidated and ascertained damages, is a question of law, to be decided by the judge, upon a consideration of the whole instrument. Where it is plain that the parties meant the sum fixed to be liquidated damages, the courts will not interfere to frustrate that intention, but, if it be doubtful, upon the whole agreement, whether the sum named was intended to be a penalty or liquidated damages, it will be construed to be a penalty, it being the tendency of the courts to consider the contract as creating a penalty to cover the damages actually sustained by a breach, rather than liquidated damages.

"Subject to the principles stated in the last section the courts have adopted certain rules of construction, in the case of contracts containing promises of this kind; which are—

"1. If the contract is for a matter of certain value and a sum is fixed to be paid on breach of it which is in excess of that value, then the sum fixed is a penalty and not liquidated damages.

"2. If the contract is for a matter of uncertain value and a sum is fixed to be paid on breach of it, the sum is recoverable as liquidated damages. There is 'nothing illegal or unreasonable in the parties, by their mutual agreement, settling the amount of damages, uncertain in their nature, at any sum upon which they may agree.'

"3. Where the contract involves several distinct matters of various kinds, and one fixed sum is stipulated to be paid for any breach, of whatever kind, it is a penalty and not liquidated damages."

53. Recovery for Imperfect or Incompleted Work.

As stated in Art. 49, recovery can be had under a contract for partial performance, when the contract may be considered as divisible or severable. That is, where a part of the agreement may be entirely fulfilled, while other portions remain unfulfilled. In this case, however, while the party in fault may recover damages for the work done, or goods delivered under the divisible contract, he is always liable for such damages as can be shown to have resulted from his failure to completely perform his agreement.

When the contract can not be considered divisible, but must be looked upon as one and entire, recovery can not be had for anything short of a substantially complete performance. That is to say, a substantially incomplete performance discharges the contract entirely, as stated in Art. 49, and even when the performance is sufficiently complete to prevent discharging the contract, so that recovery can be had for the work done, the owner may still enter a claim for damages for each and every particular in which the performance has been incomplete. Here again if the default is shown to be a willful neglect or refusal to comply, the law is construed much more severely than for mere oversights.

SPECIFIC PERFORMANCE.

54. General Rule as to Specific Performance. Suits to enforce specific performance can not always be maintained, for actions can be brought at law for such damages as may be shown to have resulted from a breach of the contract, or from a total failure to perform. It has been customary, however, to allow specific performance to be enforced in certain cases where suit is entered in a court of equity, but even in equity specific performance will not be enforced where a payment of damages will put the plaintiff in as good a position as if the agreement had been actually performed. Also if an action for damages would not lie, neither would an action for specific performance. In a case in equity, however, many considerations will be taken account of, in the way of meting out justice to the parties, which could not be considered in a case at law upon the terms of the contract itself.

Where specific performance is ordered by a court of equity, the same court will, if necessary, enforce its decree either by a mandate enforcing the performance named or by an injunction to prevent the doing of the contrary.

Since the parties to an engineering agreement can, as a rule, be fully compensated for a failure to perform on the part of either, by a recovery of damages, specific performance can not ordinarily be enforced, and hence this subject will not be further discussed in this connection.

DISCHARGE OF RIGHT OF ACTION UNDER A CONTRACT.

55. The Right of Action. Upon any breach of a contract there arises in favor of the injured party a legal right of action for compensation. "This right of action can then not be discharged by any payment or performance, or tender of payment or performance, by the promisor, without the consent and acceptance of the promisee; for the promisee, after breach, becomes entitled to the compensation or remedy provided by process of law, and is not bound to accept any tender or offer made in satisfaction of his legal rights." This right of action can only be discharged in one of the following four ways: (*a*) By a Release; (*b*) By an Accord and Satisfaction; (*c*) By a Judgment; (*d*) By Lapse of Time.

(*a*) A Release of a legal right of action consists in a voluntary agreement to discharge a claim, and is only valid when supported by a consideration or when executed under seal. Otherwise it is a mere unsupported promise which binds no one. But a voluntary delivery to the debtor of the evidence of a debt, as of a note or bond, or the destroying of the same, with the intention of discharging the debt, does operate as a release. A release of one of several debtors, jointly, or jointly and severally, liable for the same debt, releases all.

(*b*) Release by Accord and Satisfaction consists in an agreement on the part of the creditor to accept something in satisfaction of his claim, accompanied by the delivery or performance of what has been agreed upon. Here the execution of the agreement is the satisfaction referred to in the phrase "accord and satisfaction," the agreement to accept this being the accord. It should be noted that the right of action is not discharged until this agreement or "accord" is fully executed when "satisfaction" has been rendered.

(*c*) Release by a Judgment. Evidently a judgment obtained through a suit at law in favor of the plaintiff discharges all further right of action against the defendant in the case so adjudicated. His former right is now merged in what is called a contract of record, and this is discharged by the payment of the judgment, or by such satisfaction as can be obtained by process of execution. An adverse judgment against the plaintiff does not discharge the obligation or right of action unless this adverse judgment was rendered on the merits of the case. Of course any judgment may be set aside by the court in which it is rendered, or set aside by a higher court, in which case judgment may be entered in favor of the other party if so ordered, or the parties may be remitted to their original positions.

(*d*) The discharge of right of action through Lapse of Time is in virtue of certain statutory limitations providing that after the lapse of a certain period of time, which is different for different kinds of contracts, the right of action under the contract ceases to exist, and is said to have been discharged by lapse of time. Even in the absence of any statutory provision the courts will not allow a case to be opened on a contract which has long stood as a dead letter. In the common law the period of time which bars the right of action is commonly twenty years. This will apply even to sealed instruments, and for parole agreements this time will be shortened and discharged by payment presumed for shorter periods.

It must not be understood, however, that the courts will allow either party to an agreement to benefit through lapse of time from a fraudulent contract, although the lapse of an unreasonable time before suit is entered by the defrauded party will have the effect of affirming the contract. In other words the law reasonably requires that in case of either fraud or breach of contract a prompt recourse to the courts shall be had.

56. Removal of Statutory Bar to Right of Action.

While statutes of limitation are a bar to a right of action or recovery in the courts, they do not act to extinguish the claim, and hence notwithstanding the time in which suit may be entered has elapsed, the right of action may be revived by (*a*) a promise to pay the debt; (*b*) a subsequent acknowledgment of the debt; or (*c*) a part payment of the debt. In other words, any acknowledgment on the part of the debtor of the existence and legitimacy of the claim, after the right of action has been barred by the statute of limitations, serves to revive the claim for another like period. This acknowledgment of the existence of the debt, in order to serve to revive its legal status is not merely a recognition of the fact of the debt, but must consist in an agreement to pay the debt.

After such a removal of the bar to the right of action, suit may be entered upon the original contract by showing that the claim has been revived by the free act of the debtor. In other words, the debtor has here waived his legal rights of defense, and such a waiving of his rights does not require a consideration to support it, as was shown in Art. 27,

5

PART II

General Clauses in Engineering Specifications and Accompanying Documents.

57. General Considerations. Nearly all the works designed by engineers and architects are executed by other parties called contractors. The contractor usually buys all the materials and furnishes all the labor required in the execution of the work, as designed, and he agrees to do this within a stated time and for a fixed sum. To insure his doing this satisfactorily certain written documents are prepared and signed by both parties, that is to say, by the man, company, or corporation having the work done and who is to pay for the same, and by the contractor, or the man, company, or corporation who does the work and furnishes the materials.

Standing between these two parties to an agreement is the engineer or architect who has planned the work and who usually superintends its execution and assists in the final settlement between the parties to the agreement. Although paid by the party having the work done he occupies a judicial and not a partisan position and he is expected to act justly and fairly towards both parties.

In order that there shall be no misunderstanding in regard to the intentions of the designer, plans are usually drawn showing the general and detail features of the work, and accompanying these there is a written description of the work, of the materials to be used, of the time and manner of the payments, etc. This document is called the specifications. The drawings and this description are then referred to as the *plans and specifications*.

In order to get open and general competition in doing the work a date is set on which *bids* will be received, and blank *forms of proposals* are prepared by the engineer which can be filled out by the *bidders*, and *notices* or *advertisements* are inserted in the papers and in the engineering journals calling the attention of contractors to this public letting. These and other accompanying documents will be discussed in the order of their sequence in actual practice.

ADVERTISEMENTS.

58. The advertisement should be as short as possible to contain the necessary information, in order to save expense. It should usually contain the requisite information on the following subjects:

(1) A title indicating the kind of work to be done.

(2) Place, date, and hour of opening the bids.

(3) Person, company, or corporation letting the work.

(4) An adequate description of the work, with especial reference to the kind and quantity (or cost) of work to be done.

(5) Conditions of payment, if these are peculiar.

(6) Instructions as to where to obtain plans, specifications and blank forms of proposals.

(7) Statement as to amount of cash or of certified check or of bond to accompany the bid.

(8) A reservation of the right to reject any or all bids.

(9) Any other peculiar feature, as the letting of the work in parts or as a whole; bids to be received only from experienced contractors, etc.

59. The Theory of Advertisements. The object of the advertisement being to secure as large a competition as possible from responsible bidders, it follows that the information conveyed in it should be such as not only to attract the attention of such parties, but such as would enable them to decide whether or not it would be worth their while to submit a bid. A prominent title indicating the general character of the work would serve to attract the attention of contractors engaged in that line of work. It is a common practice to omit this title, with the result that one is obliged to read nearly the entire advertisement, which is usually printed in small type, before he can learn what the nature of the work is. It is usual also to announce that the proposals or bids which are to be submitted shall be sealed, with the implied understanding that these seals are not to be broken until the bids are opened at the place, date, and hour named. This is for the purpose of preventing collusion and fraud. In other words the bids are to remain secret and unknown except to the bidders themselves until the hour arrives for opening them. It is also customary to state that these bids or proposals shall be opened in the presence of the bidders, in other words at an open meeting of the board, or committee, or corporation, or council. To this meeting all persons are free to come and see the bids opened, and to hear them publicly read, with the privilege of taking down the prices named if they choose.

The description of the work included in the advertisement should be sufficient to enable the contractor to decide whether or not it was of such a character as he would be willing to undertake, and also sufficient to enable him to determine the amount of work to be done, and the time required to perform it, as well as the probable approximate cost of the same, and the amount of capital required to successfully prosecute it. The advertisement should also indicate whether or not the work would be let in parts or only as a whole. If it may be let in parts, the advertisement should indicate what the lines of division are, so that one might know what parts he was at liberty to bid upon.   It is customary to pay for contract work on monthly estimates of the engineer, reserving from each month's estimate of the worth of materials furnished and work done twenty or twenty-five per cent. until final completion.   This enables the contractor to carry out the work without having the requisite capital to complete the work with his own means.   If the conditions of payment are to be other than this, thus making them unusual and peculiar, such conditions should be stated in the advertisement.

Having drawn the attention of contractors to the work and given them the necessary information to enable them to decide whether or not they would wish to submit bids upon the same, it remains to give them such information as may be necessary to enable them to procure promptly the necessary plans and specifications, the blank forms of proposals, and information as to the amount and kind of guarantee which they must submit with their bids to have them considered.

60. The Guarantee. The object of the guarantee is always to insure that the successful bidder, or the party who is given the contract, will sign the contract for doing the work and furnish the requisite bond for faithful performance.   In other words this guarantee is simply an assurance of his good faith and honest intentions in submitting his bid, and it is customary to make it consist of cash or the equivalent of cash in the form of a bank check duly certified by the bank as being receivable for the amount stated.   This check is to be made payable to the party letting the work, or his agent, and is to be forfeited to such party in case the bidder fails or refuses to enter into a contract for the performance of the work after the award has been made to him.   The deposits made by the unsuccessful bidders are, of course, immediately returned to them, and that of the successful bidder is held until he has

entered into a contract as above stated, after which it is also returned to the owner. Sometimes it is considered a hardship for the bidders to have to make this cash deposit in submitting their bids, in which case the bidders are asked to furnish a bond or guarantee signed by parties known to be responsible, binding themselves in a stated sum, which sum they agree to pay if the bidder named therein fails to enter into a contract for the faithful performance of the work.* Some such guarantee as this should always accompany every bid received in open competition. While this might not be necessary for men of known business integrity, yet in an open competition bids will be received from strangers, and without this kind of an assurance of honest intention the successful bidder will often refuse to enter into a contract on the basis of a bid. In this case the handing in of a bid would involve no financial responsibility, and hence bidders might carelessly submit bids without having taken due precautions to determine the cost of the work, and hence might have made a price altogether too low and one which would involve serious losses on their part if they would undertake to carry out the work for the sum named. In case such a party should receive the award and then after more careful investigation learn that the work could not be performed for the price named in the bid he would decline to enter into a contract and the letting would have to be made over again. This would necessitate readvertising the work, and a considerable delay, in addition to some cost. It is desirable therefore always to require a certain guarantee of good faith which shall accompany the bid itself, and which shall involve considerable loss to the bidder if he declines to enter into a contract in case the work is awarded to him.

61. Right of Rejection. It is well always in the advertisement to reserve the right to reject any or all bids, for

---

*Or the agreement may be to pay the difference between the price named and the contract price for which the work may finally be let.

if this is not done the fair inference is that the contract will be let to the lowest bidder. In some instances, when the work is done under state or city auspices, the law may require that the contract shall be let to the lowest bidder if let at all. In this case the advertisement should state that "the right is reserved to reject all bids," since if parties should not choose to let the work to the lowest bidder their only recourse would be to reject all the bids and advertise the work again. If the parties letting the work are not bound by this legal requirement, and if they have reserved the right "to reject any or all bids" in the advertisement, then they are at liberty to let the work to any of the bidders without subjecting themselves to a charge of unfairness. It must be admitted, however, that if the work is not let to the lowest bidder, the parties letting the work subject themselves to invidious criticism, and they should have very good and satisfactory reasons which they are willing to produce in defense of their action, in order to clear themselves from blame before the various parties interested in the letting of the work.

62. Illustrative Examples. The following advertisements have been selected from the current journals as fairly embodying in suitable form the requirements as above stated. The student should note the terse and condensed style of these advertisements in which the greatest possible amount of valuable and required information is given clearly but in the least possible space:

CELINA, OHIO, WATER-WORKS—NOTICE TO CONTRACTORS.—Sealed proposals will be received by the trustees of the water-works of the village of Celina, Ohio, up to 8 o'clock P. M. of the 10th day of April, 1895, for furnishing the materials and constructing a system of water-works for said village.

There will be required about 773 tons of cast iron pipe; about 18 tons of special castings; 101 fire hydrants; 76 valves and boxes; brick pumping station and chimney; 2 pumps of a combined capacity of two million gallons per day; 2 boilers; a steel stand-pipe 16 feet in diameter and 125 feet high, etc.

Bids will be received for furnishing any of the materials above or for constructing the works complete. Proposals must be addressed to the Secretary of the Water-Works Trustees, Celina, Ohio, and must contain a certified check or its equivalent, made payable to said secretary in an amount equal to two (2) per cent. of the amount of the bid.

Plans may be seen and specifications and blank form of proposal procured at the office of the trustees, Celina, Ohio, or at the office of the engineers, —————, Buffalo, N. Y. The right is reserved to reject any and all bids.

                                        ● ———, Pres.,
                                        ———, Sec.,
                                        ———,
Trustees of the Water-Works, Celina, Ohio.
—————, Buffalo, N. Y., Engineers.

PROPOSALS FOR IRON LATHING AND AREA GRATINGS.— Office of Building of Library Congress, 145 East Capitol street, Washington, D. C., November 12, 1894.—Separate sealed proposals for furnishing, delivering, and putting in place complete the iron furring and lathing required for the ceilings, partitions, etc., in the first, second, and attic stories, and for the iron gratings and tile lights required for the areas of the Building for Library of Congress in this city, will be received at this office until 2 o'clock P. M. on Tuesday, the 27th day of November, 1894, and opened immediately thereafter in presence of bidders. Specifications, general instructions and conditions, and blank forms of proposal may be obtained on application to this office. —————————,

                                        Superintendent and Engineer.

NOTICE TO SEWER CONTRACTORS.—Sealed proposals for building about four (4) miles of pipe sewers in sections 7 and 8 of the Medford sewerage system will be received by the commissioners of sewers at their office until 4:45 P. M., Saturday, March 30, 1895. All proposals must be on forms furnished by the city and accompanied by a check of five hundred (500) dollars drawn on some national bank, and made payable to the treasurer of the city of Medford. Some approximate quantities are as follows: 20,477 lin. ft. of pipe sewer; 18,081 cu. yds. of earth excavations of all depths; 67 manholes aggregating 578.2 vert. ft. Bricks, pipe, cement, and iron work will be furnished by the city. Plans may be seen, specifications and forms of contract and proposals may be obtained at the office of the commissioners. Each bidder is required to make a statement indicating what sewer work he has done, and to give reference that will enable the board to judge of his business

standing; and no bid will be received in case the bidder has not looked the work over on the ground. The commissioners reserve the right to reject any or all bids, if they deem it to the interest of the city so to do. —————,
Chairman Commissioners of Sewers.
—————, Engineer, Medford, Mass., March 18, 1895.

The following advertisements are given as examples of extreme brevity, but since they appeal to a particular class of contractors, accustomed to do such work, they perhaps convey all the information really necessary to give in the advertisement:

To BUILDERS.—Office of the Light-House Engineer, Eighth District, New Orleans, La., March 20, 1895. Proposals will be received at this office until 2 o'clock P. M., Wednesday, the 1st day of May, 1895, for furnishing the materials and labor of all kinds necessary for the construction, erection, and delivery of the buildings for the Brazos River Light Station, Texas. Plans, specifications, forms of proposal, and other information may be obtained on application to this office. The right is reserved to reject any or all bids, and to waive any defects. —— ———, Major, Corps of Engineers, U. S. A., Light-House Engineer.

OFFICE OF ENGINEER, Ninth and Eleventh Lighthouse Districts, Detroit, Mich., March 25, 1895.—Sealed proposals will be received at this office until 3 o'clock P. M. of Monday, the 15th day of April, 1895, for furnishing seven skeleton iron towers for Hay Lake Channel, St. Mary's River, Mich. Plans, specifications, forms of proposals, and other information may be obtained on application to the undersigned. The right is reserved to reject any or all bids, and to waive any defects. —— ———, Major Corps of Engineers, U. S. A., Lighthouse Engineer.

OFFICE OF THE COMMISSIONERS, D. C., Washington, D. C., March 28, 1895.—Sealed proposals will be received at this office until 11 o'clock A. M., April 5, 1895, for grading and regulating streets and roads. Blank forms of proposals, specifications and all necessary information may be obtained at this office. —— ———, —— ———, —— ———, Commissioners, D. C.

U. S. ENGINEER OFFICE, 537 Congress street, Portland, Me., March 4, 1895.—Sealed proposals for dredging in Harrisseckit river, Me., and Bellamy river, N. H., will be received here until 3 P. M., Monday, April 15, 1895, and then publicly opened. All information furnished on application. ——— ———, Major Engineers.

PROPOSALS FOR CONSTRUCTION of dams and shore protec-tions on Upper Mississippi river, between Muscatine, Iowa, and New Boston, Ill. U. S. Engineer Office, Rock Island, Ill., March 16, 1895. Sealed proposals will be received here until 2 P. M., April 15, 1895, and then publicly opened. All information furnished on application.

U. S. ENGINEER OFFICE, Boston, Mass., Feb. 25, 1895. Sealed proposals for dredging in "The Narrows," Boston Harbor, Mass., will be received here until noon, April 2, 1895, and then publicly opened. All information furnished on appli-cation. ——— ———, Lt. Col. Eng'rs.

INSTRUCTIONS TO BIDDERS.

63. Preliminary Information. A description of many of the general conditions of the work and of the manner of letting it may well be grouped together and printed in con-nection with the blank forms of proposals. This information is usually placed under the title of "Instructions to Bidders." A fair sample of such preliminary information is given below. All of this information might be, and often is, embodied in the specifications themselves, but they are here separated for greater clearness in the analysis of the various documents involved in the letting of an engineering contract:

INSTRUCTIONS TO BIDDERS.

FOR A WATER-WORKS SYSTEM AT THE U. S. MILITARY POST AT
FORT RILEY, KAN.

1. No bids will be received for any part of the work herein described from parties who can not show a reasonable acquaintance with, and preparation for, the proper performance of the class of work for which the bid is submitted. Evidence of such competency must be furnished if desired.

2. Proposals must be made on the blank forms to be obtained at this office.

3. Bids will be received as follows:

First. On wells and connections complete to the wall of the pump pit. Bidders will state methods which they propose to use in sinking wells.

Second. On boiler, coal and dwelling house, pump pit and reservoir with roof complete.

Third. On all machinery including boilers, furnaces, stack, concrete floor in boiler and coal house, pump, connections, suction and discharge pipes to the outside of pump pit wall, benches, tools, etc. Bidders must state what kind of pump they propose to furnish.

Fourth. On the pipe system, complete with hydrants and valves, and to include the following items:

(*a*) Price per foot for eight (8) inch mains.

(*b*) Price per foot for six (6) inch mains.

(*c*) Price per foot for four (4) inch mains.

These items are introduced to cover any slight variations in lengths over or under the amounts herein specified, and contractors hereby agree to such extension or reduction at the prices named.

Bidders may make in addition to the above a bid for the entire work complete.

4. Each proposal must be accompanied by a written guaranty in the sum of $2,000 (executed strictly in accordance with the printed instructions, and upon the blank forms furnished under this circular), signed by two responsible persons, to the effect that if the proposal is accepted within sixty days from the date of the opening of the proposals, the bidder will, within ten days after being notified of such acceptance, enter into a contract and give bond with good and sufficient sureties, and that in case of failure of the bidder to enter into contract and give bond, they will pay the difference between the amount of his bid and the amount for which contract may be made with another party.*

5. The amount of the penalty of the bond to be furnished by the contractor will not be less than one tenth nor more than the full sum of the consideration of the contract.

---

*In place of this it is more common to require the bid to be accompanied by a certified check (or cash) for a specified sum, to be forfeited in case the bidder fails to enter into contract if the work is awarded to him.

FORMS OF PROPOSALS.

64. The Object of Blank Forms of Proposals. In order to insure that all the bidders shall submit their proposals on exactly the same items and estimate prices in the same units, it is necessary to prepare printed blank forms to be used by all the bidders, these forms being complete in all respects except the prices and the names of the bidders. So important is it to have the bids exactly comparable in all respects that it is customary to reject all bids not made out on these printed forms as well as all bids which, though made on the printed forms, have changed the conditions of the same in any particular, either by erasures, interlineations, or additional conditions. If the bidder desires to submit a proposition in a different way or with other conditions than those stated in the printed form he should submit his bid on the printed form without correction or change and then append to his bid an auxiliary paper embodying such changes as he would wish to make, and the price he would submit if these changes were agreed to. In this way he complies strictly with the requirements by submitting a bid which is regular in every respect, and in addition submits what is practically another bid on a modified basis. While the modified bid is, of course, irregular, and would not be considered in conjunction with the regular bids, it would give to the parties letting the contract the information which he desires them to have, and states the modifications which he would agree to if the bid were let to him on the basis of his formal and regular proposal. The work might then be let to him on the basis of his valid proposal, with the expectation of making the terms in the final contract in accordance with the bidder's amended proposition. If the parties letting the contract, however, should not choose to do this, the bidder would still be bound by his formal or regular proposal. The importance of making the bids strictly comparable in every respect is so very essential to fair and intelligent treatment of the bidders themselves, and so necessary in order to determine which is really the lowest bid, that the practice of preparing and supplying such blank forms of proposals should always be followed.

65. Manner of Letting the Work. Before such forms can be prepared, however, several questions must be decided, among which are the following:

1. Shall the work be let as a whole, or shall it be let in parts.

2. Whether let as a whole or in parts, shall bids be received for fixed sums for the whole or for the several parts, or shall they be received on a basis of certain suitable units of measurement. As, for instance, per cubic yard, as for earth work, per perch for masonry, per pound for iron work, per square yard for street paving, per mile for railroad rails and ties, or per lineal foot for water pipe or sewers, etc.

3. Shall the work be let in such a way as to involve the payment of a bonus or additional sum for the performance above that required, and a corresponding reduction in price for a failure to meet the requirements.

4. Shall the work be let for a certain price for the original construction, and a certain price per annum for maintenance for a given period.

5. Shall the contractor be required to furnish all materials and perform all the labor, or shall the principal purchase a portion or all of the material and turn it over to the contractor for use in the construction of the work.

66. Contract Let as a Whole or in Parts. Some of the considerations in favor of letting work as a whole rather than in parts are:

(*a*) By this means one man or company alone is responsible for the faithful performance of the work both as to quality and as to time. This prevents a division of responsibility which is always bad, and in the case of carrying out contract work is often the cause of failure to have the work completed within the time specified, without being able to locate the responsibility for such delay. Where there are several contractors upon the same piece of work, each may so stand in the way of another that the work may be greatly delayed, and yet each one of the several contractors may have a reasonable defense which would shield him from personal liability.

(*b*) When there is but a single contractor the business is concentrated so that the work of the engineer or of the inspectors is greatly lessened from having to deal with one man instead of many separate contractors.

(*c*) When several contractors are engaged upon the same work it is difficult for them so to plan their parts, in time, as to avoid a certain amount of delay where the work of one is dependent upon antecedent work of another. When the work is done by a single contractor he can arrange to avoid such delays as are almost necessarily incident to the working of several contractors in sequence.

(*d*) When the work is such as is commonly let in a single contract, or in other words, when bids can be received from parties who have been accustomed to carry out all parts of such a work, it is usually more economical to let the work in a single contract than it is to let it in parts. In the former case there is but one man to reap a profit from the construction, whereas if let in parts, each contractor must, of course, make his estimate in such a way as to allow himself a reasonable profit.

Some of the arguments in favor of letting the work in parts are:

(*a*) The project may involve constructions of such different kinds as to make it impracticable for one contractor to undertake the entire work. In this case the letting in parts is necessary to a skillful performance.

(*b*)   Where there are local parties who are competent to execute portions of the work, but not the whole, and who are anxious to bid upon such portion, it may be wise to let the work in parts provided it is reasonably certain that competitive bids can be received on all the parts.   Even in this case it is desirable also to receive bids upon the whole work, so that when the bids are opened it will appear which is the more economical method of letting.   Even when it is reasonably certain in advance that the contract will be let as a whole it is often wise to receive bids on the parts in order to satisfy local demands, and to avoid invidious criticism and public detraction.   This is especially true in the case of public works, if the local bidders who wish to submit proposals on parts of the work, but who would be incompetent to bid upon the whole, are shut out by receiving bids only upon the entire project.

67.   Contract Let for a Fixed Sum or per Specified Units.   When the work to be performed under a contract is perfectly definite as to quantity, it is best to let the contract for a fixed sum.   When either the quantity of work to be done or the quality or kind of material to be encountered, as in excavations, is more or less unknown and indeterminate, it is necessary to let such parts of the work at least, in terms of some suitable unit of measurement.   Thus in the case of excavation, the kind of material which will be encountered is always more or less uncertain, and the quantities to be moved are usually undetermined in advance.   In various other lines of work, also, the exact quantities are not measured or computed in advance of the construction, so that in all such cases it is necessary to let the work per unit of measure.   It is often wise, however, to assume a certain definite amount of work of each kind to be performed, and let the contract for a fixed sum on the basis of this assumption, providing for variations from these amounts in the blank form of proposal by requiring the bidder to state not only a fixed sum for the assumed total, but also a price per unit of measure, in accordance with which the quantities assumed as the basis of the bid may be either increased or diminished, it being understood, however, that the quantity stated is approximately the amount of work to be performed. In this way it becomes known in advance about what the work is to cost, and if the quantities are changed somewhat these changes do not become a source of contention between the parties.

In choosing the units of measure which shall serve as the items to which prices are to be affixed by the bidders, it is necessary to select and describe these units in such a way that they can not be misunderstood; thus in masonry it is better to use the cubic yard as a unit rather than the perch, since this latter has different values in different localities. Also it should be clearly defined in the proposal itself in what way these measurements should be taken, as, for instance, in masonry, whether all openings should be excluded, and in tunnel excavation that the measurement should include only the material excavated inside the given sectional boundaries, and in water pipe on which bids are received per foot in length for the various sizes, that the measurements should be taken on the center lines of such pipes, after they are laid, etc. Also in the case of the furnishing of materials, machinery, and appliances, the printed proposal should indicate where the material is to be delivered and whether or not the machinery is to be erected. The failure to make the proposal clear in these and other minor particulars is often the cause of serious disagreements, provoking delays, and sometimes of considerable expense.

68. Contract Involving a Specific Performance.

When machinery is purchased on the basis of a specific performance, as in the case of pumping engines, steam boilers, steam ships, and the like, where a specific performance is made the basis of the contract price, it is customary and proper to provide for specific additional sums for stated percentages of excess of performance over and above that which forms the basis of the bid, and also for stated deductions from the contract price for given percentages by which the performance fails to meet the standard. In this way the contractor is fairly paid for accomplishing more than he agreed to, and the purchaser obtains a fair reduction in price for any failure to reach the agreed standard. When a specific performance is made the basis of a contract without these agreed premiums and discounts, the purchaser is at liberty to refuse to accept the work at any price, in case of even a partial failure to meet the specified requirements; while, on the other hand, if the contractor has far exceeded the specifications, he gets no benefit whatever for the enhanced value of the product. A specific performance, therefore, when made the basis of the acceptance of a piece of contract work without these provisions for premiums and discounts is a very onesided and unfair contract, and its use should be discouraged by engineers.

In all cases where a specific performance is made the basis of a contract price, the conditions of this performance should be so clearly stated in the specifications, and the nature of the tests to determine this performance so distinctly described that no misunderstanding could arise when the time comes for making these tests. These descriptions belong in the specifications rather than to the proposals.

69. Contract Including Maintenance. In the case of street pavements, especially where the material is new or untried, it is common to require the contractor to maintain it for a given period, at a stated price per annum. In this case this maintenance price must also be provided for in the proposal, as well as the price charged for first cost.

70. Contract for the Work Only. It is often wise for the principal to purchase materials himself which shall be

G

used by the contractor in the carrying out of the work. Thus the principal may wish to use a particular kind or quality of material which he does not wish to describe specifically in the specifications, or which, if described, he can not well assure himself that the contractor will furnish. Especially is this the case with such materials as can not be clearly identified by ordinary methods of inspection; as, for instance, various kinds of paints, cement, iron and steel, paving brick, besides a great number of specialties in the line of manufactured articles and machines. Or, the contractor may not be able to purchase this material on as favorable terms as the principal, because of the greater risk involved in the sale of this material when the contractor must be looked to for payment. Thus, when the bidders are informed that the principal will furnish materials which would otherwise cost the contractor large sums of money, many contractors of small means would be encouraged to bid upon the work, who otherwise would not be able to handle it. For these and other reasons, therefore, it is frequently wise for the principal to purchase the material and turn it over to the contractor for use in the work.

71. Proposal for Building a Dam, Spillway, Levee, Outlet Tunnel, and Overflow Chamber.

To the First New Mexico Reservoir and Irrigation Co., Roswell, New Mexico.

Gentlemen:—The undersigned propose to do all the work and furnish all of the material in accordance with the printed form of contract and specifications, a copy of which is herewith annexed, and bind————, on the acceptance of this proposal, to enter into and execute a contract in the form of said enclosed specifications and contract for the execution of said work at the prices named below, to wit:

Excavation:

(a) Earth, including all forms of soil, or clay, per cubic yard————.

(b) Gravel and sand, including all forms and combinations of these materials, per cubic yard————.

(c) Loose rock in open cut, including all kinds of loose rock not requiring blasting, per cubic yard————.

(*d*) Solid rock in open cut, including all kinds of rock requiring blasting, per cubic yard———.

(*e*) Rock in tunnel, including all tunnel work to the outer line of the lining wall, if such be required, otherwise to the lines of the drawings, per cubic yard———.

Fill:

(*a*) Earth, clay, gravel or sand, not rolled, per cubic yard———.

(*b*) Same materials spread in courses and rolled dry, per cubic yard———.

(*c*) Same materials spread in courses, dampened and rolled, per cubic yard———.

(*d*) Same materials spread in courses, pulverized, harrowed, wet down and rolled thoroughly, per cubic yard———.

(*e*) Clay and gravel mixed in layers, harrowed, wet down and rolled thoroughly (clay puddle), per cubic yard———.

(*f*) Loose rock dumped or thrown in as in temporary dam, per cubic yard———.

(*g*) Rip rap laid on face of dam, per cubic yard———.

(*h*) Facing rock laid dry with close joints for distance of 4 inches from surface and rammed, per cubic yard———.

Masonry:

(*a*) Rubble masonry laid in Portland Cement Mortar, as described, per cubic yard———.

(*b*) Masonry lining of tunnel, as described, per cubic yard, actual volume———.

(*c*) Dimension stone masonry, laid in Portland Cement Mortar, as described, per cubic yard———.

Enclosed is a certified check for five hundred dollars; which sum is to be forfeited to the First New Mexico Reservoir & Irrigation Co. if the party or parties making this proposal fail to enter into contract, with approved securities, within fifteen days after the contract is awarded to said party or parties.

Respectfully,

(Signature and address of contractors.)

———, ———

———, ———

St. Louis, Mo.,———1890.

Note.—Each bid shall be placed in a sealed envelope addressed to "———, President First New Mexico Reservoir & Irrigation Co.," care of "———, Consulting Engineers, ———, St. Louis, Mo.," and shall be indorsed "Proposal for building Dam, etc."

The First New Mexico Reservoir & Irrigation Co. reserves the right to reject any or all bids.

J. & F.

72. Proposal Bond. In lieu of a cash deposit accompanying the bid as a guarantee of good faith and of intention to enter into contract, if the same be awarded to the party, a bond may be received, duly signed and certified, which will insure either the signing of the contract, or the payment of such damages as may result from a failure to sign. These damages would usually be measured by the difference between the amount named by the party furnishing the bond, and the sum for which the contract might finally be let, and this is usually named as the amount of the forfeiture under the bond. It is the usual custom of the United States Government to require a bond of this sort rather than a cash deposit. It is evidently a less hardship upon the contractor to furnish such a bond. The following is the form of this document as used by the United States Government:

PROPOSAL BOND OR GUARANTY.

We,————————, of————————, in the State of————————, and————————, of————————, in the State of————————, hereby guarantee and bind ourselves and each of us, our and each of our heirs, executors and administrators, to the effect that if the bid of———————— herewith accompanying, dated————, 1894, for furnishing all materials and labor, and constructing the Power House and Office Building for the 800-ft. Lock at St. Mary's Falls Canal, shall be accepted, in whole, or in part, within sixty (60) days from the date of the opening of proposals, the said bidder—, ————————will, within ten (10) days after being notified of such acceptance, enter into a contract with the United States in accordance with the terms and conditions of the advertisement, and will give bond with good and sufficient sureties for the faithful and proper fulfillment of the same. And in case the said bidder— shall fail to enter into contract within the said ten (10) days with the proper officer of the United States, and furnish good and sufficient bond for the faithful performance of the same according to the terms of said bid and advertisement, we and each of us hereby stipulate and guarantee, and bind ourselves and each of us, our and each of our heirs, executors and administrators, to pay unto the United States the difference in money between the amount of the bid of the said bidder—, and the amount for which the proper officer of the United States may contract with another party to furnish said materials and labor and construct the Power House and Office Building as specified, if the latter amount be in excess of the former, for the whole work covered by the proposal.

WITNESSES:

————————,   ————————[SEAL]
————————.   ————————[SEAL]

Dated————————, 1894.
[Executed in triplicate.]

JUSTIFICATION OF GUARANTORS.

STATE OF ————————
County of———————— } ss.

I,————————, one of the guarantors named in the within guaranty, do swear that I am pecuniarily worth the sum of forty thousand dollars, over and above all my debts and liabilities.

[Signature of Guarantor]————————

Before me,
[Signature of Officer administering oath, with seal, if any]

————————————

STATE OF————————
County of———————— } ss.

I,————————, one of the guarantors named in the within guaranty, do swear that I am pecuniarily worth the sum of forty thousand dollars, over and above all my debts and liabilities.

[Signature of Guarantor]———— ————————

Before me,
[Signature of officer administering oath, with seal, if any.]

————————————

CERTIFICATE.

I,————————, do hereby certify that————————, and————————, the guarantors above named, are personally known to me, and that, to the best of my knowledge and belief, each is pecuniarily worth, over and above all his debts and liabilities, the sum stated in the accompanying affidavit subscribed by him.

[Signature of certifying official.]————————

NOTE.—The certificate may be given separately as to each guar-antor, and modified accordingly.

U. S. ENG. CORPS.

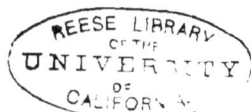

ENGINEERING SPECIFICATIONS.

73. Engineering Specifications Defined. Engineering specifications consist of a series of specific provisions each one of which defines and fixes some one element of the contract. These clauses relate, in general,

First: To the work to be done.

Second: To the business relations of the two parties to the contract.

In the first sense, the specifications supplement and explain the plans (if there be any) and define the character of the materials and the methods to be employed on the work, or if unaccompanied by plans they embody the principles and rules in accordance with which the plans must be drawn and the work executed. In this sense the specifications enable the bidder to estimate the cost of the proposed work and after the contract is let they serve as the rules of inspection and acceptance of such work.

In the second sense the specifications define the rights and duties of the two parties to the contract to each other and embody proper provisions for changes in the plans, and for the settlement of disputes which may arise; they also describe the conditions of payment, acceptance, etc., etc.

74. Classes of Specifications. There may be said to be three general classes of engineering specifications:

(*a*) Specifications accompanying complete detail plans:

(*b*) Specifications accompanying a general plan only.

(*c*) Specifications unaccompanied by any plan, and commonly known as General Specifications.

All of these classes of specifications are in common use and each has its own particular sphere of usefulness.

(*a*) Thus when the design is novel, or when the engineer wishes a particular design carried out, he usually prepares full detail plans, or drawings, showing how all parts of the proposed work shall be done.

In the case of public works, also, when the law requires the contract to be let in open competition, and also specifies that it shall be let to the lowest bidder, it is almost necessary to prepare full detail plans in order to avoid an inadequate or inferior design being put into competition with better ones, and, from its diminished cost, receiving the contract.

(*b*) If the engineer can limit the bidders to a selected class of reliable contractors, who have reputations to lose if they should do inferior work, he may prepare very general plans only and allow the contractor to make the details to suit himself, in accordance, however, with certain specific requirements as given in the specifications, and subject to the approval of the engineer.

(*c*) If the engineer is indifferent as to even the general design, provided the finished work answers equally well certain prescribed demands, as given in a set of general specifications, he may not prepare any plans whatever, but leave the contractor (who must now also be chosen by the engineer or only responsible parties allowed to bid) to use any design he may choose, such designs to be submitted, however, with his bid, and this, together with the general specifications to form the basis of the contract.

75. General and Specific Clauses. Any specification may be said to be composed of two kinds of clauses, general and specific.

All those clauses which relate to the business portion of the contract, or which go to define the relations of the parties to the civil contract as a business proposition, may be said to be the general clauses.

All those clauses which are descriptive of the engineering or structural features of the design, either as explanatory of the plans, or of the materials to be used, or of the methods to be employed, may be called the specific clauses.

Since the general clauses are common to all kinds of specifications, they will be discussed first.

THE GENERAL CLAUSES IN SPECIFICATIONS.

76. The General Clauses in Specifications may relate to any or all of the following subjects:

(1) Time of commencement, rate of progress, and time of completion of the work.

(2) As to the character of the workmen to be employed.

(3) Suitable appliances to be used.

(4) Monthly estimates of work done and payments to be made.

(5) Provision for inquiring into the correctness of the monthly estimates.

(6) Reserving a certain percentage as a repair fund, for a stated period after completion.

(7) Conditions of the final estimate.

(8) Engineer's measurements and classifications final and conclusive.

(9) Determination of damages sustained by failure to complete the work within the time agreed upon, or as extended.

(10) The discharge of unpaid claims of work men and material men.

(11) No claims for damages on account of suspension of work.

(12) No claims for damages on account of delay.

(13) No claims on account of unforeseen difficulties.

(14) Protection of finished work.

(15) Protection of property and lives.

(16) Protection against claims for the use of patents.

(17) Assignment of contract.

(18) Contractor not released by subcontracts.

(19) Abandonment of contract.

(20) Cancellation of contract for default of contractor.

(21) Workmen's quarters and other temporary buildings.

(22) Cleaning up after completion.

(23) Removal of condemned material.

(24) Relations to other contractors.

(25) Provision for drainage.

(26) Provision for public traffic.

(27) Contractor to keep foreman or head workman, and also copy of plans and specifications on the ground.

(28) Cost of examination of completed work.

(29) Faults to be corrected at any time before final acceptance.

(30) Surveys, measurements, and estimates of quantities not guaranteed to be correct.

(31) The contract subject to interpretation and change by the engineer.

(32) Settlement of disputes.

(33) Extra work.

(34) Definition of "Engineer" and "Contractor."

(35) Documents composing the contract.

(36) Meaning understood.

77. Explanatory Note. In all that follows on the subject of specifications, after explaining and discussing a given subject, one or more illustrations will be given in solid type, from actual specifications, together with the initials of the author. The full name and professional engagement of the author can then be found by referring to the *Key to Personal References*, page 5. In general the latest practice only of the engineers quoted in this way will be cited. It must also be understood that in every case the gentlemen so quoted have themselves selected the sample specifications used and have consented to such use.

78. Time of Commencement, Rate of Progress, and Time of Completion of the Work.

It is usual to make the time of commencement of the work as soon after the signing of the contract as is thought practical, as, for instance, ten, fifteen, or thirty days, depending on the character of the work.

The rate of progress is specified in order to give the engineer authority for canceling the contract if the rate of progress is such as to indicate that the contractor will certainly be unable to complete the work on time, or at all. Thus he may be obliged to abandon the work altogether, or he may choose to do so, in which case, if rate of progress is specified, the parties of the first part need not wait for the full time for completion to arrive before being able to take the work from the hands of the contractor and complete it by hiring the labor and purchasing the materials, or by reletting it to another contractor.

The time of completion is nearly always stated, and while the time allowed should be ample it should be only such as is required when a reasonable degree of diligence is exercised on the part of the contractor.

If, for any sufficient reason, the contractor is delayed in his work, for reasons beyond his control, the time of completion is usually extended by the principal by a corresponding length of time, and then this extended period fixes the required, or specified date of completion.

And the said party of the second part further agrees that he will commence the work herein contracted to be done within twenty days from the date of this contract; that the rate of progress of his work shall be such as, in the opinion of the Engineer, is necessary for completion within the time herein specified, and that he will so conduct the said work that on or before July 1, 1899, the whole work covered by this contract and specification shall be entirely completed.   A. F.

79. As to the Character of the Workmen to be Employed. In order to secure good work it is necessary to employ skilled workmen. The engineer must therefore have some control over the character of the labor employed by the contractor. This is obtained by specifying that only skilled labor shall be employed and giving to the engineer the power of discharge over any laborer, mechanic, foreman, or superintendent employed by the contractor on the work. It is also customary to provide that this power shall extend to cases of disobedience of instructions, impudence to engineer or inspectors, drunkenness, etc., as shown in the following illustration:

And the said party of the second part further agrees to employ only competent, skillful men to do the work; and that whenever the Engineer shall inform said party of the second part in writing that any man on the work is, in his opinion, incompetent, or unfaithful, or disorderly, such man shall be discharged from the work, and shall not again be employed on it.   A. F.

80. Suitable Appliances to be Used. If not prevented by a special clause in the specifications, contractors who are unprovided with suitable mechanical appliances for doing the work properly will often undertake to perform the work with cheap and inadequate means, which would necessarily result in faulty construction, or in delaying the work. It is customary, therefore, to prescribe that all appliances shall be suitable and adequate to the purpose, and subject to the approval of the engineer. It is not wise, however, to specify particular methods or means of doing the work, since if for any reason a partial failure should result, the contractor will endeavor to obtain personal release by charging failure to the specified appliances or methods. A specification like the following is therefore recommended.

The contractor is to use such methods and appliances for the performance of all the operations connected with the work embraced under this contract as will secure a satisfactory quality of work and a rate of progress which, in the opinion of the engineer, will secure the completion of the work within the time herein specified. If, at any time before the commencement, or during the progress of the work, such methods or appliances appear to the engineer to be inefficient or inappropriate for securing the quality of the work required or the said rate of progress, he may order the contractor to increase their efficiency or to improve their character, and the contractor must conform to such order; but the failure of the engineer to demand such increase of efficiency or improvement shall not relieve the contractor from his obligation to secure the quality of work and the rate of progress established in these specifications. A. F.

81. Monthly Estimates of Work Done and Payments to be Made. It is customary, in all kinds of engineering construction, for the engineer in charge to estimate at the end of each month the quantity of material furnished on the ground, and of work done. These estimates are approximate only and serve as a basis for making monthly payments to the contractor. It is customary to reserve from ten to twenty-five per cent. of these monthly estimates until the final completion of the work. By means of these monthly payments the contractor is enabled to carry on the work to final completion with a much smaller capital than would be required if no payments were made until the work was finished. The percentage reserved from the monthly payments is intended to serve as a guarantee of final completion, and as a fund to draw upon when the time of final settlement arrives, for the payment of damages resulting from the work not having been performed within the specified time, or for other purposes as indicated subsequently in these general specifications. In the matter of payment for materials furnished, but not incorporated finally into the work, it is usually considered safe to include in the monthly estimates all materials delivered, either upon the ground, that is to say along the line of the work, and subject to the inspection and control of the engineer, and also to pay for materials and machinery furnished and stored where they are under the control and subject to the inspection and approval of the engineer. Of course no material would be included in these monthly estimates which had not been duly inspected and accepted. The following is a common form for this specification.

In order to enable the said contractor to prosecute the work advantageously, the engineer shall, once a month, on or about the last day of each month, make an estimate in writing of the amount of work done, and materials delivered to be · used in the work, and of the value thereof, according to the terms of this contract. The first such estimate shall be of the amount or quantity and value of the work done and materials delivered since the party of the second part commenced the performance of this contract on his part. And every subsequent estimate (except the final one) shall be of the amount or quantity and value of the work done since the last preceding estimate was made. And such estimates of amount and quantity shall not be required to be made by strict measurement or with exactness; but they may, at the option of the engineer, be approximate only.

Upon each such estimate being made the parties of the first part will pay to the party of the second part the following proportions or percentages thereof, to wit:

85 per cent. thereof up to and until such time as· the total estimated value of the work done and materials delivered shall amount to $1,000,000.

90 per cent. thereof after the total estimated value of such work and materials delivered shall have amounted to $1,000,000, until the party of the first part shall have fully and completely performed this contract on his part.

A. F.

82. Provision for Inquiring into the Correctness of the Monthly Estimates. The monthly estimates made by the engineer acting as the agent of the party of the first part, may be held to be binding upon this party, in case he has either made a mistake in the quantity of work done, or material furnished, or has entered into collusion with the contractor, and rendered false returns. Since the engineer is the agent of the party of the first part, his acts would bind his principal, after payment had been made on the same, if it were not expressly provided that the party of the first part shall not be estopped, or prevented, from determining by other means

the amount of work done and material furnished. In other words the party of the first part should not necessarily be bound by either the monthly or final estimates rendered by his agent, and which are intended to serve as the basis of payment. It is understood, of course. that the contractor always has this privilege of inquiry and proof of the correctness of the estimates. It is customary, therefore, to insert a clause like the following.

And it is hereby expressly agreed and understood by and between the parties hereto that the said parties of the first part, their successors and assigns, shall not, nor shall any department of the City of New York, be precluded or estopped by any return or certificate made or given by any engineer, inspector, or other officer, agent or appointee of said Aqueduct Commissioners, or of said parties of the first part, under or in pursuance of anything in this agreement contained, from at any time showing the true and correct amount and character of the work which shall have been done and materials which shall have been furnished by the said party of the second part, or by any other persons under this agreement.

A. F.

83. Reserving a Certain Percentage as a Repair Fund, for a Stated Period after Completion.

In order to provide for inherent defects in the work which may not appear on the surface, or until after the construction has been in service for some time, it is often desirable to retain a portion of the total cost of the work for a specified period of time, on which sum the party of the first part is authorized under the specifications to draw for the repairing or correcting of any and all faults or defects which may become apparent by use within the specified period. It is usual, however, to give the contractor the privilege of making such repairs under the direction, and subject to the approval, of the engineer, in place of having the engineer make such repairs and charge them against the reserve fund. This clause may read as follows.

The contractor hereby further agrees to make all the needed repairs on the said work during a period of ——months after its final completion ; and he hereby further agrees that the party of the first part is authorized to retain out of the moneys payable, or to become payable, to him, under this agreement, the sum of five per cent. on the amount of the contract, and to expend the same, or so much thereof as may be required, in making the aforesaid repairs to the satisfaction of the engineer, if within three days after the delivery or mailing of a. notice in writing to the contractor, or his agent or attorney, he or they shall neglect to make the aforesaid needed repairs; and he hereby further agrees to be responsible for any accident that may occur on account of the defective condition of the work.

E. A. F.

84. Conditions of the Final Estimate. If, in the opinion of the engineer, the contractor has completed his work in all respects in accordance with the terms of the contract, he should proceed with due diligence to make the final estimate of all quantities in the several clauses, and to certify to his principal the amount of money due to the contractor, and also the amounts which should be held in reserve under the various clauses of this character in the specifications. The party of the first part thereupon should immediately pay to the contractor such moneys as are legally due him, provided this party is satisfied that the final estimates submitted by the engineer are correct. If this party should have any doubts on this point, he should be at liberty, under the specifications, to inquire further into the correctness of such estimates. This portion of the contract may be stated as follows.

It is further mutually agreed, that whenever this contract, in the opinion of the engineer, shall be completely performed on the part of the contractor, the engineer shall proceed with all reasonable diligence to measure up the work, and shall make out the final estimates for the same and shall certify the same. The party of the second part will then, excepting for the cause herein specified, pay to the contractor, within — days after the execution of said certificate the remainder which shall be found to be due, excepting therefrom such sum or sums as may be lawfully retained under any of the provisions of this contract: Provided that nothing herein contained shall be construed to affect the right hereby reserved, to reject the whole or any portion of the aforesaid work, should the said certificate be found to be inconsistent with the terms of this agreement, or otherwise improperly given.   E. A. F.

85. Engineer's Measurements and Classifications Final and Conclusive. In order to avoid disputes as to both the quantity and the quality of the work done, it is customary to specify that the measurements and classifications of the engineer shall be final and conclusive, and binding upon both parties. This is a very important provision, and places a great responsibility upon the engineer, while it binds, at the same time, the two principals to the contract, and forces them to submit to the engineer's decisions, except as some special provision such as that stated in article 81 allows one or both of the parties to examine into the correctness of the engineer's estimates. As a matter of course either party is always at liberty to examine questions of fact and, so far as it is practicable, to remeasure quantities at subsequent times. Either party would be at liberty in case of a suit at law to have such quantities remeasured to determine such question of fact, but so far as the classification of the material is a matter of opinion on the part of the engineer, and so far as measurements of quantities have become impracticable at a subsequent period, to this extent a clause such as is here proposed binds absolutely both parties to the contract. Neither party now has any release from the decision of the engineer, except on one of two grounds:

First. Either party may bring a suit in equity, in which case the terms of the contract are not made the basis of the suit; or

Second. Either party may enter a plea of fraud on the part of the engineer, which, if sustained, would of course vitiate the decisions of such engineer. Neither of these grounds offers much encouragement to either party. A case could not be sustained in equity contrary to the terms of an expressed written agreement, except it could be shown that gross and violent injustice had been worked by a strict compliance with its terms. Neither is it desirable in a civil suit to enter a plea of fraud, since this is very difficult to maintain, and can only be maintained by proving the moral depravity of the engineer.   A clause such as the following, therefore, if incorported in a contract and agreed to by both parties places both parties absolutely at the mercy of such engineer, and the contractor should never submit to it, if he has reason to suppose that the engineer is likely to act unfairly toward him under the authority thus granted to him.   As a rule, however, this confidence which is reposed in the engineer by both parties to the contract is not misplaced.   Although the engineer is paid for his services by one of the parties to the contract, he understands that his position is a judicial one, and not that of an advocate or partisan, and that it is his business to see that justice is done to both of the parties.   The clause usually reads as follows.

It shall be understood and agreed by the parties hereto, that due measurements shall be taken during the progress of the work, and the estimate of the engineer shall be final and conclusive evidence of the amount of work performed by the contractor under and by virtue of this agreement, and shall be taken as the full measure of compensation to be received by the contractor.   The aforesaid estimates shall be based upon the contract prices for the furnishing of all the different materials and labor, and the performance of all the work mentioned in this specification and agreement, and when there may be any ambiguity therein, the engineer's instructions shall be considered explanatory, and shall be of binding force.   E. A. F.

86. Determination of Damages Sustained by Failure to Complete the Work within the Time Agreed upon, or as Extended.

It is seldom that a specific performance of any contract can be enforced.   In other words, either of the parties to almost any civil contract is at liberty to break the same, or fail to carry it out, for which failure, however, the law provides that the party breaking the contract shall pay a, penalty.   The amount of this penalty usually remains to be ascertained after the contract has been broken, and when the time of settlement arrives.   The legal remedies for breach of contract are given in Arts. 49–53.   It

7

is sufficient to remark here, that in determining the amount of
the damages, the law will only allow the actual proven damages to be collected, and always discourages any constructive,
or conventional, or arbitrary estimate of such damages. In
other words, the damages are the compensation to the injured
party, requisite to repay him for his loss, which can be traced
directly to the breach of contract.

While damages to the extent of the actual injury sustained
can always be recovered, by a suit at law, in the case of a
breach of contract, it is customary in the writing of engineering
specifications to insert one or more clauses defining the amount
of the damages which it is agreed by the parties will be sustained in case of certain specific failures to carry out the contract; and since these failures are assumed to be on the part of
the contractor, and since money is usually due him from the
other party, it becomes possible, in this case, to remunerate the
injured party by withholding a certain sum of money from the
contractor who is guilty of the breach of contract. If a specific
agreement to this effect be entered into by the parties, in
advance, the compensation for the injury done because of a
specific breach of the contract, may be recovered by simply
withholding such a sum from the contractor, and paying over
to him in final settlement the remainder. Because, therefore,
of the facility with which such a settlement can be accomplished, and also to further provide against such a contingency
arising by furnishing to the contractor a sufficient motive to
prevent such specific breaches, and furthermore, in order to
avoid a suit at law for the recovery of such compensation, it
has become customary to insert what is commonly called a
"penalty clause." *

While recovery can be had by a suit at law for the actual
damages sustained for any breach of the contract, either with
or without a specific clause to this effect, the penalty or dam-

* The reader is requested to refer to Arts. 51–53 for a discussion of the legal phases
of this question.

age clause in the specifications usually refers to one or more specific kinds of breach of contract, the more common one being that of failure to complete the work within the time agreed upon. The object of a penalty clause covering this particular kind of breach of contract is rather to insure completion of the contract within the time specified than to recover damages for a failure to do so. For this reason it has been commonly supposed if a heavy penalty were provided for a failure of this kind, it would serve as a strong motive to the contractor to hasten the work. This being the object of such a clause it has been common to specify a penalty or damage of so many dollars per day, for each and every day elapsing after the date agreed upon for the completion, before the work is finally completed, the sum so named being often a very extravagant one.

There are several ways of stating this clause, some of which are very much better than others. The following are the more usual forms:

I. *Provision for a specific "penalty."* When a specific "penalty" is named for either a particular or for any breach of the contract, whether this sum named be a per diem, or a gross amount, the court will usually construe it as meaning that such a sum is a fund provided in the specifications for the purpose of meeting such damages as may result from a breach of the contract, and that only the actual damages sustained and proved in a suit at law can be recovered from such fund. In other words, a penalty clause so stated has little or no force, since the law provides exactly the same remedy for any breach of contract, without a specific agreement.

II. *The naming of a per diem, or gross sum, as being the "ascertained and liquidated damages"* which will be sustained by the injured party for a specific breach of contract therein named, this usually being for failure to complete the work within the time specified. In this case the word "penalty"

is not used, and if it can be made to appear on trial that both
parties to the agreement really intended that the sum named
should be forfeited in case of the failure therein described, and
provided further this sum is not too extravagant and unreasonable, and provided the fact of failure and consequent liability
be fully established, then and in that case the law will sustain
the damage clause, and the injured party will be allowed to
deduct it from any moneys due the contractor, or if this fund
be insufficient, he may even sue the contractor and his bondsmen and recover the remainder. The following is a good
example of this method of stating such a clause:

And the said party of the second part hereby further
agrees that the said parties of the first part shall be and they
are hereby authorized to deduct and retain out of the moneys
which may be due or become due to the said party of the second
part, under this agreement, as damages for the non-completion
of the work aforesaid within the time hereinbefore stipulated
for its completion, or within such further time as in accordance
with the provisions of this agreement shall be fixed or allowed
for such performance or completion, the sum of one hundred
dollars per day for each and every day the time employed upon
said work may exceed the time stipulated for its completion, or
such stipulated time as the same may be increased, as hereinbefore provided, which said sum of one hundred dollars per
day is hereby, in view of the difficulty of estimating such damages, agreed upon, fixed and determined by the parties hereto
as the liquidated damages that the parties of the first part will
suffer by reason of such default, and not by way of penalty.

A. P. B.

III. *An agreement that the engineer shall ascertain and
make an estimate of the actual damages sustained* by a failure
to complete the work within the time specified (or for other
specific breach), and naming some or all of the items to be
included in such estimate. In this case no effort is made in
advance to determine what the actual damages are, and the
agreement simply consists in making the engineer an arbitrator
to act for both the parties, in determining the amount of the
damage as a question of fact. This is probably the strongest

method of stating this clause, while it is also the fairest to all parties concerned.

Because of the difficulty in proving in a suit at law the actual damages sustained from the failure to complete an engineering contract within the time specified, the contractor usually pays very little attention to a penalty clause stated as described above in form I. As a rule, contractors are better informed as to the law of contracts than the engineers who write the specifications, and when this clause is stated as first described the contractor regards it lightly, well knowing that it has no particular significance. When stated in the second manner, however, provided the sum named be reasonable, the contractor will give it much greater weight, and the party paying for the work can withhold money under it with much greater assurance of being sustained by the courts. The courts, however, have a repugnance to any agreement made in advance as to questions of fact, which in the nature of things could only be adequately determined after the breach had transpired. But because of the difficulty of fixing accurately the amount of such damages, even after the breach, the law consents to a previous agreement upon a specific sum, provided this be reasonable, and provided it be so clearly stated that the parties signing the contract can not have misconstrued it. Concerning the last method given of stating this clause, the law also has a repugnance to delegating the authority of the court to a layman in the person of an arbitrator. When, however, the question at issue is a *"condition precedent"* to settlement, as in this case of fixing the amount of the damages, and when this arbitrator is the engineer in charge of the work, who is evidently the most competent person to estimate the amount of such damage, the law readily consents that he should act in such capacity, and if both parties have agreed that his decision should be final and conclusive in the premises, there would seem to be no way of evading his decision, except by proving that it was fraudulent. As fraud.

invalidates nearly all agreements, and nearly all obligations, if it can be shown that the engineer, when acting in the capacity defined in this clause, has knowingly and willfully overestimated the amount of the damage; in other words, if it can be shown that he acted dishonestly in the matter, his verdict can be set aside and the matter can come before the court. Otherwise the court will rule that his verdict must hold, and the question can not be opened. As it is very difficult to establish a question of motive, and as the burden of proof rests wholly upon the contractor, it would seem that this method of writing the damage clause had many advantages. The following is a fair example of such a clause.

In case said contractor shall fail to fully and entirely, and in conformity with the covenants, terms and agreements of this contract, perform, and complete said work, and each and every part and appurtenance thereof, within the time hereinbefore limited for such performance and completion, or within such further time as may be allowed by said Board for such performance and completion, said chief engineer shall appraise the value of the direct and computable damages caused to said city by such failure, owing to the disbursements made by said city on account of the further employment of engineers, inspectors and other employees, including all disbursements for office rent, transportation, supplies, and other matters connected with said employment; also the value of such other direct and computable damages as shall be caused by such failure; and the amount so appraised, when approved by said Board, shall be deducted by said Board out of such moneys as either may be due, or at any time thereafter become due, to said contractor under and by virtue of this contract, or any part thereof; and in case said appraised value shall exceed the amount of said moneys, then said contractor will pay the amount of such excess to said city, on notice from said Board of the excess so due; and it is hereby agreed that the decision of said chief engineer as to the said appraisal, when approved by said Board,* shall be final and binding on both parties to this contract.

E. K.

87. The Discharge of Unpaid Claims of Workmen and Material men.

The laws of many states provide that persons, who supply either labor or material to any con-

---

*It may or may not be wise to make the verdict of the engineer subject to the approval of his principal.

tractor or other person, to be used in the construction of any building or other permanent work, if not paid by such party, may file a lien upon such completed or uncompleted work, this serving as a kind of first mortgage upon the property, under which the property can be sold and the claim satisfied. When such a law obtains, the only safe course, for the person paying for the work, is to satisfy himself before he fully pays for the work that all such claims have been liquidated, or he may if he choose, require the contractor to furnish a bond which may be sued upon, either by himself, or by such material man or laboring man as may have such a claim. This bond to be large enough to cover all such liabilities.

When the party paying for the work desires to satisfy himself that such claims have all been discharged by the contractor, the clause may be written as follows:

Said contractor further agrees that he will pay punctually the workmen who shall be employed on the aforesaid work, and the persons who shall furnish material thereunder, and will furnish said Board with satisfactory evidence that all persons who have done work or furnished materials under this contract and shall have filed any account of such claims with said Board, have been fully paid, or are not entitled to any lien under the laws of this state; and in case such evidence be not furnished as aforesaid, such amount as said Board may consider necessary to meet the lawful claims of the persons aforesaid, shall be deducted from the moneys due said contractor under this contract, and shall not be allowed until the liabilities aforesaid shall have been fully discharged and the evidence thereof furnished said Board; and if such evidence is not furnished before the final payment under this contract falls due, said Board may pay such claims in whole or in part to the person or persons, firm or corporation claiming the same, and charge the amount thus paid to said contractor, who shall accept the same as payment to the amount thereof upon this contract.

E. K.

When the party paying for the work does not care to put himself to the trouble of obtaining the information as to the discharge of all such claims by the contractor, he may so frame the wording of the bond that it will cover this case satisfactorily. In this case this portion of the bond may read as follows:*

The said————as principal, and————and———— as securities, hereby bind themselves and their respective heirs, executors or administrators, unto the City of St. Louis, in the penal sum of——dollars, lawful money of the United States, conditioned that in the event the said————shall faithfully and properly perform the foregoing contract according to all the terms thereof, and shall, as soon as the work contemplated by said contract is completed, pay to the proper parties all amounts due for material and labor used and employed in the performance thereof, then this obligation to be void, otherwise of full force and effect, and the same may be sued on at the instance of a material man, laboring man, or mechanic, for any breach of the condition hereof; provided, that no such suit shall be instituted after the expiration of ninety days from the completion of the above contract.

88. No Claims for Damages on Account of Suspension of Work.

When the work contracted for is of a public character, as for a city, or for the United States Government, and when it is expected to continue for a considerable period, and be paid for by appropriations from time to time, and also in other like contingencies, it is common to insert a clause to the effect that the contractor shall make no claim for damages for necessary delays he may experience in carrying out the work, when these delays are caused by the failure of appropriations, or by legal proceedings, and the like.

On ten days notice the work under this contract may, without cost or claims against the party of the first part, be suspended by them for want of funds, or for other substantial cause. Upon receipt by the contractor of the order for the suspension of the work, all the materials shall be piled up compactly, so as not to impede travel on the sidewalk or carriageway, or the use of fire plugs, gas or water stops, and all surplus material and rubbish shall be removed immediately from the street. When the party of the first part shall order the work to be resumed the contractor shall complete the same upon the terms and conditions of this contract.

E. A. F.

*This is the form universally adopted in all contracts made by the City of St. Louis. If not specifically so stated the material man, or the laboring man could not sue on the bond.

89.   No Claims for Damages on Account of Delay.

In order that the party of the first part shall be freed from all claims which may be set up by the contractor for damages on account of various delays and hindrances which he may have experienced in carrying out the work, and which he may make appear to have been caused directly or indirectly by the party having the work done, or by other contractors upon the work, the following clause is often inserted:

The contractor shall not be entitled to any claims for damages for any hindrance or delay from any cause whatever in the progress of the work, or any portion thereof, but said hindrance may entitle said contractor to such extension of time for completing the contract as may be determined by the engineer, provided, he shall have given notice in writing of the cause of the detention.   E. A. F.

90.   No Claims on Account of Unforeseen Difficulties.

In case it is the purpose of the contract to place upon the contractor all the responsibility for contingencies which may arise in the prosecution of the work, for which greater risk the party having the work done will, of course, pay in the increased price made by the contractor to cover such risk, the clause may be written as follows:

The contractor agrees that he will sustain all losses or damages arising from the action of the elements, the nature of the work to be done under the specifications, or from any unforeseen obstructions or encumbrances on the line of the work which may be encountered in the prosecution of the same.

E. A. F.

91.   Protection of Finished Work.

It is usually customary to hold the contractor responsible for the protection and care of the work until it is all finally completed and accepted. Even such portions of the work as have been completed and provisionally accepted and payments made upon the same, should be taken care of and fully protected by the contractor, until the entire work has been turned over. This often entails considerable expense upon the contractor, and when disputes on this question are liable to arise, it is well to insert such a clause as the following:

Contractors will be held responsible for any and all materials or work to the full amount of payments made thereon, and they will be required to make good, at their own cost, any injury or damage which said materials or work may sustain from any source or cause whatever, before final acceptance thereof.

O. M. P.

92. Protection of Property and Lives. It is always understood that the contractor shall he held responsible for all damages to property which may arise from any fault of his, or from any accident which may occur during the performance of the work. He is also held responsible for all losses of life or limb, and for all personal damages which may be sustained either by his own workmen or by the public, by or on account of the works he has under construction. In other words, it is made his duty to protect both life and property, so far as possible, from all damage, so far as these may be traceable to the works themselves. If this responsibility were not specifically placed upon the contractor, the party having the work done would often be obliged to sustain the loss, since he authorizes the execution of the work, and the contractor is his employee or agent. This clause is often written as two separate clauses, one referring to the damage to property, and the other to the damage to persons.

Furthermore the wording of the bond is usually so made as to cover both of these items, so that in case the damage or loss is greater than could be repaid by the amount of money at any time due the contractor when the accident occurs, suit may be brought upon the bond against the bondsmen to recover the remainder.

Inasmuch as claims for damages, either to person or property, usually manifest themselves in the form of suits at law against the party authorizing the work and paying for the same, and not against the contractor himself, it is common to assume that this will be the case in all claims for damages, and to word the clause accordingly. The following clause covers all of the above contingencies in an acceptable manner:

Said contractor further agrees that he will indemnify and save harmless said City and Board, and the officers and agents thereof, from all claims, suits, actions, and proceedings of every name and description, which may be brought against said City or Board, or the officers and agents thereof, for or on account of any injuries or damages to persons or property received or sustained by any person or persons, firm or corporation, by or from said contractor, or by or in consequence of any materials or explosives used on said work, or by or on account of any improper material or workmanship in its construction, or by or on account of any accident, or of any other act or omission of said contractor, or his agents, or servants, and said contractor also agrees that so much of the money due, or to become due, to him under this contract as shall be considered necessary by said Board, may be retained by said Board until all such suits or claims for damages, or otherwise, as aforesaid, shall have been finally settled and determined, and evidence to that effect furnished to the satisfaction of said Board.

E. K.

The following is a common method of wording this clause, which defines the contractor's responsibility without referring to suits at law:

The contractor shall put up and maintain such barriers and red lights as will effectually prevent any accident in consequence of his work, for which the city might be liable, and he shall be liable for all damages occasioned in any way by his acts or neglect, or that of his agents, employees, or workmen.

E. A. F.

93. Protection against Claims for the Use of Patents.

When it is anticipated that patented appliances or methods may be used either by the contractor in prosecuting the work, or as forming a part of the completed work itself, in order that the party authorizing the work may be able to collect from the contractor such fees as he may be forced to pay therefor, a special clause in the specifications may be written to cover this case. This clause may be as follows:

All fees for any patented invention, article or arrangements that may be used upon or in any manner connected with the construction, erection, or maintenance of the work, or any part thereof embraced in these specifications, shall be included in the price mentioned in the contract, and the contractor shall protect and hold harmless the party of the first part against any and all demands for such fees or claims, and before the final payment or settlement is made on account of the contract, the contractor must furnish acceptable proof of a proper and satisfactory release from all such claims.

E. A. F.

94. Assignment of the Contract.* If it is the intention of the party letting the work that the person or persons who take the contract shall perform the work themselves, without subletting it, it is necessary to prescribe that this shall be done in order to insure that it may not be sublet. One great objection to the subletting of contracts is that the subcontractor can not be held directly by the principal, since these two have not entered into contract. The principal can only hold the original contractor, and all dealings with the subcontractor must be through him. This gives rise to delays and unsatisfactory performance, and is usually prohibited by the specifications. The following form is adequate to this purpose:

Said contractor further agrees that he will give personal attention constantly to the faithful prosecution of the work, and will not assign or sublet the work or any part thereof, or any of the moneys or orders payable under the contract, without the previous written consent of said board endorsed on this contract, but will keep the same under his personal control; that no right under this contract, nor to any moneys or orders due or to become due hereunder, shall be asserted against said city or board, or any department, officer, or officers thereof, by reason of any so-called assignment, in law or equity, of this contract, or any part thereof, or of any moneys or orders payable thereunder, unless such assignment shall have been authorized by the written consent of said board endorsed on this contract; that no person other than said contractor now has any claim thereunder, and that no claim shall be made excepting under this specific clause of this contract, and under that clause relating to claim of workmen and materialmen.   E. K.

95. Contractor not Released by Subcontracts.
When it is anticipated that a portion at least of the work

*See Article 30.

will be sublet to other contractors, and when in the nature of things this is advisable, it may be specified that such subletting of all or of any portion of the work in no wise releases the contractor from full and faithful performance. The following specification would then hold:

No subcontract shall under any circumstances relieve the contractor of his liabilities and obligations under his contract; should any subcontractor fail to perform the work undertaken by him in a satisfactory manner, and should this provision be violated, the party of the first part may at their option end and terminate such contract. E. A. F.

96. Abandonment of Contract. In most large engineering contracts it is wise to provide for the emergency of abandonment. This term is here used to include not only deliberate and acknowledged abandonment of the work on the part of the contractor, but also such violations of the contract, either in the letter or in the spirit, or such unnecessary delay in its execution as may be construed as a virtual abandonment of the contract, so far as its express fulfillment is concerned. In such cases it may become necessary or desirable to take the work out of the hands of the contractor altogether, and to hire the necessary labor and purchase the necessary material, and complete the work under the direct superintendence of the engineer, charging all such items of expense against the contractor, and providing for the payment of the same, even though they should exceed all moneys due the contractor on the completion of the work. While the common law would warrant the party paying for the work in assuming the control of it, and charging the cost of the same against the contractor, in case of his express and acknowledged abandonment, it would not authorize the engineer in assuming control of the work because of delay or other violations of the terms of the contract. A provision such as the following may therefore be inserted:

Said contractor further agrees that if the work to be done under this contract shall be abandoned, or if this contract shall be sublet or assigned by said contractor, or any of the moneys or orders payable thereunder shall be assigned, otherwise than as herein provided, or if at any time said chief engineer shall be of the opinion, and shall so certify in writing to said board, that the said work is unnecessarily or unreasonably delayed, or that said contractor is willfully violating any of the terms, covenants and agreements of this contract, or is not executing this contract in good faith, or is not making such progress in the execution of said work as to indicate its completion within the required time, said board shall have the power and right to notify said contractor to discontinue all work or any part thereof under this contract, and upon such notification said contractor shall discontinue said work, or such parts thereof as said board may designate; and said board shall thereupon have the power to employ by contract, or otherwise, and in such manner and at such prices as it may determine, any persons and obtain any animals, carts, wagons, appliances, implements, tools, and other means of construction, which it may deem necessary to work at and be used to complete the work herein described, or such part thereof as said board may have designated; also, the power to use such appliances, implements, tools, and materials and means of construction of every description as may be found upon the line of said work, both such as enter into the completed work, and such as are necessarily used in and about the same in the course of construction, and to procure other proper materials for the completion of the same; also to charge the expense of all of said labor, materials, animals, carts, wagons, appliances, implements, tools and means of construction to said contractor; and the expense so charged shall be deducted and paid by said board out of such moneys as may be due or become due at any time thereafter, to said contractor under this contract, or any part thereof. In case such expense is less than the sum which would have been payable under this contract if the same had been completed by said contractor, it is agreed that said contractor shall be entitled to receive the difference; and in case such expense shall exceed the sum which would have been payable under this contract if the same had been completed by said contractor, then said contractor shall pay the amount of such excess to said city, on notice from said board of the excess so due. It is further agreed that neither an extension of time, for any reason, beyond that fixed herein for the completion of the work; nor the performance and the acceptance of any part of the work called for by this contract, shall be deemed to be a waiver by said city of the right to assume control of this contract for the reasons and in the manner hereinbefore provided. E. K.

97. Cancellation of Contract for Default of Contractor.

In the previous case it was provided that under certain contingencies the party of the first part would be warranted in assuming entire control of the work, and completing it under the contract, and for the contractor. His agency in the matter being displaced by that of the engineer, because either of gross violation of the contract, or for incompetency or unwillingness to carry it out. That clause provided, therefore, that the engineer should under such contingencies be appointed to carry out the contract with the party of the second part, in his stead, the contract itself, however, still remaining in force, and the final settlement to be made in accordance with its terms.

For a similar set of contingencies as above described, the party of the first part may prefer to cancel the contract altogether, and instead of completing the work under the supervision of the engineer, he may prefer to let a new contract for the carrying on of the work. To do this, the contract itself must be rescinded or canceled, and in order to give the party of the first part the legal authority for doing this, a clause such as the following may be inserted. Here all moneys due upon the contract at the time the contract is canceled will be forfeited to the first party. See article 49.

In lieu of the exercise of the power hereinbefore given, in case of said contractor's default, to employ workmen, purchase tools and materials, and complete the work, said board reserves the right and option, instead thereof, to annul and cancel this contract and relet the work, or any part thereof, and said contractor shall not be entitled to any claim for damages on account of such annulment, nor shall such annulment affect the right of said city to recover damages which may arise from such failure on the part of said contractor to fulfill the terms of this contract. And in case of such annulment all moneys due said contractor, or retained under the terms of this contract, shall be forfeited to said city, and be paid to the credit of the fund for extending water pipe in said city; but such forfeiture shall, however, not release said contractor, or his sureties for the fulfillment of this contract, and said contractor and his sureties shall be credited with the amount of the moneys so forfeited toward any greater sum that they may become liable for to said city on account of the default of said contractor.

E. K.

98. Workmen's Quarters and Other Temporary Buildings. It is usually necessary for the contractor to erect temporary buildings for the protection of his tools and machinery, or for office purposes, and sometimes, when the work is at a distance from boarding house facilities, it is necessary for him to provide temporary quarters for his labor. The location, erection, and removal of such temporary structures should also be subject to the approval of the engineer in charge. If temporary quarters for workmen are not really necessary, it is best to prohibit them, at least to prohibit their erection on the property belonging to the party of the first part. The following is an example of such a clause:

> The contractor may build such sheds, storehouses, etc., as are necessary for the work, but the location of such sheds, etc., must be such as will not interfere with the work of other contractors, and must be approved by the water commissioner. No buildings, sheds, or tents to be used as quarters for workmen or teams will be allowed on the city property.*

M. L. H.

99. Cleaning up after Completion. In nearly all kinds of engineering construction the grounds surrounding or along the line of the work are necessarily more or less defaced and encumbered by various disturbances of the surface, or by refuse and waste material, temporary buildings, etc., and it is usually made the business of the contractor on the completion of the work to clear up the grounds, and to put them in as presentable a condition as practicable. This does not involve any grading or removal of earth, unless it be the excess or waste which remains on the natural surface from his own excavations. It does, however, include the cleaning up of his own work, whether it be buildings, foundations, masonry, conduits, pits, etc. The following is such a clause written to cover the case of waterworks engine pits:

> When the work is completed, all pits, pipes, chambers, conduits, etc., shall be carefully cleaned out. The surround-

---

* To which might be added the following: Suitable privy conveniences shall be erected, as directed by the engineer, for the use of the workmen, and their use is made obligatory. The committing of nuisances is prohibited on all parts of the premises.

ing grounds shall be cleared of all rubbish caused by construction, all sheds, etc., and left in a neat and presentable condition.   M. L. H.

100. Removal of Condemned Material. Whenever any material which has been brought upon the ground by the contractor has been inspected and rejected by the engineer, or his assistants, it should at once be removed from the line of the work, in order to prevent its use when the engineer or his inspectors are not present. To further insure against the use of condemned material by the contractor, it is sometimes specified that all such material shall be stored by the contractor in a specified place, where it shall be kept under lock and key, and under the control of the engineer only. In case the contractor declines to remove such material from the line of the work, or declines to take out any defective work, there should be a provision authorizing the engineer to do this at the contractor's expense. The following clause may be used:

Defective work and material may be condemned by the engineer at any time before the final acceptance of the work, and when such work has been condemned it shall be immediately taken down by the contractor, and rebuilt in accordance with the plans and specifications. When defective material has been condemned, it shall be at once removed from the line of the work, and stored as directed by the engineer, or otherwise disposed of to his satisfaction. In case the contractor shall neglect or refuse to remove or replace any rejected work or material after a written notice, within the time designated by the engineer, such work or material shall be removed or replaced by the engineer at the contractor's expense.

M. L. H.

101. Relations to Other Contractors. Where more than one contractor is expected to be engaged simultaneously upon the same work, it is well to insert a clause in the specifications defining the obligation of each of these contractors to the others in certain particulars as follows:

The contractor is required so far as possible to so arrange his work and to so dispose of his materials as will not interfere with the work or storage of materials of other contractors

8

engaged upon the work. He is also required to join his work to that of others in a proper manner, and in accordance with the spirit of the plans and specifications, and to perform his work in the proper sequence in relation to that of other contractors, and as may be directed by the engineer.

M. L. H.

102. Provision for Drainage. Where the natural surface drainage is likely to be interfered with by the work of the contractor, it may be specified that he shall maintain provision for such surface drainage during the progress of the work, and that he will be held liable for all damages from his neglect to comply with this provision. The clause may read as follows:

If it is necessary in the prosecution of the work to interrupt or obstruct the natural drainage of the surface, or the flow of artificial drains, the contractor shall provide for the same during the progress of the work in such a way that no damage shall result to either public or private interests. For any neglect to so provide for either natural or artificial drainage which he may have interrupted, he shall be held liable for all damages which may result therefrom during the progress of the work.

103. Provision for Public Traffic. If it becomes necessary in the prosecution of the work to obstruct the public streets or sidewalks, and if it is practicable to carry on the work without closing these streets against all traffic it should be specified that

The contractor shall make suitable and adequate provision for the safe and free passage of persons and vehicles by, over, or under the work, while in progress. Such provision to be made to the satisfaction of the engineer.

E. A. F.

104. Contractor to Keep Foreman or Head Workman, and also Copy of Plans and Specifications on the Ground. Whenever work is visited by the engineer or his assistants or inspectors, the plans and specifications should be available for examination and if instructions are to be given for the further prosecution of the work or for any changes or corrections, some responsible person should always be present who is authorized to receive such instructions for the contractor, as his agent. In this case the instructions given to this agent have all the legal force which they would have if given directly to the contractor. This clause may read as follows:

> At all times when work is in progress, there shall be a foreman or head workman on the grounds, and also copies of the plans and specifications. Instructions given to such foreman or head workman shall be considered as having been given to the contractor.

E. A. F.

105. Cost of Examination of Completed Work.

Whenever the engineer desires to examine work which has been completed in whole or in part, this examination involving the tearing down of some portion of the work, and a corresponding expense both in taking down and in reconstructing it, it is only fair to provide that in case the work should be found to have been performed in accordance with the contract, the cost of tearing down and rebuilding should be paid by the party of the first part; but if it should be found that the work had not been constructed according to the contract, this cost should fall upon the contractor. The following is such a clause:

> Whenever required by the water commissioner, the contractor shall furnish all tools and labor necessary to make an examination of any work completed or in progress under this contract. If the work so examined is found to be defective in any respect, or not in accordance with this contract and specifications, the contractor shall bear all expenses of such examination and of satisfactory reconstruction.
>
> If the work so examined is found to be in accordance with the specifications and contract, the expense of examination and reconstruction will be estimated to the contractor at a fair price to be determined by the water commissioner.

M. L. H.

106. Faults to be Corrected at any Time before Final Acceptance.

It should usually be understood between the parties that no act of the engineer or of the inspectors should be construed as final acceptance of any portion of the work, unless it is specifically so declared in writing by the engineer.

Also that any failure to detect faulty or incomplete performance before the time of final acceptance should not be construed as an acceptance of the work. After the final acceptance by the engineer, the contract is no longer binding on the contractor in the way of requiring specific performance, but a reservation may be entered in the contract in accordance with which, if any defect or fault should subsequently appear which was undetected before the time of final acceptance, the party of the first part should have the right to recover damages for such fault or defect. A clause to this latter effect is not usually inserted, but it is legitimate if the circumstances should seem to require it. The circumstances might require it when the work is of such a character that faults could not readily be detected until the works had been put in operation. The following is an example of such a combined clause:

> Failure or neglect on the part of the engineer or any of his authorized agents to condemn or reject bad or inferior work or materials, shall not be construed to imply an acceptance of such work or materials if it becomes evident at any time prior to the final acceptance of the work and release of the contractor by the party of the first part; neither shall it be construed as barring the party of the first part, at any subsequent time, from the recovery of damages or of such a sum of money as may be needed to build anew all portions of the work in which fraud was practiced or improper material hidden, whenever found.

107. Surveys, Measurements, and Estimates of Quantities not Guaranteed to be Correct.

It is not usually possible to give in advance complete measurements, dimensions, and estimates for all parts of the work. Especially is this true of the more detailed dimensions. It should always be understood, therefore, that the contractor must be responsible for the proper adjustment of the dimensions and details of the different parts of the work to each other and that the dimensions and figures given on the plans and specifications are always subject to changes during the progress of the work.

The following clause refers especially to the construction of a steel viaduct:*

Contractors are also required to check all leading dimensions and clearances as a whole and in detail, the fitting of all details, and to become responsible for the exact position and elevation of all parts of the work, which will only primarily be located by the engineer of the department of public parks. They will maintain their own field engineering, that of the city being for the purposes of original lay-out, inspection, and checking. The contractor must provide and maintain such facilities for the engineer or his assistants as he may require for the convenient examination and inspection of the work in progress. He will pay the cost of testing all material in laboratories or shops, and the cost of such mill and shop inspection as he may be called upon to perform in addition to that furnished by the engineer, the selection of such laboratory or inspectors being dictated by the engineer, to whom they will report. He will furnish such monthly progress photographs as may be required to maintain the record. A. P. B.

108. The Contract Subject to Interpretation and Change by the Engineer in the Following Particulars:†

(*a*) Where meaning is obscure and uncertain.

(*b*) As to what is implied beyond that which is specifically described.

(*c*) In case of discrepancies between plans and specifications.

(*d*) In case changes of plans or methods of work are afterwards decided upon.

Since the engineer is the author of the specifications, he evidently is the proper party to interpret their meaning. It goes without saying that the specifications and plans should be as clear and definite as possible in all particulars, but it is quite impossible to free language from many inherent defects, neither is it practicable to describe minutely and in detail all

*In this case both foundations and superstructure formed one contract. If the owner should prepare the foundations he must guarantee his surveys and locations to be as shown on the drawings, or as described in the specifications. In this case provision must also be made for a comparison of the standards of length used by the owner and by the contractor.

† See Arts. 32, 33, 34, 38 and 39.

parts of the work. There will, therefore, usually be some
uncertainty as to the real meaning of the words used in the
specifications, or even of the drawings themselves, and many
of the details of the work must be understood by implication,
rather than described in either the specifications or the plans.

Occasionally also by some oversight the plans and specifications will not agree. This usually results from changes of
plan after one or the other has been drawn, such changes being
made in the one place and not in the other. As a rule the
specifications control, rather than the plans, and the figures on
the plans control, rather than the actual dimensions of the
drawings when taken to scale. The engineer should, however,
be at liberty to determine what the real meaning was intended
to be in all cases of discrepancies.

Very few contracts for large works are carried out from
beginning to completion without changes being introduced in
both the plans and in the specifications during the progress of
the work. These changes may arise from a newly devised
method or plan which may be considered superior, or from
unlooked for obstacles met with in the work, or from suggestions on the part of the contractor himself. They also are
frequently made in order to reduce the cost of the work, and
on the other hand are sometimes made in order to improve its
character. It should be understood, therefore, that the engineer
has the privilege of making such changes in the plans and
specifications at any time.

So far as the engineer may add to the plans or specifications by way of interpretation of their true meaning, as in (*a*)
and (*b*), such supplementary and explanatory matter should
not involve any change in the contract price.

In the matter of discrepancies, however, between plans
and specifications, if the contractor can show that he based his
estimate on one of these to the exclusion of the other, and when
interpreted by the engineer, he finds he had estimated on a plan

materially cheaper than that now required, it would be but just and right to allow him the difference in the cost, since he had the right to suppose that the plans and specifications were in accord.

When changes are introduced in the plans or specifications after the contract is let, such changes create a new contract, and as a matter of course there must be a new agreement as to compensation. Without a special clause authorizing such changes neither party could change the terms of the contract against the will of the other without breaking it. Furthermore without some understanding as to how the compensation should be determined for such change in plans or specifications, the party of the first part would be at the mercy of the contractor in this matter, and he could charge an extravagant price for such changes, and there would be no remedy. The following is a suitable clause, covering all these matters:

Said contractor also agrees that said chief engineer shall decide as to the meaning and intent of any portion of the foregoing specifications, or of the plans, where the same may be found obscure or in dispute; and said chief engineer shall have the right to correct any errors or omissions therein, when such corrections are necessary to the proper fulfillment of the intention of said plans and specifications; the action of such correction to date from the time said chief engineer gives due notice thereof. And it is also agreed by said contractor that said board may, at any time, make any changes in the location, form, dimensions, grades, and alignments, and may make any variations in the quantity of the work to be done, as exhibited in the advertisement or notice of letting hereto attached, or in the form of proposal or bid for said work, and may entirely exclude any of the items of work relating to said quantities at any time, either before the commencement of the work, or during its progress, without thereby altering or invalidating any of the prices herein named, or this contract in any other respect; should such action diminish the amount of work that would otherwise be done, no claim shall be made for damages on the ground of loss of anticipated profits on work so dispensed with; and should such action be taken after the commencement of any particular piece of work, and result thereby in extra cost to said contractor, said chief engineer shall certify to said board the amount to be allowed therefor, which he shall consider fair and equitable, as between the parties, and his decision, when approved by said board,* shall be final and conclusive.   E. K.

109. Settlement of Disputes.† While the contract lies between the party paying for the work, being the party of the first part, and the contractor who does the work, being the party of the second part, the contract itself is administered and enforced by the engineer, who is usually employed by the party of the first part. It is well understood also that the engineer is supposed to act always in a strictly professional and administrative capacity, that he has no personal interest in favor of, or against either party, and that his sole object is to see that the contract is faithfully carried out in accordance with its express terms and real meaning. It is also recognized that he is the most competent person to determine all differences and disputes, where these arise between the parties to the contract, or between two or more contractors engaged upon the same work. It is proper and right, therefore, that he should be made the referee in all cases of dispute or misunderstanding, and that his position as arbitrator should be made final and conclusive in the premises. If it be expressly agreed upon between the parties themselves that the engineer shall act in this capacity, then his decision does become binding and final upon the parties, even to the exclusion of the action of the courts, unless it can be shown that the engineer acted through prejudice, or ignorance, or fraud. As it is usually very difficult to establish a case against the engineer on either of these grounds, a clause such as the following usually acts to settle all disputes and to keep such controversies out of the courts. Honesty and fairness is also so common a characteristic of engineers that a clause such as the following is nearly always acceptable to both parties, and very seldom results in injustice being done to either party.

* This decision of the engineer is usually made final and conclusive without approval by his principal.

† See articles 12 and 13.

To prevent all disputes and litigation, it is further agreed by and between the parties to this contract, that said chief engineer shall be the referee, in all cases, to determine the amount, quality, acceptability, and fitness of the several kinds of work which are to be paid for under this contract, and to decide upon all questions which may arise as to the fulfillment of said contract on the part of said contractor, and his decision and determination, when approved by said board* shall be final and conclusive. Said contractor shall also afford all reasonable facilities for access to his work to any other parties or contractors who may be doing extra work or be working on a section of the conduit adjacent to his own, and any difference which may arise between two contractors in regard° to their adjoining work is to be adjusted by said chief engineer, whose decision in the matter shall be final and binding upon both parties.†   E. K.

110, Extra Work.‡ While all changes in plans and specifications have been provided for in section 51, it is well to insert a special clause on the subject of extra work. It is common for contractors, on the completion of a piece of work, to bring in a bill of extras, which they claim represents work which they were asked to perform, and which was not included in the plans or specifications, and which was not specially provided for by particular agreement with the engineer with the corresponding compensation to be paid for it. What the contractor's ideas or intentions may be on this subject does not usually develop until the work has been fully completed and the time of final settlement has arrived. In many instances it is then too late to determine the exact facts concerning this extra work, either because of the incompleteness of the records,

* See footnote, page —.

† In the opinion of the author of this work it is doubtful if a clause such as is here given will always stand in a court of law. The reader is referred on this subject to Arts. 12 and 13 in the Synopsis of the Law of Contracts. In accordance with the principles there laid down it would seem that the courts will only sustain such a clause as the above when it can be shown that the acts of the engineer taken under it have been such as a court could properly refer to an expert referee, or to a person presumably more competent than the court to determine. In general such questions would be such as might be called "*Conditions Precedent*" to a legal settlement. Such "conditions precedent" would include all questions, such as the value of extra work, the amount of damages actually sustained from any breach of the contract, the extent of any failures to comply with the contract, and all matters which are not so much questions of fact as questions of quantitative and qualitative value, which can only be estimated, and which the engineer is presumably competent to evaluate. See also Art. 86.

‡ See Articles 3S and 39.

or because of the impracticability of making the necessary
measurements. Such a bill of extras, therefore, brought in at
the time of settlement is always the source of a certain amount
of difficulty and irritation, and when the piece of work extends
over a considerable length of time, such a contingency as above
described should be prevented by requiring all such bills of
extras to be presented from month to month. Furthermore, it
is desirable also that the contractor should reveal to the engineer his intentions in regard to claims for extras, before such
extra work is executed. In this case if he will not accept of the
price fixed by the engineer for doing 'such work the engineer
should have the privilege of letting this extra work to another
party. In this way extravagant prices for such work could be
prevented, and disputes avoided, and the following is given as
a good example of such a clause, on a piece of work which
extended over a considerable period of time :

No claim for extra work shall be considered or allowed,
unless such extra work shall have been previously ordered by
said engineer, in writing. The claims for such extra work,
when so ordered, shall be presented to such board on or before
the 15th day of the month following that in which said extra
work was done, otherwise such claims during that month will
be forfeited and waived. In case any extra work shall be
required in the proper performance of the work contemplated
to be done under this contract, it is understood that said board
reserves the right to have such extra work done by any other
person, firm, or corporation than the said contractor, unless an
agreement upon the prices to be paid for such extra work can
promptly be reached between said board and contractor. Should
said extra work be let to any other person, firm, or corporation
than said contractor, said contractor further agrees that he will
not, in any way, interfere with, or molest such person, firm, or
corporation, and that said contractor will suspend such part of
the work herein specified, or will carry on the same in such a
manner as he may be ordered by said engineer, so as to afford
all reasonable facilities for doing such extra work; but said
contractor agrees to make no claim for damages, or for any
privileges or rights, other than expressed by this contract, by
reason of the suspension and the doing of such extra work,
except for an extension of time to perform this contract, as may
be certified to said board in writing by said chief engineer, and
approved by said board.   E. K.

111. Definition of "Engineer" and "Contractor."

While it is not at all necessary as a rule to define the terms "Engineer," "Contractor," "Board," etc., it is usually well to insert such a definition, to prevent any legal quibble in case suit is brought by either of the parties to the contract. In this definition also the agency of persons acting for either of the principals or for the engineer is also defined.

Wherever the word "engineer" is used herein, it shall be and is mutually understood to refer to————and to his properly authorized agents, limited by the particular duties entrusted to them.

Wherever the word "contractor" is used herein, it shall be and is mutually understood to refer to the party or parties contracting to perform the work to be done under this contract, or the legal representatives of such party or parties.

E. A. F.

112. Documents Composing the Contract.

While in common law all the documents, acts, agreements, public advertisements, etc., which relate to or serve to explain the full meaning and intent of the contract, are made portions of such contract, it is well also to specify particularly what documents combine to make what is understood by the parties as "the contract." This clause is frequently inserted in the enacting agreement, which may or may not precede the specifications proper. It is here inserted as a clause in the specifications, but perhaps more properly belongs in what is sometimes designated more specifically as "the contract." The clause may read as follows:

It is understood by the contracting parties that the following documents are essential portions of the complete contract: The advertisement, the instructions to bidders, the proposal, all drawings, maps, and plans, hereto attached or herein described, the specifications, specific contract, and the contractor's bond.

113. Meaning Understood.

It is not unusual for contractors to enter a plea, either during construction or on final settlement, that such and such parts of the specifications were not understood, and that their bids were made under a misapprehension.   To prevent the making of such a claim the following may be inserted:

Said contractor hereby admits that he has read each and every clause in this contract, and fully understands the meaning of the same, and hereby agrees that he will comply with all the terms, covenants and agreements herein set forth.

E. K.

114.   The Use of General Clauses in Engineering Specifications. While the general clauses here described with illustrative examples may appear to the reader unnecessarily voluminous, their purpose and effect is to clearly define the business relations of the parties, and to prevent injustice being done to either party.   They are also calculated to prevent litigation and delay in the final settlement, and if they are able to effect these ends they are well worth inserting, even where the work to be done is relatively small, and unimportant.   The engineer should be careful, however, that all such clauses are consistent as between themselves, and it is best also to make them mutually exclusive.   In other words, the same thing should not be described or defined in more than one clause, as repetitions only weaken the document.   Furthermore, *no condition or limitation should be inserted without a full intention of strict compliance.*   If the engineer begins to relax in his requirements in one particular, the contractor will not be slow to take advantage of such precedents, and to claim similar privileges in other directions.   If the engineer could know in advance who the contractor was to be, many of the clauses here offered might be dispensed with, in case the contractor was known to be thoroughly honest and competent.   The specifications are prepared in advance, however, and it is wise to assume that the contractor will be a more or less irresponsible party, without reputation to sustain, and whose sole object is personal gain.   It must also be understood that the clauses here given are offered only as illustrative examples, and not to be blindly copied.   The engineer in writing the specifications should have clearly in mind what the business relations are intended to be, and make his general clauses consistent with that conception. He could probably consult other specifications, or the clauses as given above, as suggestions and to enable him to avoid omitting some essential condition which he wishes to insert. It is believed, by the author, that the clauses here quoted have the support of the leading members of the profession in this country, and that they are well adapted to determine the conditions which the engineers who use them desired to impose. It must not be supposed, however, that all these general clauses would ever be embodied in any one specification.

PART III.

Specific Descriptive or Technical Clauses in Specifications.

115. Essential Features of Good Specifications.
We now come to consider that portion of any given set of specifications which relates to and describes the work itself. In writing specifications of this kind, the following requirements should be complied with:

(*a*) The work should be described first as a whole, and then in detail.

(*b*) Every portion and detail of the work should be described in clear and simple language which will be understood by the contractors who are supposed to bid on the work. These descriptions should have reference to the ultimate end to be accomplished rather than to the means and methods to be employed. It is usually not wise to specify methods unless in the opinion of the engineer some particular method is far preferable to any other.

(*c*) The clauses in the specifications should be made so far as possible mutually exclusive. That is to say, no part of the work should be specifically described in more than one place. Repetition of descriptions tends to weaken the document.

(*d*) The specifications should be clear in the matter of indicating what is absolutely required without any alternative, and what is named as indicating in general the character of the product, and in which alternative materials, methods, or results will be allowed. If the engineer anticipates that some clauses

126

in the specifications are to be rigidly insisted upon, while others will not be specifically enforced, this intention or state of mind of the engineer should be revealed in the specifications themselves. In other words, the contractor should know in advance how the specifications are to be interpreted, so far as it is possible to give this information in the specifications themselves.

(*e*) The last named requirement demands that the engineer should be so familiar with all the details of the work described, from actual experience, that he is able to know in advance what his decision will be in the various contingencies which may arise during the progress of the work. His foresight in this particular must be complete and distinct, which can only be the case when the engineer who writes the specifications has had considerable experience in carrying out practically the same kind of work.

(*f*) In choosing units of measure, in describing the work, to be used for determining compensation, only specific and definite units should be chosen, or they should be so defined as to admit of no double meaning. For instance, to say that mortar shall be composed of one part cement to two parts of sand comes very far from defining a particular ratio of ingredients. If the words "by weight" be added, it still fails to define, inasmuch as wet sand is much heavier than dry; or if the words "by measure" be used, this also fails to define, since cement may be measured in the original package, where it is thoroughly compacted, or it may be dumped and measured in a fluffy condition, in which it occupies some 50 per cent. more volume.

(*g*) The engineer should be familiar with the ordinary methods employed by different kinds of mechanics and should so design his work as to obtain satisfactory results without requiring a much higher grade of work than is customary by the mechanics who will be called upon to execute it. It is practically impossible with the most thorough supervision and inspection to get mechanics to vary their ordinary practice materially. The failure to recognize this fact often leads engineers to specify methods or results which are practically unattainable, and this leads either to continual violation of the specifications and its accompanying irritations and delays, or to an abandonment on the part of the engineer of the strict interpretation of his own specifications.

(h)   In the matter of materials it is customary to specify not the very highest and best the market affords, but such a grade of material as would be satisfactory in service, and which can be supplied by the standard manufacturers of that particular product. In this way the engineer gets the benefit of a wide competition, and of a correspondingly low price. The minimum requirements for materials which serves as a criterion of rejection determines very largely the cost of the work. If, therefore, the engineer in preparing his design bases his calculations upon what might be commonly known as good or first-class materials, with a minimum limit fairly below this generally recognized first-class grade, he will usually obtain a material practically as good as the market affords, without being obliged to pay an extravagant price for it, and without suffering from the delays and troubles caused by the rejection of a large proportion of the material furnished. To base a contract on the very highest tests known of a given material, and to require this extraordinary quality for all the material furnished is extremely unwise.

(i)   If possible to avoid it, it is best not to specify a particular manufactured product or proprietary article by name. If this is done at all, more than one such name should be given if possible, and others admitted if shown to be as good as these to the satisfaction of the engineer. To limit the materials or articles specified to that of a single manufacturer subjects the engineer to invidious criticism and suspicion, and it is always wise to avoid even the appearance of evil.

(*j*) It is not uncommon to specify that the materials furnished shall be of well known brands, or the products of factories or works of established reputation. Similarly it is sometimes specified that the contractor himself must show a familiarity with the work he proposes to perform. Also in the matter of stone, for instance, that it shall be taken from quarries which have been in long use, the stone from which by actual use is known to have good weathering qualities.

(*k*) Before writing the specification, the engineer must also have a clearly defined notion as to the amount of responsibility which is to be placed upon the contractor. If the engineer prescribes the plan, the materials, and the methods to be employed, he, of course, assumes all responsibility so far as these are concerned, and he can not.in justice make the contractor responsible for his own faults. He should always be ready and willing to take upon himself such responsibilities pertaining to the design as properly belong to the designer. For instance, if the contractor is held responsible for the results, he must be given a considerable latitude as to methods, and if the engineer prescribes the plan, the materials, and all the methods to be employed, he can not hold the contractor for the results, or for the successful operation of the project, beyond the simple faithful performance of the work prescribed. In general the entire responsibility for the successful operation of the work should rest upon the engineer. It is only in case of extra-hazardous undertakings, which are largely of the nature of an experiment, in which no well defined plan is outlined, that the contractor is left comparatively free both as to plan and execution. It is only in such cases that the engineer is warranted in relieving himself of all responsibility, and placing the same wholly upon the contractor. There are cases in which contractors have, by experience, acquired peculiar ability to perform certain kinds of hazardous work, and who are willing to undertake the same under a guarantee of suc-

9

cessful execution, under which circumstances, any engineer, even of high repute in his profession, would be warranted in letting a contract and putting the entire responsibility upon the contractor.

(*l*) A strong reason for enforcing specifications literally and rigidly, instead of accepting some other material or method which possibly may be "just as good" is to be found in the relation the engineer and owner hold to other contractors who bid upon the work. It is to be presumed that these other parties have based their estimates on a strict compliance with the specifications, and it is possible that the lowest bidder has presumed on his being able to substitute cheaper materials or methods for those specified. If he is allowed to do this after the contract is let, it is evident that the other bidders have been discriminated against to their disadvantage and under a species of fraud, which should not receive the encouragement of either the engineer or of the owner. It is difficult, therefore, to see how a cheaper grade of work can properly be accepted in lieu of that specified, even though it be "just as good," without encouraging this practice of presuming upon a cheaper fulfillment, and also without treating the other bidders unfairly. Other things being equal, therefore, it is best to rigidly enforce a contract, even though a cheaper material or method might, in the opinion of the engineer, be employed with equally good results. Or, if a cheaper compliance is allowed, a corresponding reduction in price should be insisted upon.

The above are some of the numerous controlling ideas which the engineer should have clearly in mind in the writing of a set of engineering specifications. He must know in the first place exactly what he wants, and then try to so describe it that others can not mistake his meaning. The general and detail plans are usually made before the specifications are written, and the engineer has these before him in writing the specifications, and makes liberal reference to them. Since they are also a part of the specifications, he has the advantage of a double language in which to present his ideas, and, if he does not succeed in making clear to the proposed contractors exactly what is to be done, he should feel that he alone is to blame for any misunderstanding.

116. Specifications Accompanying Complete Detail Plans. As described in article 17, we have in general three classes of engineering specifications, describing the work itself, namely:

Specifications accompanying complete detail plans.

Specifications accompanying a general plan only.

Specifications unaccompanied by any plan, and commonly known as General Specifications.

When the specifications are accompanied by complete detail plans, these plans are prepared before the specifications are drawn, and in this case the specific descriptive specifications are largely composed of descriptions of the materials to be used, the methods of manufacture and erection, and of the results to be accomplished, or of the tests to which the finished product is to be subjected. Such complete detail plans are always necessary when a particular and definite plan is to be carried out. When the work is to be let in open competition, and to the lowest bidder it is also usually best to prepare complete detail plans in order to insure that the bidders will all estimate on exactly the same thing, and further, to insure that the final product will be satisfactory. In this case the detail plans must be duplicated by some of the various methods of copying drawings, and in very large and important work frequently by photo-lithographing, and these copies of the plans submitted along with the specifications to the various parties wishing to bid upon the work. If the bidders are expected to be those residing in the immediate vicinity, it is not so necessary to duplicate these drawings, all the bidders being asked to examine the drawings in the office of the

engineer. In this case the drawings usually are self-explana-
tory even to the extent of indicating the materials to be used,
so that the written specifications need not describe the work as
to its form and dimensions, but are only descriptive of the
work in a general way.

**117. Specifications Accompanying a General
Plan Only.** In this case the bidder is asked to submit detail
plans for the work in submitting his bid, the specifications
being so framed, however, as to reduce all designs which fully
comply with these requirements to a common standard of
value. To accomplish this purpose the engineer must be able
to foresee practically all the various designs which may be
submitted, and to anticipate all the advantages in economy
which are likely to control the preparation of these designs,
and to make such requirements in the specifications as, when
complied with, will in his opinion give products of equal efficiency, capacity, and permanency. In this way the engineer
gets the advantage of the experience and of the inventive genius
of all the persons bidding upon the work, with the chance
that he may secure a better design than he himself would
have been able to prepare, and also one which may cost materially less than his own design. As an illustration of this
kind of specification, we may have a bridge or roof truss of a
particular general design, the outline drawing showing simply
the general dimensions, and the location of the members, this
fixing the general style of truss, which for some particular
reason the engineer wishes complied with. The contractor is
then asked to build a truss on these general lines in compliance with the accompanying written specifications.

**118. Specifications Unaccompanied by Plans,
Known Commonly as General Specifications.** In
this case no plans whatever are submitted, but only the most
general requirements to be satisfied. As, for instance, in the
case of a bridge, the total span, the loads to be carried, the

character of the stream to be bridged (this determining the character of false works required, and often determining the character of the superstructure itself), the kinds of materials to be used in its construction, the maximum unit stress allowed in the various parts, etc. Or if the contract is for machinery, the specifications may define the amount and kind of work to be done, which fixes the capacity of the apparatus. They would also define the efficiency or economy of operation of the plant, and make various requirements in regard to the material used, and methods of construction which would be supposed to govern its permanency. In this case, as in the one preceding, the engineer must have constantly in mind in writing the specifications the possibility of complying with them with some kind of cheap product which, while fulfilling the letter of the requirements of the specifications, would not be at all what he hopes to obtain, or what would be consistent with the more standard forms of construction.

ILLUSTRATIVE SPECIFICATIONS OF VARIOUS ELEMENTARY PORTIONS OF ENGINEERING WORK.

119. Scope and Purpose. It is proposed in the following articles to give illustrative examples of what is considered good practice jn describing many of the more common elements of engineering construction. The practice pursued in discussing the general clauses in engineering specifications, of first considering the purpose of the clause, and then giving an illustrative example, will be followed here. Frequently, however, more than one illustration will be given, and the sources from which they are obtained will also be indicated. It is thought this arrangement will serve a better purpose than to give a series of complete specifications of various kinds of engineering work, without a detailed discussion of the several clauses.   By the arrangement here used duplication of parts is avoided, and the reasons for the particular descriptions can be given in their proper connection.   These reasons will perhaps be more useful to the young practitioner than the particular specifications themselves.

SPECIFICATIONS ON EXCAVATIONS AND EMBANKMENTS.

120.   Earth Work, Excavation, and Grading. A specification for excavation, or grading, should satisfactorily cover the following ground:

(*a*)   Location and general description of the work.

(*b*)   As full a description of the character of the materials to be excavated as can be furnished.

(*c*)   A classification of the materials which will be employed, and the methods of measurement.

(*d*)   A description of the lines of limits of excavation and fill, including borrow pits, and waste banks.

(*e*)   The disposition to be made of the excavated materials.

(*f*)   The distance to which the material is to be transported, commonly included under the general head of "Haul."

It is always wise for the engineer to determine in advance approximately the character of the material to be encountered and to give to the contractor the benefit of such information. It is not wise, however, for him to guarantee the quality of the material to be as indicated, since this furnishes to the contractor grounds for claims which the engineer will find it difficult to adjust.   The character of the material is usually learned approximately by borings along the line of the work.   In the case of railroad work it is not customary to do this, but from the general knowledge of the geologic formations, the various kinds of materials can be fairly anticipated.

The specifications should be very explicit and clear beyond all possibility of doubt in the matter of grades or classification of materials to be encountered, and the methods which will be used in measuring the quantities. Innumerable difficulties are constantly arising in the carrying out of engineering specifications from misunderstandings on these points. Materials encountered in excavation are not only various as to quality, but all gradations are found, so that it is practically impossible to determine where one classification or kind of material ends and another begins. In fact two engineers will often classify the same materials differently under the same specifications, and there is no possibility of determining such questions, except by agreeing to abide by the decision of the engineer. It is best not to have too many classes of materials, and all kinds of materials are usually grouped under three general heads, namely: *Earth, loose rock,* and *solid rock.**

121. Grading. The following specifications for railroad grading are used by the Pennsylvania Railroad:

Under this head will be included all excavations and embankments required for the formation of the roadbed; cutting all ditches or drains about or contiguous to the road; the foundations of culverts and bridges, or walls; the excavations and embankments necessary for reconstructing turnpike or common roads, in cases where they are destroyed or interfered with in the formation of the railroad; and all other excavations or embankments connected with or incident to the construction of said railroad.

All cuttings shall be measured in the excavations, and estimated by the cubic yard, under the follow heads, viz: *Earth, Loose Rock, Solid Rock.*

Earth—will include clay, sand, loam, gravel and all other earthy matter, or earth containing loose stone or boulders intermixed, which do not exceed in size three cubic feet.

* In the specifications used on the Chicago Drainage Canal but two classes of materials were named, "Solid rock" and "Glacial drift." Where no solid rock was anticipated, but one class was named and this was called "Excavation." Sometimes more than three classes are recognized and provided for when the grades are distinct and well marked.

Loose Rock—shall include all stone and detached rock lying in separate and contiguous masses containing not over three cubic yards; also, all slate or other rock that can be quarried without blasting, although blasting may be occasionally resorted to.

Solid Rock—Includes all rock occurring in masses exceeding three cubic yards, which can not be removed without blasting.

The roadbed will be graded twenty feet wide in earth cuttings and fifteen feet in fillings, except where otherwise directed by the engineer, with side slopes of such inclination as the engineer shall in each case designate, and in conformity to such depth of cuttings and fillings as may have been or may hereafter be determined upon by said engineer.

Earth, gravel and other materials taken from excavations (except when otherwise directed by the engineer), shall be deposited in embankments, the cost of removing which will be included in the price paid for excavation. All material necessarily procured from without the road and deposited in the embankments will be paid for as excavation only. In procuring materials for embankment from without the line of the road, the place will be designated by the engineer in charge of the work; and in excavating and removing it, care *must* be taken to injure or *disfigure* the land as little as possible. The embankments will be formed in layers of such depth (generally one foot), and the materials disposed and distributed in such manner as the engineer may direct, the required allowance for settling being added.

No borrow pits will be opened nearer than four feet from base of embankment slope, and will receive same slope as corresponding embankment. All borrow pits will be excavated in a regular manner and so as to leave no holes for standing water, generally with a descent at bottom to allow free passage of water.

Wherever the excavations furnish more material than is required for embankments, the surplus will be used to increase width of embankment, or deposited in spoil banks or waste piles, as and where the chief engineer may direct.

The roadbed, in cuts and on banks, to be made in a workmanlike manner; to be perfectly even and regular according to grade stakes as set from time to time by the engineer in charge, and to be exactly of the width directed.

All slopes to be formed even and straight, according to slope stakes, and to such incline as directed in each case.

All ditches in cuts or along banks to be made of such width and grade as the engineer in charge may direct.

If the contractor shall make excavations or embankments in excess of the directed width, then such excess shall not be paid for.

Over culverts and behind bridge abutments the embankments shall be formed carefully, so as to avoid damage to, or bulging of, the masonry. Only the best materials will be used for this purpose, the same to be deposited in layers of not over ten inches thick. The contractor to be responsible for any damage to the masonry.

Contractors, when directed by the engineer in charge of the work, will deposit on the side of the road, or at such convenient point as may be designated, any stone or rock that they may excavate; and if, in so doing, they should deposit material required for bank, the additional cost, if any, of procuring other materials from without the road will be allowed. All stone or rock excavated and deposited as above will be considered the property of the railroad company, and the contractors upon the respective sections will be responsible for its safe keeping until removed by said company, or until the work is finished.

The line of road, or the gradients, may be changed if the engineer shall consider such change necessary or expedient; and for any considerable alterations. the injury or advantage to the contractor will be estimated, and such allowance or deduction made in the prices as the engineer may deem just and equitable; but no claim for an increase in prices of excavation or embankment on the part of the contractor will be allowed or considered, unless made in writing before the work in that part of the section where the alteration has been made shall have been commenced. The engineer may also, on the conditions last recited, increase or diminish the length of any section for the purpose of equalizing or balancing the excavations and embankments.

Whenever the route of the railroad is traversed by public or private roads, commodious passing places must be kept open and in safe condition for use; and in passing through farms the contractor must also keep up such temporary fences as will be necessary for the preservation of the crops.

The above specification makes no provision for what is commonly called "haul." By "haul" is meant an additional compensation for carrying the excavated material beyond a certain limiting distance. Such a provision usually accompanies a specification for excavation, and may be given in the following language:

The price paid for "excavation" in all the several classes thereof will be understood to cover and pay for the entire expense of its removal by any method whatever, including loading, unloading, transportation and deposit in the manner prescribed in these specifications, in the places designated by the engineer, provided the haul of the material so transported does not exceed —————— ( ) feet, and beyond that distance —— per cubic yard per one hundred (100) feet will be allowed and paid, for such extra haul, in addition to the price paid for excavation.

122. Excavations Under Water. Excavations made under water are usually for the purpose of securing a channel for the passage of boats. In many cases the character of the material is quite various, and largely unknown. It is proper for the engineer to make such investigations as are practicable to discover what these materials are, and to about what depth the cutting will have to be made, and to give the contractor the benefit of such information. It would not be well for him, however, to make positive statements as to the character of the material, and he should relieve himself in the specifications of all responsibility for the information given. The contractor on the other hand should inform himself of the nature of the work so far as possible, both by personal examination, and by availing himself of the investigations of the engineering department. The following paragraphs concerning excavations under water are taken from the standard specifications used by Col. O. M. Poe, of the corps of engineers, U. S. army.

All available information in the possession of the United States will be given upon application. The United States will not guarantee the correctness of its information and will not be responsible for the safety of the employees, plant or materials used by the contractor, nor for any damage done by or to them from any source or any cause. Bidders are expected to satisfy themselves as to the nature of the work to be done, and it will be assumed that proposals are based upon a thorough understanding of its character. Intending bidders are urged to visit the localities of the work, and, by personal inspection and inquiry, fully inform themselves as to the present and probable future conditions. Navigation shall not be obstructed, and no allowance or concession will be made for any lack of information on the part of the contractor regarding the work. The price bid shall be full compensation for furnishing all necessary labor, materials, and appliances of every description, and for doing all work herein specified to the satisfaction of the engineer officer in charge, and shall include all risks and delays of whatever nature attending the execution of the work.

The work comprises the improvement of two shoals in St. Mary's River, Michigan, above the canal. The upper shoal lies northwesterly, and the lower shoal northeasterly from old Round Island Light House.

The work to be done consists in excavating a channel within the side and end lines prescribed by the U. S. Agent in charge, said channel to have a bottom width of 300 feet, and a total length of about 3,000 feet, the estimated excavation being 90,000 cubic yards, bank measure, more or less. The greatest distance to the dumping ground will not exceed two miles, and the average distance will not exceed one mile.

· The material to be removed consists of boulders, clay, sand, gravel, and possibly hard pan, all in unknown proportions.*

The bottom of the completed cut shall be in a plane 25.32 feet below the upper surface of the coping of the lock of 1881 in St. Mary's Falls Canal, the face of excavation probably varying from scraping to 5 feet.

No payment will be made for excavation below 22 feet depth of water. All excavations within the specified side slopes and between 21 and 22 feet depth of water will be paid for at half rates in the final estimates.

Proposals will be received for bank measure only. For the monthly estimates 30 per cent. will be deducted from scow measurements, but the total of the estimates made for the upper shoal will never exceed the total of the estimates made for the lower shoal, until all the required excavation has been finished at one of the shoals, and an equal amount of excavation has been done at the other.† · O. M. P.

*In letting this work all materials were entered in one class, even though it were solid rock. In this particular instance no solid rock was anticipated, and hence it is not mentioned in the list of materials to be encountered, but in other sections solid rock was anticipated, and was so entered in the list of materials, but the proportion in each class was left for the contractor to estimate as best he could from the investigations made by the engineering department, and such additional borings as he might make himself, there still being but one price per cubic yard for all materials. The responsibility of the contractor in this connection is fully defined, however, in the extract given above.

†This work was let by the cubic yard as measured in bank. That is from soundings taken both before and after the excavation between the side lines indicated, and above the depth of 21 feet. Material excavated below the depth of 21 feet and above the depth of 22 feet was paid for at half rates. Any excavation below the depth of 22 feet was not paid for at all. Neither were the excavations made outside the side boundaries. In other words the material was paid for in cut as it would be described in earthwork out of water. The reason for requiring the work on one shoal to be fully completed before an equivalent sum is paid for the other is to insure the completion of the work. If a small amount of material had been left above the 21 foot line in various places, the excavation of these scattered portions would be very expensive per cubic yard, and if the contractor had already received the major part of his compensation it might be difficult to induce him to finish up the work properly.

123. Specifications for Measuring Quantities Excavated Under Water by Weight and Displacement.

The following specification is taken from the U. S. Engineer Corps specifications for excavation in the James River, and they illustrate another method of determining amount of material excavated. In this case the soundings in the river are only used for determining when the excavation has been completed to the proper depth, and to prevent excavation beyond a limit of over-depth.

CLASSIFICATION.—Mud, sand, clay, and gravel, under the class "*earth excavation*," will include those materials, however hard or compact they may be, the gravel to include pebbles up to three inches in diameter. "*Cobble*" will include stones up to three cubic feet. "*Soft Rock*" will include any rock in place which can be removed by dredges without blasting, although, for economy in removal, blasting is resorted to, and will include the disintegrated rock found for two or three miles below Richmond. "*Solid Rock*" will include rock which rings under the hammer, boulders measuring over three cubic feet, and other rock which, in the opinion of the Engineer officer in charge, can not be removed without blasting. When soft and solid rock occur together, in strata or otherwise, the amount of each will be ascertained by the inspector after the material is placed on lighters. Mixed classes of material must be separated by the contractor, at his expense, for measurement by displacement as the lighters are unloaded at the place of deposit.

UNCLASSIFIED ROCK.—In place of bidding on soft and solid rock separately, an alternate bid may be submitted for *unclassified rock*, which will include all material described in these specifications as rock, *whether soft or solid*. The bid will be for the ton of 2,000 pounds, and the weight will be ascertained by displacement, as hereinafter described. The bidder on this class will specify a price for extra haul *per ton of 2,000 pounds*.

Should the bid for unclassified rock be accepted on any section, the price per ton will be stated in the contract in lieu of those for soft and solid rock. On such section the contractor must begin the excavation of areas known to be of solid rock in part, with other areas, as directed by the Engineer, and continue their excavation, or accept an equivalent reservation in tons, according to the relative cost of excavating soft and solid rock, to be determined by the Engineer, until such solid rock excavation is made.

MEASUREMENT OF EXCAVATION.—All materials excavated in the two and a half miles below Richmond and at Drewry Bluff will be measured on deck lighters by displacement of water taken at 62 pounds per cubic foot. The basis of measurement for "solid rock" will be 155 pounds and of soft rock 120 pounds, and of other classes 120 pounds per cubic foot, until otherwise ascertained at the instance of either the Engineer or the contractor. These "other classes" will then be loaded on deck lighters (with gauges attached) whose cubic feet of submergence has been ascertained for each tenth of a foot. The load will be put on in shape suitable for accurate measurement by cross sections, at the expense of the contractor, and in such manner that the plane of submergence will be parallel to the deck of the lighter, as near as may be. The difference in cubic feet of submerged section, light and loaded, at 62 pounds to the cubic foot, will be considered the weight of the measured load. The lighters must be kept bailed out, and will be considered loaded with the same weight when afterward sunk to the same gauge readings. The proportion of a partial to a full load will be determined by the ratio of the displaced volumes of water.

The Engineer may measure the displacement of empty lighters at his discretion, and the contractor must free them from water at his own expense whenever requested by the inspector for the purpose of determining their loads.

DUMPING LIGHTERS will not be allowed in transporting materials excavated at Drewry Bluff or above, but can be used for material excavated at Harrison's Bar and Goose Hill Flats. Where dumping lighters are used measurements will be made by the capacity of the pockets.

DECISION AS TO QUANTITIES.—The duty of determining the quantity of material carried in or on lighters will be performed by inspectors appointed by the Engineer in charge, and the decision of such inspectors, acting under the orders of the Engineer, as to the amount of material excavated and removed, as well as to its place and manner of deposit, shall be final and without appeal on the part of the contractor.

OVER-DEPTH.—No allowance will be made for dredging more than twelve inches below the required depth or outside the limits of the channel as marked by stakes or ranges, unless such additional depth is necessary to break up strata of rock or cemented earth. Increased depth will be allowed in such cases if authorized by the inspector, acting under the direction of the Engineer. If a deficiency in depth is found the contractor must re-excavate the bottom until .the required depth is obtained.

DUMPING.—Excavated material will be the property of the United States and be disposed of strictly in accordance with instructions from the inspector.   W. P. C,

124. Specifications for an Earthen Dam. The following specification for an earthen dam across a stream for a storage reservoir for irrigation purposes is a fair illustration of a method of writing the specifications for such a purpose. It is sufficiently elastic to allow the engineer great discretion in the matter of adapting the methods to the particular material encountered, while at the same time it gives to the contractor a fair idea as to the amount of work which will be required of him, and therefore as to its approximate cost. This is really all he cares to know at the time he makes his bid.

The object in the mind of the engineer in writing this specification was to make as impervious a dam as possible with the materials which were known to exist in the immediate vicinity, without making the cost extravagant. The earthwork was well insured against overflow by having a very long spillway across the dividing ridge between this and an adjacent drainage basin some distance above the dam.

SPECIFICATION FOR AN EARTHEN DAM:—After the work has been staked out, the top soil and all vegetable matter shall be wholly removed from the area of the work. This area shall be cleared to the satisfaction of the engineer, and no allowance will be made therefor. The material removed, which is sufficiently free from vegetable matter, may afterwards be used for building the outer toe of the dam.

The sides of the valley shall be terraced or stepped with risers one foot high, over the whole area of contact with the dam. These terraces shall also have offsets horizontally as shown in the sketch. They shall be cut to the depth indicated by the engineer, and the material excavated will be paid for as excavation.

A trench shall be opened from the surface to and into the gravel stratum, of the dimensions as shown on the drawings, being about 16 feet wide at the surface, and about 6 feet wide at the bottom. This trench extends the entire length of the dam and up to the side slopes; it extends to the rock after the gravel stratum is passed. This work will be paid for as excavation.

A puddle core wall extends the whole length of the dam, being 6 feet wide at the bottom of the trench, which averages about 20 feet below the surface of the ground, averages about 16 feet wide at the surface of the ground, and is 6 feet wide at a point 4 feet below the top of the dam.

This puddle core is to be a mixture of clay and the gravelly sand such as found in the gravel stratum (all of the gravel to be such as will pass through a 2 inch ring), in about equal proportions. These two kinds of materials will be spread in alternate courses about 3 inches thick. The clay courses will be harrowed sufficiently to pulverize the hard clods, using a disc harrow, or such other machine as may be satisfactory to the engineer. A gravel course will then be laid and this harrowed until it is thoroughly incorporated and mixed with the clay. It shall then be wet down and allowed to stand until it is in a proper condition for compacting, when it shall be rolled thoroughly, to the satisfaction of the engineer, with a grooved roller weighing not less than 150 pounds to the lineal inch along the axis of the roller.

On this shall then be laid another 3 inch layer of clay, which shall be pulverized by harrowing, to be followed by a 3 inch layer of gravel, harrowed, wet down and rolled, as before described, and so on. If, in the opinion of the engineer, it is deemed advisable, or desirable, the clay for the puddle shall be pulverized dry before it is spread on the dam.

The above describes, in a general way, the amount of work which will be required to be done on the puddle core of the dam, but the particular operations are made subject to any changes which the engineer may choose to make, in the interest of a more effectual mixing or compacting of the materials. The general drawings show the probable depth to which the puddle wall will be carried, but the depth shown is not to be considered as exact nor final, as it may be varied according to the character of the ground developed by the excavation.

All that part of the dam on the up-stream side of the core wall shall be made of such clay material as the engineer may select from that overlying the gravel stratum in the immediate vicinity below the dam. It is thought that at least half of this overlying clay is suitable for this purpose. It shall be thoroughly incorporated with the upper clay underlying the dam by first plowing up the surface, after the top soil has been removed, and rolling down again as described below. This plowed upper surface, and also the clay fill, shall be pulverized in courses not over 6 inches thick, by harrowing or rolling, or both, before wetting down. It shall then be wet down thoroughly and allowed to stand until it is in a proper condition for most effectual compacting. It shall then be rolled to the satisfaction of the engineer, and in accordance with his directions, with a grooved roller weighing not less than 150 pounds per lineal inch of roller, measured along its axis.

All that part of the dam below the puddle core wall shall be filled with the surface clay in the immediate locality below the dam. It shall be laid in courses not over 6 inches in thickness, harrowed and rolled dry to the satisfaction of the engineer. It is the intention to make as compact a mass of this portion as possible without wetting down.

J. & F.

125. Specifications for Coffer-Dams. The following specification for the construction of coffer-dams, and for the methods of paying for the same is taken from a recent U. S. Engineer Corps specification for the building of a navigation lock and dam on the Great Kanawha river, West Virginia. It illustrates how such specifications may be drawn and the work paid for in an equitable manner, and how bids may be made up upon such work without assuming extraordinary risk, and without knowing much of the nature of the material to be excavated or of the depths to which the construction will extend.

How Built.—The coffer-dams will be built as shown generally by the drawings exhibited and as directed by the engineer. They will be formed of cribs sunk to hard pan, sheathed with plank and filled with heavy-dredged river-bed material not liable to wash. They will be thoroughly banked on the outside with clay puddling, or like material, of quality and quantity to make them sufficiently water-tight to be pumped out. The crib-filling and the banking outside will be protected to such an extent, as directed, by a top layer of loose stone.

As the work within the different sections of the coffer-dams for the dam is finished, the ends of the next section of coffer will, when required, be made of square sawed timber, rods, upright plank and puddle, built across and near the end of the finished part. Similar timber and plank bulkheads will, also, if ordered, be built by the contractor between the coffer-dam for the lock and the lock wall, to form part of the first section of coffer-dam for the navigation pass and elsewhere in making coffer connections as the engineer may require.

How Paid For.—The United States will pay the contractor for the dredging and excavation for the site of the coffer-dams his contract price for "Excavation." For logs and sheathing used, he will be paid his contract prices for "Crib

Logs in Coffer-dam" and "Sheathing." For material used to fill the coffer-dam cribs, he will be paid his contract price for "Coffer-dam Filling." The plank and sawed timber used in the coffer-ends or bulkheads will be paid for as "Sheathing," and the puddle in same as "Coffer-dam Filling." It is understood that all labor, all banking, puddling and stone used on the outside of the cribs, the spikes and bolts and all material not mentioned, required in the construction of the coffer dams, shall be furnished by the contractor without cost to the United States. The coffer-dams must be promptly banked to full height. No payment will be made for logs or filling above the level of the top of the lowest part of unfinished banking. No payment will be made for any coffer-dam materials carried off by the river or lost in any manner during construction. All repairs to the coffer-dams or their adjuncts must be borne by the contractor.

Removal.—The contractor will be required to remove the coffer-dams and their belongings at his own cost. The time and manner of the removal of the coffer-dams, or any parts of them, and the place to deposit the materials, to be prescribed by the engineer.

Ownership.—It is understood and agreed that the payments for excavation, logs, sheathing and filling, as provided for above, shall cover the entire cost of the coffer-dams to the United States, and by virtue thereof they shall become the property of the United States, in case of the failure or annulment of this contract.

Dredges and Pumps.—In building the coffer-dams the contractor will be required to employ, at the same time, not less than two suitable steam dredges at excavating and filling; and for pumping the coffer-dams he must keep at least three good sufficient pumping outfits, with pumps, engines and boats complete, in, or always ready for, operation. The dredges must be equipped to do effective work to a depth of 28 feet.

W. P. C.

126. Specifications for Protective Work. It is
customary for engineers to require the contractor to protect his work from all causes, such as landslides, rainfall, floods, ground water, quicksand, etc., without any special compensation therefor. In all such cases the contractor, of course, will build his temporary protective works as cheaply as possible, and often will not provide that degree of protection which the engineer may think is necessary and wise. A fair division of the responsibility and cost of such protective works between the

10

contractor and owner is therefore desirable, and the following specifications taken from those used on the new Croton Dam of New York City, in 1892, are offered as a very excellent solution of this problem.

It will be noted that the cost of such protective works as the engineer might consider necessary is to be paid for by the city at the standard prices per unit of measure named in the bid, and the engineer is given control of the design and general character of such works. The responsibility for the efficient execution of such works, however, is made to rest upon the contractor by holding him responsible for all damages caused by their inefficiency, except in one instance of an unprecedented flood in the Croton river. In which particular case the damages are to be duly appraised and paid for by the city.

It will also be noted that the taking care of the ground water by pumping rests wholly upon the contractor without special compensation.

The contractor shall do all other work needed to protect his work from water; he shall erect all temporary dams, cofferdams, sheet piling and other devices, take care of the river, and shall be responsible for all damage that may be caused by the action of water, whether from negligence or any other cause. Such damage is to be repaired, and the work must be restored and maintained at his cost.

All earth and rock excavation, masonry, timber and other work, temporary or permanent, for the purpose of protecting the work from the river, provided that they are ordered or approved by the engineer, are to be paid for at the prices stipulated in this contract. All work of this character is to be removed by the contractor at his own expense, if so ordered by the engineer.

The responsibility of the contractor as to damage caused by the inefficiency of the protective work shall cease, however, if such damage is caused by the river at a time when the flow of the river attains such volume as will cause it to rise to a height of more than eighty-one inches above the stone crest of the present Croton dam, such height being the greatest recorded by the city authorities.

Such damage as may be caused under the circumstances just stated shall be repaired by the contractor as soon as practicable, under the direction of the engineer, who shall appraise the cost of such work of repairs, and the amount of the same shall be paid to the contractor on the certificate of the engineer that the work has been completed to his satisfaction; and, after such certificate shall have been issued, the contractor shall again become responsible for all damage that may be caused by the action of the water, in the same manner as is specified above. .

If such appraisal of the engineer is not satisfactory to the contractor, the said contractor shall so state in writing to the aqueduct commissioners, and, thereupon, a board of arbitration, composed, first, of the chief engineer, or of such other person that the aqueduct commissioners may designate; second, of a person selected by the contractor; third, of another person to be designated by the other two, shall proceed to appraise the cost of such damage, and their decision shall be final and binding on both parties, provided it is the unanimous decision of the three members of the said board; but if the said decision is not unanimous, the appraisal of the chief engineer shall stand and become final and binding to both parties.* And, on the certificate of the aqueduct commissioners that the said appraisal has been made in accordance with the stipulations of this agreement, the amount of said appraisal shall be paid to the contractor. And the said appraisal, whether made by the chief engineer or by the said board of arbitration, shall include only the cost of the actual work done to repair the damage, and shall not include any alleged loss of profit or other loss due to the delay caused by such repairs, but an extension of time shall be granted to the contractor for the performance of his contract, equivalent, in the opinion of the engineer, to the loss of time due to the interruption of the operations of construction on account of the said work of repairs.

The contractor is to do all the draining and pumping which shall be necessary for keeping the work free from water, and if at any time the engineer is of the opinion that, in order to maintain the slopes and sides of the excavations in proper order, it is necessary to remove the water from the ground outside of the limits of the excavations, the contractor shall, at his request, sink the necessary pipes or wells to intercept the water, and place, maintain and work such pumping or other exhausting apparatus as shall be sufficient to properly maintain the said slopes and sides.

The cost of furnishing the necessary appliances and machinery, of working them, and of doing all the work connected with draining and pumping operations, is to be included in the prices bid for the various kinds of work which the draining and pumping operations are intended to protect.

A. F.

* This is a new departure in arbitration proceedings, but it has many things to recommend it.—AUTHOR.

SPECIFICATIONS FOR CEMENT MORTAR, CON-
CRETE, AND MASONRY.

127. Cement Mortar.* There are in general two
kinds of cement in common use in America, namly, Portland
or artificial cement, and Natural cement. Portland cement is
an artificial mixture of lime and clay properly burned and
ground. Natural cement is made by burning the natural rock
which contains approximately the proper ingredients, and
grinding the calcined product. Portland cements are known
by their various manufacturer's names or brands, and are mostly
imported from Germany, France, and England. Recently a
number of manufactories have been established in America.
Natural cements are usually known under a geographical name,
indicating their place of manufacture, as Rosendale cement,
made on the Hudson river; Louisville cement, made on the Ohio
river in the vicinity of Louisville; Utica cement, made at Utica
in the northern part of the state of Illinois; Milwaukee cement,
etc. In general the Portland cement costs about three times as
much as the natural cements and it has three or four times the
strength of these. It is common to require a tensile strength
of from 300 to 400 pounds per square inch for Portland
cements, which have hardened one day in the air and six days
in water, and about 100 pounds per square inch tensile strength
for natural cements, similarly treated. The Louisville cement
is quick setting, and a very fair test may be obtained of its
strength in twenty-four hours, in which case a tensile strength
of from 60 to 80 pounds per square inch may be specified, the
briquettes being allowed to remain one hour in the air, and
twenty-three hours in water.

*The reader is referred to a small book of 100 pages by F. P. Spalding assistant
professor of civil engineering, Cornell University, Ithaca, New York, on "The Testing
and use of Hydraulic Cement" (1893). This is a most excellent hand-book for the
engineer as well as for the student; it fully describes the characteristics of both natural
and Portland cements, as well as the most approved methods of testing cements and of
mixing mortars.

The strength of cement and cement mortar depends greatly on the fineness of the cement. This is usually tested by passing it through a sieve of from 50 to 100 meshes per lineal inch, having from 2,500 to 10,000 meshes per square inch. The 100 mesh sieve is much to be preferred, and is usually specified in the case of Portland cement, since probably only the particles which would pass through such a sieve' are really efficient or active in the process of hardening, the coarser parts being inert, or as so much sand.

A cement mortar is a thorough mixture of sand with cement, first in a dry state, usually in the proportion of one of cement to two of sand by measure, with natural cements, and one of cement to three or four of sand when Portland cement is used. After these ingredients have been effectually mixed, sufficient water is added to reduce the composition to the desired consistency. It is important that the sand should be clean, or free from all earthy ingredients. It is common also to specify that it shall be sharp; that is to say, the grains should not be too much rounded. Ocean beach sand is apt to be very much worn, and not sharp in this sense. River or bank sand is usually preferred on this account.

In specifying the proportions of sand and cement to be used in making up a cement mortar, it is customary simply to name so many parts of sand to one part of cement, by measure. It would, as a rule, be inconvenient to determine this ratio by weight, but a determination by measure is subject to serious objections. For instance, a barrel or original package of cement, when dumped or turned out upon a mixing platform in a loose and fluffy condition will have nearly 50 per cent. more volume than it had in the original package. It is necessary, therefore, in order that the meaning of the specifications shall be clear, to indicate whether the proportions by volume shall be taken with the cement in the original package, or in a loose state, after having been emptied from such package. It is perhaps more convenient to measure the cement after it has been emptied from the original package. In any case the engineer should decide which method he proposes to adopt, and reveal this decision in the specifications themselves. The following specification for the making of cement mortar is satisfactory in every respect, except that it does not indicate whether the cement is to be measured in the original package, or in a loose condition. Since all kinds of cement set or harden by a kind of crystallizing process, and in some cases this action begins almost immediately after the cement has been wet down, it is necessary to use the mortar as quickly as possible after it has been mixed. In the case of Portland cement, and some of the slow setting natural cements, an hour or two may be allowed to elapse between the wetting of the mortar and its final use upon the work, but, in the case of such quick setting cements as the Louisville, the mortar should be finally placed in the work within thirty minutes of the time it is first wet. If cement mortar is used after it has begun to set, it will be permanently weakened, and if used very long after it has begun to set, it will remain inert and will never hard·n.

Mortar shall be composed of one measure of cement and two measures of sand, and shall be mixed on a t'ght platform as follows: One measure of sand shall be evenly distributed on the platform; and one measure of cement shall be distributed on the sand, and a second measure of sand shall be distributed on the cement.   The sand and cement shall then be thoroughly mixed in a dry state, being turned over with shovels until this is accomplished.   Water shall then be added in a sufficient quantity to convert the sand and cement into a mortar which will stand in a pile and not be fluid enough to flow.   During the application of the water the mass must be constantly turned with shovels, so that the mortar will be of uniform consistency.

O. B.

128.   Cement Concrete.   Cement concrete is usually composed of cement mortar as described in the previous article, mixed with broken stone.   It may, however, be composed of cement mortar mixed with gravel.   If gravel can be procured free from earthy matter, varying in size from coarse sand to

stones not more than about two inches in diameter, it would
serve a better purpose in the manufacture of concrete than does
broken stone. Experiments have shown also, that when stone
is broken in a stone crusher and not screened, so that all the
finer parts remain in, including the stone dust, a stronger concrete results than with the use of the same quantity of screened
stone.

The ideal cement concrete is such a mixture of material of
graded size, from the largest used down to the finest sand, as
will make a nearly solid mass, when properly mixed. This
may then be solidified by uniting with it such an amount of
finely ground cement as will serve to completely coat each and
every particle of sand, gravel, or stone, and fill the small voids
remaining after the graded materials have been thoroughly and
uniformly mixed. Since crushed rock is always angular it will
be often impossible to make as solid a concrete mass with it as
can be made by the use of gravel. When gravel is used it is
best to have it screened to a series of regularly graded sizes,
and then such proportions of each successive smaller size used
as will serve to fill the voids in the larger size. The cement
finally fills the voids between the small sand grains.

The sand and cement should always be very thoroughly
mixed dry, then the coarser material should be thoroughly wet
and the excess of water drained off, after which the mixed sand
and cement should be incorporated with the moistened gravel
or rock, and a sufficient amount of water added while the mixing is in progress as will reduce the entire mass to the proper
consistency. The most effectual mixing can be done by
machinery, but it is more commonly done by hand. Perhaps
the best cement mixer is a cubical box mounted on trunions at
its diagonally opposite corners into which the proper proportions of the constituent parts, including the water, are placed
and the whole given a certain number of revolutions. There
are various kinds of continuous mixtures into which the proper

proportions of the several ingredients are thrown somewhat at random, and from which the concrete is supposed to continuously fall upon the work in a properly mixed condition. This method is probably fully equal to hand mixing, but is not as satisfactory as the use of the cubical box above described.

Concrete should always be laid in courses of from six to nine inches in depth, and thoroughly rammed in place in order to compact it effectually. If several courses are to be laid in order to obtain the required depth of concrete, *one course should follow another as rapidly as possible*, in order that they may become effectually joined, and form finally one monolithic mass. The courses already laid, however, should not be disturbed, after the concrete has begun to take a set. When the work is interrupted at the end of a day, and other courses of concrete are to be laid the following day, and especially when Sunday intervenes, the top of the concrete should be covered and kept wet, and when the next course is laid the top surface of the former should be thoroughly water soaked, and all earthy matter removed from it.

Masonry or other heavy weights should not be laid upon concrete until it has been allowed to harden, usually as much as twenty-four hours. In the case of quick setting natural cements, however, twelve hours may be sufficient.

Since successive freezing and thawing will prevent the ultimate hardening of cement mortar, it is customary to prescribe that no masonry or concrete in which cement mortar is employed shall be laid in freezing weather. It is a well established fact, however, that Portland cement mortar is not injured by freezing if it remains in a frozen condition for a considerable length of time. Again, when the temperature is not too low, but below freezing, freezing of the mortar may be prevented by adding salt to the water in making the mortar, or the ingredients of concrete may be heated so that the concrete will have set before freezing can take place.

The following specification for cement concrete includes as a constituent part of it the specification for cement mortar in the previous article. That which is given below is supposed to follow directly upon the previous quotation, the whole constituting a specification for cement concrete.

The broken stone shall be wetted down and then thoroughly mixed with the mortar by turning it over with shovels; no more stone shall be used than can be covered on all surfaces with mortar, and the proportion of broken stone in the concrete must not exceed five measures of stone to one measure of cement. All material must be actually measured in bulk.

Concrete must be mixed in small and convenient quantities and immediately deposited in the work. It must be carefully placed, and not dropped from any height. It shall be laid in sections, and in horizontal layers not exceeding nine (9) inches in thickness, and it must be thoroughly rammed until the stone is covered with mortar and a film of water appears on the surface. In no case shall concrete be permitted to remain in the work if it has begun to set before the rammi g is completed. When concrete is properly made the whole mass becomes one stone when it has set. and it is very important that it shall be deposited continuously in the work. All surfaces upon which concrete is to be laid must be wetted before the concrete is deposited. Plank or timber forms must be provided when necessary to confine the concrete to the shape and dimensions shown on the plans.

Before any weight is placed on concrete it shall have as much time to set as can conveniently be allowed, and in no case less than twelve (12) hours.

In cold weather material for concrete shall be heated as directed by the engineer.

The engineer will issue special instructions for concrete which is to be deposited under water.   O. B.

The following method* of making concrete by using sea washed gravel of standard sizes as obtained from graduated screens has given most excellent results. In this mixture there were three grades of sand and gravel employed, namely, fine sand, coarse sand, and small gravel stones up to one fourth of an inch in diameter, and large gravel from one half to two inches in diameter. The proportions were one part cement, two parts fine sand, four parts coarse sand and small gravel,

*See article by C. H. Platte, C. E. in Engineering News of February 21, 1895.

and eight parts of the larger gravel, making in all one part of cement to fourteen parts of sand and gravel, by measure. The cement and fine sand were mixed dry. The two grades of gravel were then thoroughly mixed and saturated with water, the surplus water being allowed to drain off. The dry mixture of cement and sand was then uniformly spread over the wet gravel and thoroughly mixed with it. The water which remained adhering to the gravel was found sufficient to moisten the cement, and also to insure a uniform distribution of such water through the mass. The mixture was then deposited in place and thoroughly rammed, and it was found to give a very solid and strong concrete. It was found that three and one half barrels of cement were used for each four and onehalf cubic yards of concrete in place. It is said that the concrete was equal in every particular to that made of one part of cement, three parts sand, and five parts broken lime stone. This species of concrete was used in the foundations of the New York and Brooklyn bridge, and also on some of the New York City cable railways. This mixture comes very near being the ideal concrete for both solidity and economy.

The following complete specification for cement mortar and concrete indicates the present practice of one of the leading civil engineers of this country, and has many excellent qualities. These specifications were used in the construction of the 155th street bridge over the Harlem river, New York City.

All mortar used, except for pointing, shall be composed of one part Portland cement mixed in a dry state, with two parts clean, sharp sand, free from loam; the entire quantity of water to be added, and the whole mass thoroughly incorporated. Mortar shall only be made in batches for immediate consumption, and none used that has commenced to set. Pointing mortar shall be composed of cement and sand in equal parts, made in very small quantities, and mixed with just enough water to bind the grains.

The cement used will be Rosendale or its equivalent, and Portland. The Rosendale shall be of the best quality of established brands approved by the engineer, and showing a minimum breaking strength of seventy pounds per square inch on the twenty-four hour test, viz.: Set one hour in air, remainder of time in water. Portland cement must be of a well-established brand, English or German manufacture, and approved by the engineer. It should be a moderately slow-setting cement, weighing one hundred and six to one hundred and twelve pounds per struck bushel, not less than ninety-eight per cent. fine, using a two thousand five hundred mesh sieve, and to test neat in seven days after two hours in air,* remainder of time in water, three hundred and eighty, to four hundred and twenty-five pounds per square inch of section. When tested in cakes, all cement must be free from cracking on the edges.

In general, all cements must be subject to the usual standard of inspection and testing, as recommended by the American Society of Civil Engineers. The contractor must keep on hand, under proper shelter, and convenient of access, sufficient stock of cement ahead of his wants to afford a reasonable time for its proper examination and testing.

The sand used must be clean, sharp sand, not too fine, free from pebbles and must not soil white paper.

The water used must be clean, fresh water.

The concrete will be mixed either by hand or machine, according to quantity needed and rapidity of requirements. Any machine used must be one of proved efficiency and reliability for uniform product. All component parts of a concrete batch must be accurately gauged as to relative volume. Cement and sand must be thoroughly mixed dry, and the thoroughly wetted broken stone incorporated therewith, and only enough water added by sprinkling to uniformly dampen the mass without wetting it, as may be directed. The amount of water required will depend upon atmospheric conditions of heat and moisture, and due allowance must be made therefor. In hand mixing, the broken stone will be spread out upon a tight platform, within a movable frame of gauged dimensions, holding an exact manageable batch of stone, nine to twelve inches deep. The stone to be flushed off uniform with edge of frame, and thoroughly wetted, allowance being made when stone has become highly heated from the sun, when the water quickly evaporates. The sand and cement mixed dry on an adjacent platform to be then uniformly spread over the stone prepared as above. The frame will then be lifted off, and the mass shoveled over in rows, the men working from opposite sides towards each other, care being exercised not to heap the mass, but simply turn it over, keeping the original thickness. During this process the sprinkling to be kept up to obtain the

*This is commonly given as 24 hours in air, but kept moist.

required dampness. The operation to be repeated rapidly and only so often as may be necessary to obtain a uniform homogeneous mass of concrete, which must be used as rapidly as possible. All concrete must be thoroughly rammed in layers, as may be directed, with metal rammers of an approved pattern, weighing from thirty to thirty-five pounds each. Corners inaccessible to direct long-handled ramming, as in air chambers of caissons, to have special devices provided to insure solid contact work. All concrete stone to be trap-rock, or a dense fine grained blue limestone, or may be a clean screened coarse gravel, when used with Rosendale cement. All concrete stone must be clean and free from dust. There will be three grades of concrete required designated by numbers.

No. 1. Concrete mass on roof of caissons and filling of air chambers—one part Portland cement, two parts sand and four parts of one-half-inch clean, broken stone.*

No. 2. Concrete for footings of Pier III, and hearting of all piers—one part Portland cement, two parts sand and four and one half parts broken stone, miscellaneous sizes, but none over two inches in any direction.

No. 3. Common concrete in Pier II—one part Rosendale cement, two parts sand and five parts gravel or broken stone.   A. P. B.

When concrete is used for the purpose of making a wall impervious to water, it must be made of small gravel or small broken stone, and it must be unusually rich. That is to say a large excess of mortar must be employed. With these precautions, with proper care in laying, it is possible to make a practically water-tight wall of cement concrete. Such a wall or partition may be constructed between two rubble stone walls, the concrete core being relied on to make the wall practically water-tight.

129. Specifications for Stone. The following specifications for stone to be used for various purposes are those in use by the Chicago, Milwaukee and St. Paul Railway. While certain qualities of stone are here specified, no method is prescribed for determining these qualities. The qualities of building stone are often examined by means of laboratory tests for strength, specific gravity, and for effect of freezing, and also

*That is to say, crushed stone which will pass through a sieve of ¾ inch mesh and be retained on a sieve of ¼ inch mesh.

by chemical and microscopic tests to determine composition
an l structure.   While such tests have considerable value in the
absence of any knowledge from experience, they do not take
the place of that kind of knowledge which is obtained from
having observed the strength and weathering qualities in actual
structures which have been long exposed to the action of the
elements.   It is always desirable, therefore, to have stone from
quarries of established reputation, the products of which have
long been upon the market.   For this reason where stone specifications are prepared for a given locality, the engineer may
inform himself of the most available kind of stone to be used
at that place, and may specify two or three alternative varieties,
by naming the quarries.   Evidently this would not be practicable where general specifications are prepared for an entire
railway system of such large extent as that of the Chicago,
Milwaukee and St. Paul.   For this system the specifications
read as follows:

Stone.

Bridge, block rubble and common rubble stone must be of
sound and durable quality, free from flint seams, powder
cracks, dry and incipient cracks, flaws and other imperfections,
and of such character as will resist the action of the weather
without injury to the masonry in the climates traversed by the
railway company's lines.

. All stone, except riprap, shall have its top and bottom
beds approximately parallel to each other and to the natural
quarry beds, and shall be approximately rectangular in shape
with sides perpendicular to its beds.

Bridge stone shall be from 14 inches to 24 inches thick,
from 4 feet to 7 feet long, and from 2 feet to 5 feet wide; but
in no case shall its length be less than two and one half times
its thickness, nor its width be less than one and one half times
its thickness.

Block rubble stone shall be from 8 inches to 14 inches
thick, 2 feet to 5 feet long, and not less than 18 inches wide.

Common rubble stone shall not be less than 6 inches thick,
16 inches long, and 10 inches wide.

Riprap must be of sound stone of such quality that will
·not disintegrate under the action of the weather.   It shall be
of random size and shape, none to be less than 20 pounds in
weight, and the majority such as can be handled by one man,

but no stone to be larger than can be handled by two men without the use of a bar.

The engineer reserves the right to specify the quarry and the particular ledge in the quarry from which the stone shall be supplied.

The stone may be inspected before or after shipment from the quarry, at the option of the railway company, and in the former case the contractor shall furnish the inspector with full facilities for examination of the stone.

The engineer reserves the right to accept or reject any or all of the stone for want of conformity with these specifications at any time previous to its being paid for in full by the railway company, notwithstanding that it may have been previously passed upon by the inspector, and in case of such rejection the title to the stone shall be in the contractor, and he shall be charged freight on the same at regular tariff rates.   O. B.

130. Stone Masonry. It is not safe for the engineer to undertake to designate a particular class of masonry by a particular name, without entering in the specifications a full description of the same. The names of classes of masonry are too indefinite and are used in too many senses to make it safe to pursue such a course. The engineer should, therefore, describe in considerable detail exactly the kind of masonry construction he desires, and he need not give to such masonry any particular class name. If he does use class names, he should define them clearly in the body of the specifications. Specifications will be given below for several different kinds of masonry.

In laying masonry and in writing the specifications for the same, three particular ends should be constantly in mind. These are: (*a*) evenness and equality of bearing in supporting the superimposed load; (*b*) so far as possible an entire absence of voids or openings in the body of the work; and (*c*) so effectual a bonding of the mass as to cause it to act so far as possible as a monolithic structure. If the masonry occupies a prominent situation so that its appearance is a matter of importance, the exterior surfaces may be made to conform to any desired plan. The following specifications are thought to be self explanatory. They are the general specifications for bridge masonry used by the Chicago, Milwaukee & St. Paul Railway. While these specifications do not require a very expensive grade of work, if fairly carried out they will produce permanent monolithic structures of great strength, provided a good quality of stone and cement have been employed, the specifications for which are issued separately.

Bridge Masonry.

All masonry shall be built according to the plans and instructions furnished by the engineer, and when built by contract will be measured, estimated, and·paid for by the cubic yard, and only to the amount of cubical contents of the same as planned and laid out.

All masonry built by contract shall be subject to the supervision of an inspector whose duties it shall be to see that the requirements of these specifications are complied with, but his presence shall in no way or in any degree lessen the responsibility of the contractor or his obligations.

The stone used in bridge masonry shall be of the quality and dimensions described and known as bridge stone in this company's specifications for stone.

The stone shall be carefully cut and dressed, forming headers and stretchers, which must be laid in regular horizontal courses in good cement mortar, with beds and builds level, the end and side joints vertical and broken at least fifteen (15) inches.

All foundation or footing courses must be made of select large stones not less than eighteen (18) inches in thickness and having a superficial area of at least fifteen (15) square feet.

No course of stone shall be less than fourteen (14) nor more than twenty-four (24) inches in thickness and each course shall be continuous around and through the wall, the courses decreasing, when at all, regularly in thickness from the bottom to the top of the wall.

Face stones shall be composed of headers and stretchers, and each stone in any course shall be of the exact thickness of the one adjoining it. The outer surfaces are to be rock face, but the edges shall be brought to lines corresponding to the finished dimensions of the masonry, and there shall be no projections of over four (4) inches beyond these lines.

The beds and joints of face stone shall be dressed back at least twelve (12) inches from the face of the wall and must be brought to a joint of not more than one half (½) an inch when laid. The under bed must extend to the extreme back of the stone; NO OVERHANG whatever will be allowed.

Stretchers shall not be less in length than two and one half (2½) times their height, and no stone shall have a less width than one and one half (1½) times its thickness.

Headers at least four (4) feet long, when the thickness of the wall will permit, shall be put in frequently to bond the wall, and they shall be so arranged that the headers of any course shall fall between the headers of the course immediately below it. There shall be one header to every two (2) stretchers, and they shall, as far as practicable, hold the size back into the heart of the wall that they show in the face.

When the walls do not exceed four (4) feet in thickness headers must run entirely through the wall, and in pier work a number of them shall extend through, even though the walls are of a greater thickness than this.

When walls exceed four (4) feet in thickness, there shall be as many headers of the same size in the back of the wall as in the face, and so arranged that a header in the rear of the wall shall be between two headers in the front.

The backing and interior of the walls shall be of large, well shaped stone of a thickness equal to that of the corresponding face stone. No voids over six (6) inches in width shall be left between these stone, and all such void must be filled with small stones and spawls thoroughly bedded in cement mortar or grouted. When the masonry is completed, it must contain no voids, and must be, as nearly as practicable impervious to water. When weep holes are necessary, they will be ordered by the engineer.

All stones shall be prepared by dressing and hammering before they are brought on the wall, and must be so shaped that their bearing beds will be parallel to their natural beds. No heavy hammering will be allowed on the wall after a course is set, and, should any irregularities occur, they must be carefully pointed off.

Each stone must be laid on its broadest bed without the use of chips, pinners or levelers, in a full bed of mortar, so that no stone shall bear upon another stone at any point without a mortar joint intervening.

Care must be taken not to injure the joints of stone already laid. Should a stone be moved or the joint be broken the stone must be taken out, the mortar thoroughly cleaned from both the stone and the masonry and the stone then reset.

The stones in each course shall be so arranged as to form a proper bond with the stones of the course immediately beneath it, and in no case shall this bond be less than fifteen (15) inches.

· Both the stone and the masonry must be kept free from all dirt that will interfere with the adhesion of the mortar or cement to the stone, and in warm weather the stone and the masonry must be wet with clean water just before laying.

When masonry is built in freezing weather, the masonry and stone must be thoroughly freed from ice or frost by using salt and hot water, and where practicable, the stone must be held over a fire just before being set.

The top surfaces of coping stones of abutments and piers are to be rough cut to a true plane, and the surfaces where the bed plates of iron bridges rest shall be bush hammered and made level. When the track is on a grade or curve, the elevation both for the curvature and grade will be provided for in the ironwork. Under no circumstances will the masonry be cut on an incline for this purpose.

The front face and top of all mud walls shall be rough cut to a true plane.

Whenever it may be necessary to remove any part of the present masonry in extending abutments or piers for second track work, it shall be stepped back so as to insure a sufficient bond between the new and the old work, so as to break joints nowhere less than twelve (12) inches.   O. B.

The following specifications for different classes of masonry are taken from the standard specifications used by the Pennsylvania Railroad Company. In these specifications three separate classes of masonry are recognized, and for any particular piece of work, it becomes necessary to specify only the class of masonry which shall be used in these general specifications:

Detailed plans will be prepared by the engineer for each structure, and copies of the same furnished to the contractor before the commencement of the work. All stone used for the different classes of masonry must be sound, durable and not liable to be affected by the weather, and shall be subject to the approval of the engineer.

Masonry will be classified as follows:

First-class masonry shall consist of ranged rock work of the best description. The face stones shall be accurately squared, jointed and bedded, and laid in regular horizontal courses, not less than twelve inches in thickness, decreasing regularly from bottom to top of the walls. They shall consist of headers and st.etchers, and there shall be at least one header to every two stretchers, and they shall be so laid that, as nearly as practicable, the headers in each course shall divide equally, or nearly

11

so, the spaces between the headers in the course immediately below. Stretchers shall be not less than three feet long and sixteen inches in width. Headers shall not be less than three feet in length and eighteen inches in width, and shall hold the size back into the heart of the wall that they show in the face.

When the walls do not exceed four feet in thickness the headers shall run entirely through, and when they exceed that thickness there shall be as many headers of the same size in the rear as in the front of the wall, so arranged that a header in the rear of the wall shall be between two headers in the front.

Every stone must be laid on its natural bed, and all stones must have their beds well dressed and made always as large as the stone will admit of. Mortar joints shall not exceed one quarter inch in width; the vertical joints of the face must be in contact at least four inches, measured in from the face, and as much more as the stone will admit of. The stone will be cut with pitched edges, but all corners, batir lines, steps and copings must be run with a neat chisel draft of one and one half inches on each corner, and the projections of the rock face must not exceed three inches beyond the face of the pitch or draft lines of the stones. The stones of each course shall be so arranged as to form a proper bond with the stones of the underlying course, and the bond shall in no case measure less than one foot. Stretchers shall in no case have less than sixteen inches bed for a twelve inch course, and for all courses above sixteen inches in thickness, at least as much bed as face. The whole of the masonry shall be laid in cement mortar, each stone being carefully cleaned and dampened before setting and each course shall be thoroughly cemented before the succeeding course is laid. No hammering on the wall will be allowed after the course is set; if any irregularities occur they must be carefully pointed off. The backing shall consist of stones with beds dressed to one half inch, and of a thickness equal to that of the corresponding face stones; they shall be laid in full cement mortar beds, so as to break joints and thoroughly bond the work in all directions, and on the completion of each course the space between the large backing stones (none of which spaces will be over six inches wide) shall be filled with small stones and spalls, thoroughly bedded in cement mortar or grouted.

All foundation courses must be laid with select large stones not less than eighteen inches in thickness, nor of less superficial surface than fifteen square feet. All bridge seats, steps and tops of walls should be finished with a coping course of such dimensions and projection as may be ordered by the engineer, dressed and cut to a true surface on top and on the showing faces and in conformity with diagrams for the same, which shall be furnished by the engineer. If required, all copings shall be fastened together with clamps of iron.

First-class arch masonry shall be built in all respects in accordance with the above specifications for first-class bridge masonry. The ring stones shall be dressed to such size and shape as the engineer may determine, and of the thickness shown on the plans. The joints must be made on true radial lines, and the face of the sheeting stones must be dressed to make close joints. The ring stones and arch sheeting stones shall break joints not less than one foot. The wing walls shall be neatly stepped in accordance with the drawings furnished, with selected stones the full width of the wing and not less than fourteen inches thick, and no stone shall be covered less than twelve inches by the one next above it.

The parapets shall be finished with a coping course of full width of parapet, with such projection as may be directed by the engineer; the coping to be not less than fourteen inches thick and to be fastened together with wrought iron clamps.

Second-class bridge masonry shall consist of broken or random range work of the best description. The face stones shall be dressed to a uniform thickness throughout before being laid, but not hammered, and shall be laid with, horizontal beds and vertical joints on the face. No stone shall be less than eight inches in thickness, unless otherwise ordered by the engineer. There shall be at least one header to every three stretchers, and both headers and stretchers shall be of similar size, when the thickness of the wall will admit, but neither shall be less than three feet in length and fifteen inches in width. The same arrangement of headers shall be required as is specified for first-class bridge masonry. Mortar joints shall not exceed one half inch in thickness. All corners and quoins shall have hammer dressed beds and joints. All corners and batir lines shall be run with an inch and a half chisel draft. The vertical joints of the face must be in contact at least four inches, measured from the face, and as much more as the stone will admit of. The work need not be laid up in regular courses, but shall be well bonded. The stones shall be cleaned and dampened before setting and shall be laid in cement mortar. The backing shall consist of stones of the same thickness as the adjacent face stone, laid in full cement mortar beds with good joints and bonds, and the spaces filled with spalls, thoroughly bedded in cement mortar, or grouted, as specified for first-class bridge masonry. Bridge seats, steps and tops of walls shall be coped in the same manner as specified for first-class masonry. Stones in foundation courses shall be of not less than twelve inches thickness and ten square feet of surface.

Second-class arch masonry shall be laid in cement mortar, and shall be of the same general character and description as second-class bridge masonry, with the exception of the arch sheeting, for which proper stones shall be selected that sall have a good bearing throughout the thickness of the arch, and shall be well bonded and be of the full depth of the arch. No stone shall be less than six inches in thickness on the intrados of the arch. The ring stones of all arches shall conform to the specifications for first-class arch masonry.

Third-class masonry shall be laid dry, or in lime or cement mortar as may be directed by the engineer. It shall be formed of good quarry stones, laid upon their natural beds, and roughly squared on the joints, beds and faces, the stones breaking joints at least six inches, and with at least one header for every three stretchers. No stone shall be used in the face of the wall less than six inches in thickness, or less than twelve inches on the least horizontal dimensions. Headers shall be at least three feet long, or extend entirely through the wall. The ends of all walls shall be dressed and finished in accordance with the plans. The stones in the foundations must not be less than ten inches in thickness, and shall contain not less than ten square feet surface, and each shall be firmly, solidly and carefully laid.

In box culverts the top courses of the side walls shall extend entirely across the walls, and the covering stones shall have a bearing of at least one foot on each wall. The thickness of covering stones shall not be less than ten inches for two feet openings; not less than twelve inches for three feet openings, and not less than fifteen inches for four feet openings. Unless built on timber foundations reaching entirely across the opening. the space between side walls of box culverts must be paved with stone, set on edge, not less than eight inches deep, and well secured at the ends with deep curbing.   P. Ry.

131. Specifications for Stone Masonry for a Large Stone Dam.

The following specifications for stone masonry are those which were used in the construction of the new Croton dam, New York City, 1892. They are commended especially for their securing a most efficient, solid, and impervious grade of work, at a minimum cost. Thus the body of the dam, composed of rubble stone masonry laid in cement mortar, thoroughly bonded, and made entirely solid, cost from $3.40 to $4.00 per cubic yard, the cement mortar being one of Rosendale cement to two of sand, the stone having to be hauled about one mile.

Another significant feature of these specifications is the paying for the face dressing per unit of surface in addition to the standard price per cubic yard, the matter of this face dressing being left until the work is executed. In this way such small details need not be determined in advance and indicated upon the drawings.

Stone Masonry.

All stone masonry is to be built of sound, clean quarry stone of quality and size satisfactory to the engineer; all joints to be full of mortar, unless otherwise specified.

Dry rubble masonry and paving are to be laid without mortar, and are to be used for walls, for the slopes of the dam embankments, and at any other place that may be designated.

This class of masonry is to be of stone of suitable size and quality, laid closely by hand with as few spawls as practicable, in such manner as to present a smooth and true surface. The work is to be measured in accordance with the lines shown on the drawings or ordered during the progress of the work. The stones used must be roughly rectangular; all irregular projection and feather edges must be hammered off. No stone will be accepted which has less than the depth represented on the plans or ordered. Each stone used for paving must be set solid on the foundation of broken stone or earth and no interstices must be left.

In the dry rubble masonry walls, large stones must be used, especially for the faces, and the walls must be bonded with frequent headers, of such frequency and sizes as shall be approved by the engineer.

Riprap may be used in connection with the protective work, and wherever the engineer may order it. It shall be made of stone of such size and quality and in such manner as he shall direct, and must be laid by hand.

After the slopes which are to receive the paving have been dressed, a layer of broken stone is to be spread as a foundation for the paving, wherever ordered. The broken stones must be sound and hard, not exceeding two inches at their greatest diameter. Broken stone, not exceeding one inch in diameter, may be used for forming roadways; it is to be spread to such thickness as ordered and heavily rolled or rammed. Broken stones may be used also wherever the engineer may direct, rolled if so directed, and paid for under this head, except the broken stone used for making concrete, the cost of which is included in the price hereinbefore stipulated for concrete laid.

Rubble stone masonry is to be used for the central part of the dam, for the overflow, for the center walls of the earth embankments, for most of the structures and appurtenances of the dam, and wherever ordered by the engineer.

Rubble stone masonry shall be made of sound, clean stone of suitable size, quality and shape for the work in hand, and presenting good beds for materials of that class. Especial care must be taken to have the beds and joints full of mortar, and no grouting or filling of joints after the stones are in place will be allowed. The work must be thoroughly bonded. The faces of the rubble stone masonry, especially the up-stream face of the walls, shall be closely inspected after they are built, and if any mortar joints are not full and flush, they shall be taken out to a depth of no less than three inches or more, if so ordered, and repointed properly.

A large quantity of rubble stone masonry in mortar is to be used in the construction of the central part of the dam and of the center wall and overflow.

The stones used therein must be sound and durable; they must have roughly rectangular forms, and all irregular projections and feather edges must be hammered off. Their beds, especially, must be good for materials of that class, and present such even surfaces that, when lowering a stone on the level surface prepared to receive it, there can be no doubt that the mortar will fill all spaces. After the bed joints are thus secured, a moderate quantity of spawls can be used in the preparation of suitable surfaces for receiving other stones. All other joints must be equally well filled with mortar.

The quality of the beds is to regulate, to a large extent, the size of the stones used, as the difficulty of forming a good bed joint increases with the size of the stones.

Various sizes must be used, and regular coursing must be avoided, in order to obtain vertical as well as horizontal bonding.

The sizes of the stones used will vary also with the character of the quarries, but, especially in the places where the thickness of masonry is great, a considerable proportion of large stones is to be used. If the size and character of the stones, in the opinion of the engineer, shall admit of it, the joints (except the beds), instead of being filled with mortar, may, at his request or on his approval, be filled with concrete made as hereinbefore specified, with the exception that the component materials be mixed in the proportion of one part of cement to three parts of small stone or gravel of such size as the engineer shall direct, and thoroughly rammed, care being taken to use a moderate amount of water only which must be brought to the surface by ramming, such filling of joints with concrete to leave no vacancies and to be thoroughly made. If concrete is so used, the spaces left between the stones should not be less than six inches, in order that proper ramming can be obtained.

No extra compensation shall be paid to the contractor for the use of such concrete, the cost of which is to be included in the price herein stipulated for the masonry in connection with which it is used.

The exposed faces of the main wing wall, of road culverts, of some of the walls and of any other rubble work that the engineer may designate, are to be made of broken ashlar with joints not exceeding one half inch in thickness; the stones not to be less than 24 inches deep from the, face, and to present frequent headers. This face work to be equal in quality and appearance to the face of the breast wall in front of the new gate house at Croton dam (Section 1), and to be well pointed with Portland cement. This face work is to be paid for by the square foot of the superficial area for which it is ordered, in addition to the price paid per cubic yard of rubble stone masonry.

Block stone masonry is to be composed mainly of large blocks and is to be used for the steps of the overfall or for other steps, or whenever and wherever ordered by the engineer. It is to be laid in Portland cement mortar, well pointed, or may be ordered laid dry at the price stipulated in clause O, item (o).

This stone, which is to receive the shock of water and ice, is to be especially sound, hard and compact, and of a durable character; it is to be prepared to the dimensions given so that no joint will in any place be more than one inch wide. The outside arrises must be pitched to a true line.

The outer faces of the masonry dam and of its gate chambers, of the overflow (except steps), and of any other piece of masonry that may be designated, are to be made of range stones, as shown on the plans, the stone to be of unobjectionable quality, sound and durable, free from all seams, discoloration and other defects, and of such kind as shall be approved by the engineer.

All beds, builds and joints are to be cut true to a depth of not more than 4 inches, and not less than 3 inches from the faces to surfaces allowing of one half inch joints at most; the joints for the remaining part of the stones not to exceed two inches in thickness at any point.

All cut arrises to be true, well defined and sharp.

Where this class of masonry joins with granite dimension stone masonry the courses must correspond, and the joining with arches and other dimension stone masonry must be accurate and workmanlike.

Each course to be composed of two stretchers and one header alternately, the stretchers not to be less than 3 feet long nor more than 7 feet long, and the headers of each successive course to alternate approximately in vertical position.

The rise of the courses may vary from bottom to top from 30 inches to 15 inches in approximate vertical progression, and the width of bed of the stretchers is not to be at any point less than 28 inches. The headers are not to be less·than 4 feet in length.

This class of masonry, for the faces of the dam and gate chamber, including the headers, is to be estimated at 30 inches thick throughout. At other places that may be designed by the Engineer, the size of the stones is to be established by him, and the facing stone masonry is to be estimated according to the lines ordered or shown on the plans. In no case are the tails of the headers to be estimated.

The work to be equal in quality and appearance to the facing stone masonry work built by the aqueduct commissioners for their masonry dam across the east branch of the Croton River near Brewster.

All copings that may be ordered and the heads of the arches of the highway culverts, will be classed as facing stone masonry.

The price herein stipulated for facing stone masonry is to cover the cost of pointing, of cutting chisel drafts at all corners of the gate-house dam and other corners, and of preparing the rock faces; but if any six-cut or rough-pointed work is ordered in connection with this class of masonry it shall be paid for at he prices therein stipulated for such work.

The face bond must not show less than 12 inches lap, unless otherwise permitted.

The pointing of the faces to be thoroughly made with pure Portland cement after the whole structure is completed; unless otherwise permitted, every joint to be raked out therefor to a depth of at least two inches, and, if the engineer is satisfied that the pointing at any place is not properly made, it must be taken out and made over again.

Granite dimension stone masonry must be made of first-class granite of uniform color, free from all seams, discoloration and other defects, and satisfactory to the chief engineer.

It is to be used for the gate openings in the gate chamber, for the coping of the dam, for the gate-house superstructures and for the crest and first step of the overflow, and at any other place that may be designated by the engineer.

The stones shall be cut to exact dimensions, and all angles and arrises shall be true, well defined and sharp.

All beds, builds and joints are to be dressed, for the full depth of the stone, to surfaces, allowing of one quarter ($\frac{1}{4}$) inch joint at most. No plug hole of more than 6 inches across or nearer than 3 inches from an arris is to be allowed, and in no case must the aggregate area of the plug-hole in any one joint exceed onequarter of its whole area.

The stone shall be laid with one quarter ($\frac{1}{4}$) inch joints, and all face joints shall be pointed with mortar made of clear Portland cement, applied before its first setting. All joints to be raked out to a depth of two inches before pointing.

The pointing of all masonry, including the faces of the main body of the dam and of the center walls which are below the ground, is to be done thoroughly with Portland cement mortar, mixed clear where used for all exposed faces of brick and cut stone masonry of all kinds (including the rubble facing); and mixed for other work in such proportion as the engineer shall determine. The cost of all pointing is to be included in the price stipulated for the masonry to which it is applied.

The exposed faces of the cut stone are to be finished in various ways, in accordance with the various positions in which they are placed. They shall be either left with a rock or quarry face, rough-pointed, or fine hammered (six-cut work).

The various classes of face dressing must be equal in quality and appearance to those on the sample in the office of the chief engineer.

In rock face work the arrises of the stones inclosing the rock face must be pitched to true lines; the face projections to be bold, and from 3 to 5 inches beyond the arrises. The angles of all walls on structures having rock faces are to be defined by a chisel draft not less than $1\frac{1}{2}$ inches wide on each face.

In rough-pointed work, the stones shall at all points be full to the true plane of the face, and at no point shall project beyond more than $\frac{1}{4}$ inch, the arrises to be sharp and well defined. Each stone to have its arrises well defined by a chisel draft, which is included in the price for rough-pointed dressing.

In fine hammered work the face of the stones must be brought to a true plane and fine dressed, with a hammer having six blades to the inch.

In measuring cut stone masonry, when the stones are not rectangular, the dimensions taken for each stone will be those of a rectangular, cubical form which will just inclose the neat lines of the same. The price herein stipulated for granite dimension stone masonry is to cover the cost of preparing the rock faces, of making the chisel drafts, and of preparing all holes and recesses and grooves.

No payment will be made for cutting grooves and recesses other than the price paid for the dressing of their surfaces, which are to be fine hammered.

For rough-pointed and fine hammered (six-cut) dressing, a price per square foot of dressing will be paid in addition to the price per cubic yard of masonry, viz.:

For rough-pointed dressing, the price stipulated in clause O, item (*t*), and for fine hammered (six-cut) dressing, the price stipulated in clause O, item (*s*).

The exposed parts of the cut stone are generally to be prepared with rock face.

The inside surfaces and copings are generally to be rough-pointed.

All the gateways, grooves, sills, floors, and all other surfaces designated by the engineer are to be fine hammered.

A. F.

132. Specification for First-class Bridge Masonry.

The following specification for first-class bridge masonry represents the current practice of one of the leading American engineers:

Masonry—The stone to be used shall be the best quality of limestone from the quarries, except the nose stones for the curved upstream starlings of Piers I, II and III, one stone to each course shall be granite.

The stone must be strong, compact and of uniform quality and appearance and free from any defects which in the judgment of the engineer may impair its strength or durability.

No course shall be less than 16 inches in thickness and no course, except the coping, shall be thicker than the one beneath it.

Each bed of every stone shall measure at least thirty-six inches in each direction, except that where the thickness of the course is less than twenty-four inches the bed need not exceed one and one half times the thickness of the stone.

The bottom bed shall always be the full size of the stone and no stone shall have an overhanging top bed.

Stretchers shall not be less than four feet nor more than seven feet long and stretchers of the same width shall not be placed together vertically, but this shall not apply to the ends of stretchers where headers come centrally between stretchers.

Headers shall be at least four feet long and shall be at least three quarters their full width for the whole length. There shall be at least two headers on each side of every course between the shoulders.

At least two stones on one side of each course shall reach through to the stone on the other side.

The joints of the face stones shall be cut twelve inches back from the face.

The beds and the joints shall be cut to within one quarter of an inch of a true plane. The depressions below this plane shall not be more than one tenth of the whole surface. The vertical joints shall not average more than three eighths of an inch and shall not exceed one half. Thin horizontal mortar joints will not be insisted on, but every stone shall be set in a full bed of mortar and brought to a proper bearing with wooden mauls, no levelers being allowed.

All stones must be carefully cleaned and wet before setting, and no mortar beds shall be laid until the course below has been cleaned and wet.

The beds and joints of all face stones shall be cut to true pitch lines.

The face of the upstream cutwaters of Piers 1 and II shall be fine pointed work with no projections exceeding one half inch.

The coping shall be two feet thick with a bush-hammerep face throughout and be cut with a wash as shown on the plans.

A four inch draft line shall be cut on all vertical angles.

All other portions of the piers shall have a rough quarry face with no projections exceeding three inches, the quarry face to average at least one and one half inches from the pitch line of the joints and never be run back of such pitch line.

No grab holes shall be made in the face of the coping or on the pointed work of the cutwater.

The stones of the curved upstream starlings of Piers II and III shall be doweled into those of the course below with one and one eighth inch steel dowels extending six inches into each course, these dowels to be placed about ten inches back from the face and seven inches on each side of each joint. The stones of the upper course shall be drilled through before setting after which the hole shall be extended six inches into the lower course, a small quantity of mortar shall be put into the hole, the dowel dropped in and the hole filled with mortar and well rammed.

The coping shall be cut with close joints throughout the whole pier and each stone under the bridge seat shall go completely across the pier.

The joints in the two courses below the coping in all piers shall be cramped with cramps of one inch round iron, sixteen inches long, the ends four inches into each stone.

The backing shall be composed of stones of the same thickness as the face stones and with beds cut in the same manner as required for the face stones. The spaces between the large stones shall not occupy more than one fifth of the entire area of the pier inside of the face stones and these spaces shall be filled with good rubble masonry carefully laid up on full mortar beds and well rammed.

All face stones shall be laid in Portland cement mortar two parts of sand to one of cement. The backing shall be laid in natural cement mortar, two parts of sand to one part of cement.

The Portland cement shall be an imported cement, equal in quality to O. F. Alsen & Sons' best quality, natural cement shall be equal to the best grades of Louisville cement. At least 88 per cent. shall pass through a sieve of 350 meshes per square inch. Each carload shall be treated as a separate lot. Samples shall be mixed separately and formed into briquettes which shall be kept one day in air and six days in water and then broken. When broken they shall have an average tensile strength of at least eighty pounds per square inch and a minimum of at least seventy pounds. None of the briquettes shall crack or blow while in water. The tests shall be made in the manner usual on works under charge of the same chief engineer.

When the masonry is laid up in freezing weather the backing shall be laid in Portland cement, three parts of sand and one part of cement, and such other precautions taken against freezing as the engineer may direct.

The joints of the face stones shall be picked out and pointed in mild weather with one part of sand and one part of Portland cement which shall be driven in with a calking iron.

The masonry of the pivot pier will be built hollow, the central opening to be of cylindrical shape and 24 feet in diameter, the interior face of the masonry to be rough and every fourth stone to project at least one foot from the general interior surface. This central space shall be filled with concrete.

The upper three feet of the filling of the pivot pier shall be made of Portland cement concrete, the center of the pier to finish six inches above the masonry coping, the top to be finished with a granitoid surface two inches thick. The anchor bolts to secure the center casting of the turntable shall be set before the concrete is completed, and the concrete filled around it.   G. S. M.

SPECIFICATIONS FOR STREET PAVEMENTS AND MATERIALS.

133. Specifications for Paving Brick. The essential properties of a good paving brick are: (*a*) Strength to resist cross breaking; (*b*) strength to resist crushing; (*c*) toughness or strength to resist shocks and blows; (*d*) it must be comparatively non-absorbent. Any brick which possesses these qualities in a high degree will also resist abrasion or wear satisfactorily.

The above four qualities can be determined very accurately by laboratory tests. The author has had a large experience in testing paving brick for these qualities, and offers the following specification as one which can be readily fulfilled, without limiting the competition overmuch, and, therefore, without adding appreciably to the cost.

Paving Brick.

Bidders must submit not less than 20 samples of the brick which they propose to use upon the work, at least ten days previous to the letting of the contract. Ten of these brick will be submitted to the following tests at the expense of the bidder, the remaining ten supplied by the successful bidder will be retained as samples for comparison until the work is completed.

The brick shall be what is known as paving brick, made from a suitable quality of shale, and burned to a uniform consistency throughout, but not to the degree of actual vitrifaction.* The tests to which the brick shall be subjected are as follows:

(*a*) *Cross breaking.* Five brick will have their top and bottom faces, as laid edgewise upon the street, ground to perfectly true and parallel planes. They shall then be supported edgewise on rounded knife-edges seven inches apart, and loaded in the center to rupture. The total breaking load in this test, divided by the thickness of the brick (top width as laid in pavement) shall give a result of not less than 3000 pounds, this being the strength of the brick per inch in width when laid in the pavement.†

* Here might follow any desired description of the size, dimensions, rounding of edges, etc.

† This method of reducing cross breaking strength to strength per inch in width gives a great advantage to the deeper grades of brick, but this is an advantage which such bricks are properly entitled to. It evidently would not be fair to compute from this test the tensile strength of the brick per square inch, that is to say to obtain the cross breaking modulus of rupture by the ordinary formula, and use this function in making the comparison, since in this case no advantage would accrue to the deeper brick. The shallow brick being able to show as great a strength per square inch as the deeper brick. It is true the brick as tested after having been ground down have not their normal width, but it is probable that on the average one set of samples of brick will have been reduced in width about the same as the average of any other sample in the process of grinding down. Therefore it is thought the specification here given is both fair and adequate to determine the strongest brick against cross breaking when used in a pavement.

(*b*) *Crushing.* One end of each brick, after having been broken as above, shall then be tested in crushing. For this purpose it shall be dressed approximately to two inches square and a height equal to the width of the brick, which is approximately four inches. This makes a crushing specimen two inches square, of four inches cross-section, and nearly four inches high, which is tested with the load parallel to the longer longer dimension, these top and bottom surfaces having been ground to true planes as described above. In other words, the brick is tested for crushing by applying the load edgewise upon the brick, or as it will be applied in the pavement. When tested in this manner the brick shall show a crushing strength of not less than 12,000 pounds per square inch.*

(*c*) *Impact Test.* For this test five entire bricks of each sample shall be placed in a tumbler or rattler, along with ten cast iron brick with rounded edges, weighing six pounds each. These materials shall not be accompanied by any kind of cushioning material or any other matter whatever. The rattler shall then be revolved for thirty minutes at a speed of thirty revolutions per minute. The bricks are weighed before they are put in and after they have been removed, and the loss in weight found. This last divided by the original weight gives the percentage of loss. The average percentage of loss of the five brick shall not exceed twenty per cent.†

(*d*) *Absorption.* The five brick which have been tested for resistance to impact shall be broken across and then dried for four hours at a temperature of not less than 212 ° F. and then weighed. They shall then be soaked in water for twenty-four hours (or forty-eight hours) and when removed from the

---

*The author has tested paving brick in this way that has had a strength of over 25,000 pounds per square inch, and the 12,000 pound minimum limit is readily complied with by many makers of paving brick. A crushing test of paving brick by applying the load to the side of the brick, or testing the brick flatwise, is very deceptive, inasmuch as the specimen is too short in the direction of the stress to allow failure to occur in a normal manner. Even cubical forms are theoretically much too short. A crushing test specimen should always have the length in the direction of the load nearly twice that of its least lateral dimension, in order to obtain a normal failure. This height is necessary, since the theoretical angle of rupture of such materials when friction is taken into account, is 45 degrees plus one half the angle of repose, the tangent of which is the coefficient of friction. This makes the angle of rupture about 58 degrees to 60 degrees with the horizontal, instead of 45 degrees, as is commonly stated in the text books, where friction is neglected.

†This test is a very severe one, but one of the most valuable to which paving brick can be subjected. If the brick resists this test satisfactorily, it may be relied on to resist the abusive action it will receive in a street pavement. This should not be called an abrasion test, since the loss here suffered is not due to wear, but is wholly due to chipping and breaking off under impact. It, therefore, can not be compared with the ordinary abrasion tests of brick where small castings are used and the treatment continued for a long period. In this latter case the loss of weight is due to abrasion proper. But since a brick pavement does not wear out in this way, but chips off and breaks up, it is thought the proper test is an impact test, such as above described, rather than an abrasion test.

water they shall be wiped dry with a towel, and again weighed. The increase in weight divided by the original or dry weight gives the percentage of water absorbed.   This test shall indicate an absorption of less than two per cent. (or three per cent.*

Average Results.   It is to be understood that in each of the tests described above the standard of requirement is to be applied to the average result from five specimens, these averages being within the requirements.   It is also to be understood that no one specimen shall fall without the requirement by more than 40 per cent., and in case either the average does not come up to the requirement, or any one test fails of the requirement, by more than 40 per cent., the entire sample of twenty brick shall be rejected and not allowed to enter the competition; or the bidder will be allowed to submit other twenty samples at his option.

The average result of the four tests made on each sample will be indicated by the sum of four characteristic numbers, corresponding to their respective tests, these four characteristic numbers being derived as follows:

1.   The characteristic number for the cross breaking test shall be the cross breaking strength in pounds per inch in width divided by 1,000.

2.   The characteristic number for the crushing test, shall be the crushing strength in pounds per square inch divided by 4,000.

3.   The characteristic number for the impact test shall be twenty-five minus the percentage of loss of weight sustained in this test.

4.   The characteristic number for the absorption test shall be three minus the percentage of absorption.

In case the characteristic numbers found for the impact and absorption tests prove to be negative, they are to be subtracted from the sum of the other characteristic numbers, or added to them negatively.   In other words the algebraic sum of these four characteristic numbers shall form the basis for determining which brick has sustained the best test.

Thus, if the average cross breaking strength were 3,500 pounds per inch in width, its characteristic number would be 3.50.   If the average crushing strength were 15,000 pounds

*While the test for absorption is not very important, and while the maximum limit is not well defined, the precentage of absorption does indicate something of the general character of the brick, and, therefore should be made.   It should not, however, have equal weight with the other tests.   Probably any absorption less than two or three per cent. would never prove a source of weakness in any brick.   While, therefore, the two per cent. limit is specified above, probably three per cent. would not be objectionable. In making the absorption test, it is best to use the brick which have already been tested for impact.   These have their outside faces largely removed, and will allow water free access to the interior portion of the substance of the brick.   It is also well to break these brick in two before subjecting them to the moisture test.

per square inch, its characteristic number would be 3.75. If the average loss in weight in the impact test was 18.25 per cent. the characteristic number of this test would be 6.75. If the average percentage of absorption was 1.50, the characteristic number for this test would be 1.50*, the sum of these characteristic numbers being 15.50. This becomes the total characteristic number, indicating the average value of this sample of brick. The sample giving the highest total characteristic number as determined by the above rules, to be declared the brick which sustained the best test.

While the particular numerical quantities chosen in the rules given above may not be the best to use, the writer believes these rules fairly represent the proper method to be employed in arriving at the best average tests.

134. Specifications for Brick Paving. The specifications in this and the following articles for various kinds of wearing surfaces of street pavements are taken from the standard specifications used in the city of St. Louis. In these specifications all the general clauses and also all detailed description of the grading, curb, gutter, and foundation will be omitted, since it is the intention to include in them only that portion of the specification describing the wearing surface.

In this specification for brick pavement, after describing the curbing, preparation of the roadbed, which involves a thorough rolling with a steel roller, weighing not less than ten tons, or three hundred pounds per lineal inch of roller; also the concrete foundation of six inches in depth, the following specification is given for

WEARING SURFACE.

Upon the foundation of concrete shall be laid a bed of coarse sharp sand free from loam or vegetable matter one and one half (1½) inches in thickness when compacted to serve as a bed for the bricks, which will be laid directly upon and imbedded in it, with close end and side joints. Upon this base of sand a pavement of the best quality of vitrified paving brick shall be laid. The brick shall not be less than seven and one

*This indicates how the absorption test can be made to enter into the final total with a small weight, since it is not so important as the other tests. The impact test, on the other hand, is given a relatively large weight as it is of the greatest importance.

half inches, nor more than ———— inches long, not less than
two and one eighth inches, nor more than two and one half
inches wide, not less than three and one half inches, nor more
than four and one fourth inches deep. All paving brick must
be homogeneous and compact in structure, free from loose
lumps of uncrushed clay or from laminations caused by the
process of manufacture or fire cracks or checks of more than
superficial character or extent. All brick so distorted in burning as to lay unevenly in the pavement shall be rejected. All
brick shall be free from lime or magnesia in the form of pebbles and shall show no signs of cracking or spalling on
remaining in water ninety-six (96) hours.

The brick shall have a specific gravity of not less than 2.
They shall not absorb more than 3 per cent. of water when
dried at 212 degrees Fahrenheit, and immersed for twenty-four
hours in water.

The bidders shall submit twenty-five samples of the brick
they propose using. A portion of these bricks shall be subjected to such physical tests as the board of public improvements shall deem necessary,* and the remainder shall be retained
as samples of the material to be furnished and used. Any brick
which does not stand the tests satisfactorily will be rejected, and
no bid contemplating the use of the rejected brick shall be
entertained. Samples may be submitted by manufacturers, in
which case the bidder proposing to use brick of such manufacture will not be required to submit samples. The quality of
the brick furnished must conform to the samples presented by
the manufacturers and kept in the office of the street commissioner.

All brick may have a proper shrinkage but shall not differ
materially in size from the accepted samples of the same make,
nor shall they differ greatly in color from the natural color of
the well burned brick of its class and manufacture.

No bats or broken brick shall be used except at the curbs
where nothing less than a half brick shall be used to break
joints. The bricks to be laid in straight lines and all joints
broken by a lap of at least two inches, to be set upon the sand
in a perfectly upright manner as closely and compactly together
as possible, and at right angles with the line of the curb, except
at street intersections where they are to be laid as the street
commissioner may direct.

The pavement to be thoroughly rammed two or three
times, as may be directed by the street commissioner, with a
paver's rammer, weighing not less than seventy-five pounds, or
a roadroller weighing not less than three nor more than six
tons.

* Tests similar to those described in the previous article are required.

12

All the joints in the pavement shall be completely filled with clean coarse river sand, and an additional layer of sand not less than one inch in depth shall be spread uniformly over the whole surface of the pavement. The joints may be filled with hot pitch or asphaltum, with some sand or gravel at bottom, or with cement grout, this latter being now(1895)recommended in order to prevent the brick from chipping on the edges. On steep grades, however, this would give too smooth a surface. St. L.

Following the above specification is a "maintenance clause," similar to that given in the following article, providing for the maintenance of the pavement in good repair for a period of nine years. The contract price provided also for an annual sum to be paid for maintenance, and the bond given by the contractor covered the maintenance, as well as the original construction.

The tests to which the brick are submitted under this specification are the same as those given in Art. 133.

135.   Specification for Asphaltum Pavement.

After describing the preparation of the roadbed, curbing, concrete foundation, having a depth of five inches, etc., the following specifications of the asphaltum body and wearing surface are employed.

BINDER.

The second or binder course will consist of a fine bituminous concrete composed of clean broken stone, slag or gravel, not exceeding one and one half ($1\frac{1}{2}$) inches in their largest dimensions, thoroughly screened, and asphaltic cement made from lake asphalt, as below described. The stone, slag or gravel, will be heated by passing through revolving heaters and thoroughly mixed by machinery with the asphaltic cement in the proportion of not less than fifteen gallons of the asphaltic cement to one (1) cubic yard of stone, slag or gravel. The mixture will be so made that the resulting binder has life and gloss without an excess of cement. Should it appear dull from over heating or lack of cement it will be rejected. This binder will be hauled to the work and spread on the base with hot iron rakes, and immediately rammed and rolled with hand and steam rollers while in a hot and plastic condition, until it has a thickness of one and one half ($1\frac{1}{2}$) inches. The upper surface will be made exactly parallel with surface of the pavement to be laid.

WEARING SURFACE.

Upon this binder course thus prepared shall be laid a wearing surface or pavement proper, the basis of which shall be composed of lake asphalt unmixed with any of the products of coal tar, of a nature and quality proved to be durable and proper by having been in successful use in roadway pavements in one or more cities of the United States for a period of at least two years and in an amount greater than five thousand square yards in each of said cities.

The wearing surface shall be composed of—

1st.   Refined lake asphaltum.
2d.   Heavy petroleum oil.
3d.   Clean sharp sand.
4th.   Fine powder of carbonate of lime.

Refined asphalt shall be smooth and free from lumps of unmelted pitch or organic matter not bituminous. It shall not at any time reach a temperature over 375 degrees Fahrenheit. The asphaltic cement shall be prepared from such refined asphalt as may be approved by the street commissioner, and suitable heavy petroleum oil or other approved solvent.

The heavy petroleum oil, which may be the residum by distillation of the petroleum oils as found in the market, generally contains water, light oils, coke, and a gummy substance soluble in water. This petroleum oil is freed from all impurities and brought to a specific gravity of from 18 degrees to 22 degrees Beaume, and a fire test of 250 degrees Fahrenheit.

To the melted asphalt, at a temperature of not over 325 degrees Fahrenheit, the oil, after having been heated to at least 150 degrees Fahrenheit, is to be added in suitable proportions to produce an asphalt cement. To accomplish this, from 15 to 21 pounds of oil per 100 of refined asphalt will be required. As soon as the oil has begun to be added, suitable agitation, be means of an air blast or other acceptable appliances, will commence and be continued till a homogeneous cement is produced. The appliances for agitation shall be such as to accomplish this in at least ten hours, during which the temperature shall be kept at from 290 degrees to 325 Fahrenheit, and no higher. If the cement then appears homogeneous and free from lumps and from inequalities, as shown by samples from different parts of the still, it may be used. Should it not prove homogenous, such deficiencies as may exist shall be corrected by the addition of hot oil or melted asphalt, in the necessary proportion.

They shall be mixed in the following proportions by weight:

Pure asphalt .... 100 parts.
Heavy petroleum oil .... 15 to 20 parts.

The asphaltic cement being made in the manner above described, the pavement mixture shall be formed of the following materials, and in proportion stated:

Asphaltic cement	from 12 to 15
Sand..	" 83 " 70
Pulverized carbonate of lime	" 5 " 15
	100 100

Limestone dust shall be an impalpable powder of carbonate of lime, the whole of which will pass a 30-mesh screen, and at least 75 per cent. pass a 100-mesh screen.

The sand and asphaltic cement are heated separately to about three hundred degrees Fahrenheit. The pulverized carbonate of lime, while cold, is mixed with the hot sand in the required proportions, and is then mixed with the asphaltic cement at the required temperature. and in the proper proportion, is a suitable apparatus, which will effect a perfect mixture.

The pavement mixture, prepared in the manner thus indicated, shall be laid on the foundation. It shall then be carefully spread, by means of hot iron rakes, in such manner as to give a uniform and regular grade, and to such depth that after having received its ultimate compression, it shall have a thickness of two inches. The surface shall then be compressed by rollers; after which a small amount of hydraulic cement shall be swept over it, and it shall then be thoroughly compressed by a steam roller, weighing not less than ten (10) tons, in order to get a thoroughly compressed wearing surface, the rolling being continued as long as it makes an impression on the surface.

The powered carbonate of lime shall be of such degree of fineness that 5 to 15 per centum by weight of the entire mixture for the pavement shall be an impalpable powder of limestone, and the whole of it shall pass a No. 26 screen. The sand shall be of such size that none of it shall pass a No. 80 screen, and the whole of it pass a No. 10 screen.

In order to make the gutters, which are consolidated but little by traffic, entirely impervious to water, a width of twelve inches next the curb shall be coated with hot pure asphalt and smooth with hot smoothing irons, in order to saturate the pavement to a certain depth with an excess of asphalt.

TOOLS AND SAMPLES OF MATERIALS.

The contractor shall furnish and have on the line of work at all times, a complete and sufficient plant of tools, rollers, carts, etc., as may be determined by the street commissioner, to carry on the work in an expeditious and workmanlike manner, also furnish samples of the crude lake asphalt to be used in the work, properly labeled, also samples of the wearing surfaces as prepared for use, and the statement of the amount of each material used in making up the pavement mixtures, when called for by the street commissioner.

In order that the asphalt may be fully tested, each bidder must deposit with the street commissioner, at least three days before making his bid, samples ot materials he intends to use, together with certificates and statements as follows:

1st.   A specimen of the crude asphaltum not less than five (5) pounds in weight with a certificate stating the place from whence the asphaltum was taken.

2d.   A specimen of the asphaltic cement not less than five (5) pounds in weight with a statement of its composition, and also a statement of the composition of the proposed wearing surface.

3d.   A sample of the pavement surface showing the Asphalt after two years' actual use in a street, said sample be not less than one foot square and to be accompanied by a certificate from the proper city official showing the time during which said pavement has been in use on the street on which it was laid, and the certificate shall further show that the pavement from which the sample is taken, or similar pavement, has been in successful use on one or more roadways in said city for a period longer than two years, and in an amount greater than five thousand (5000) square yards.

4th.   A statement of the location and the capacity in square yards per day of the works or factory where the paving material is to be prepared.

Specimens must be furnished to the street department as often as may be required during the progress of the work.

MAINTENANCE.

The said ————, party of the first part, expressly guarantees to maintain at grade and surface in good order the aforesaid work of reconstruction throughout and at the end of the full period of nine years, commencing one year after the said work of reconstruction is completed and accepted, and binds himself, his heirs and assigns to make all repairs which may from any imperfection in said work or materials or from any rotting, crumbling or disintegration of the materials, become necessary within that time; and the party of the first part shall, whenever notified by the street commissioner that repairs are required, at once make such repairs at his own expense, and if they are not made within the proper time, the street commissioner shall have power to cause such repairs to be made, and the cost thereof shall be paid out of the fund provided for the payment of contracts for street maintenance, and the amount shall be deducted from any money then due under the contract, or which may thereafter become due.   At the end of the nine-year period the street commissioner must determine whether or not the street is in good order at grade and surface, and the principal and his sureties under this contract shall not be discharged from liability on their maintenance bond hereunder until the street commissioner shall so determine and certify thereto in writing to the principal under this contract. And it is further expressly agreed, that if any time during the term for which the contract for the maintenance of the above street is in force, the pavement of said street, or any part thereof, has deteriorated to such an extent as to require, in the opinion of the board of public improvements, reconstruction, the street commissioner shall, with the approval of the board of public improvements and of the mayor, notify the contractor that reconstruction is necessary, and the contractor shall, within three months after receiving such notice, reconstruct the whole or such part of the pavement with the same kind of material as here tofore applied, or with some other material approved by the board of public improvements. And if the contractor fails to reconstruct the street within three months after having been notified, the board of public improvements may, with the approval of the mayor, cancel the contract and relet the work of reconstructing the pavement, and that the cost of such reconstruction shall be paid by the city and the amount collected by suit from the contractor or his sureties, not to exceed fifteen dollars per square of pavement, included in the contract.

And it is further agreed that whenever any repairs of the street are made necessary from the construction of sewers, the laying of pipes or telegraph wires, or from any other disturbance of the pavement by parties acting under permits issued by the city, the contractor shall, on notification from the street commissioner, immediately make all necessary repairs in conformity with the specifications for this class of work. The cost of all such repairs, exclusive of trenching and back filling, which shall be done by the parties who hold the permits, and in the same manner as now required by existing ordinances, shall be paid for at the full contract price for a superficial square of new pavement out of the fund set apart for the payment of contracts for the maintenance of streets, and the amount shall be certified by the street commissioner to the auditor, who shall reimburse, by transfer, the aforesaid fund from the funds of the proper department, if the repairs were made necessary by the construction of any public improvement; and out of the funds to be deposited by persons obtaining permits for opening streets before such permits are granted, if the repairs are made necessary by work done under such permits. And it is agreed that the contractor shall have the right to make all repairs which become necessary by the construction of any public improvement or work done by private parties under permits given by the city.   St. L.

136. Specification for Asphalt Pavement. The following specification for asphalt pavement was prepared in 1892, by Mr. A. P. Boller, of New York City, for such a pavement upon the new Harlem river bridge at 155th st., New York. It probably embodies the latest and most approved methods of making such a pavement, and so far as it is applicable to ordinary street pavements, it might be followed with advantage.

The sub-surface must then be brought to a uniform grade and cross-section not to exceed a crown of three inches in width of roadway by filling all depressions with a fine bituminous concrete or binder, to be composed of clean, broken stone not exceeding one inch in their largest dimensions, thoroughly screened, and coal tar residuum, commonly known as No. 4 paving composition.

If required by the department of public parks, clean, sharp sand may replace a portion of the broken stone.

The stone or stone and sand must be heated by passing through revolving heaters, and thoroughly mixed by machinery with the paving composition in the proportion of one (1) gallon of paving composition to one (1) cubic foot of stone.

This binder must be hauled to the work and spread with hot iron rakes in all holes or inequalities and depressions below the true grade of the pavement, to such thickness that after being thoroughly compacted by tamping and hand rolling the surface shall have a uniform grade and cross-section, and the thickness of the binder at any point shall be not less than three quarters of an inch.

The upper surface shall be exactly parallel with the surface of the pavement to be laid.

Upon this foundation must be laid the wearing surface or paving proper, the basis of which or paving cement must be pure asphaltum, unmixed with any of the products of coal tar.

The wearing surface must be composed of:—

1. Refined asphaltum.
2. Heavy petroleum oil.
3. Fine sand, containing not more than one per centum of hydro-silicate of alumina.
4. Fine powder of carbonate of lime.

The asphaltum must be specially refined and brought to a uniform standard of purity and gravity of a quality to be approved by the engineer.

The heavy petroleum oil must be freed from all impurities and brought to a specific gravity of from eighteen to twenty two degrees Beaume, and a fire test of two hundred and fifty degrees Fahrenheit.

From these two hydro-carbons shall be manufactured an asphaltic cement which shall have a fire test of two hundred and fifty degrees Fahrenheit, and at a temperature of sixty degrees Fahrenheit shall have a specific gravity of 1.19, said cement to be composed of one hundred parts of pure asphalt and from fifteen to twenty parts of heavy petroleum oil.

The asphaltic cement being made in the manner above described, the pavement mixture will be formed of the following materials, and in the proportions stated:

Asphaltic cement ....from 12 to 15
Sand ....from 83 to 70
Pulverized carbonate of lime...from 5 to 15

The sand and asphaltic cement are to be heated separately to about three hundred degrees Fahrenheit. The pulverized carbonate of lime, while cold, shall be mixed with the hot sand in the required proportions, and then mixed with the asphaltic cement at the required temperature, and in the proper proportion, in a suitable apparatus, which will effect a perfect mixture.

The pavement mixture prepared in the manner thus indicated must be brought to the ground in carts at a temperature of about two hundred and fifty degrees Fahrenheit, and if the temperature of the air is less than fifty degrees, iron carts, with heating apparatus, must be used in order to maintain the proper temperature of the mixture; it shall then be carefully spread by means of hot iron rakes, in such manner as to give a uniform and regular grade, and to such depth that after having received its ultimate compression, it will have a thickness of two inches at crown of roadway, tapering off, if required, to about one inch at gutters. The surface shall then be compressed by hand rollers, after which a small amount of hydraulic cement shall be swept over it, and it shall then be thoroughly compressed by a steam roller weighing not less than two hundred and fifty pounds to the inch run; the rolling to be continued for not less than five hours for every one thousand yards of surface.

The powdered carbonate of lime must be of such degree of fineness that five to fifteen per centum by weight of the entire mixture for the pavement shall be of an impalpable powder of limestone, and the whole of it shall pass a No. 26 screen. The sand must be of such size that none of it shall pass a No. 80 screen, and the whole of it must pass a No. 10 screen.

In order to make the gutters, which are consolidated but little by traffic, entirely impervious to water, a width of twelve inches next the curb must be coated with hot, pure asphalt and smoothed with hot smoothing irons in order to saturate the pavement to a certain depth with an excess of asphalt.

If rock asphalt be used, it must be natural bituminous limestone rock: (1) from the Sicilian mines at Ragusa, equal in quality and composition to that mined by the United Limmer and Ver Wohle Rock Asphalte Company, Limited; (2) from the Swiss mines at Val de Travers, equal in quality and composition to that mined by the Neuchatel Rock and Asphalte Company, Limited, or (3) from the French mines at Seyssel, equal in quality and composition to that mined by the Compagnie Generale des Asphaltes de France, and it shall be prepared and laid as follows:

(1) The lumps of rock shall be finely crushed and pulverized, the powder shall then be passed through a fine sieve. Nothing whatever shall be added to or taken from the powder obtained by grinding the bituminous rock. The powder shall contain from nine to twelve per cent. natural bitumen, eighty eight to ninety one per cent. pure carbonate of lime, and must be free from quartz, sulphates, iron pyrites, or aluminum. (2) This powder shall be heated in a suitable apparatus to two hundred or two hundred and fifty degrees Fahrenheit, and must be brought to the ground at such temperature in carts made for the purpose, and then carefully spread on the foundation previously prepared, to such depth that, after having received its ultimate compression, it will have a thickness of two inches. (3) It shall be skillfully compressed by heated rammers and rolled until it shall have the required thickness of two inches. (4) The surface to be rendered perfectly even by heated smoothers, and to be rolled with a steam roller weighing not less than two hundred and fifty pounds to the inch run, the rolling to continue for not less than five hours for each one thousand yards of surface.   A. P. B.

137. Specification for Granite Pavement. The following specification for granite pavement is that used in the city of Milwaukee so far as the granite paving is concerned. These granite blocks are laid upon a concrete foundation six inches thick, and this latter upon a carefully prepared surface which has been thoroughly rolled with a heavy roller. The concrete is made of natural cement one part, sand two parts, and broken stone five parts. On this is spread a sand cushion two inches thick when compacted, on which the granite blocks are laid.

Granite Block Paving. The blocks must consist of a hard granite uniform in grain and texture, without lamination or stratification and free from excess of mica or feldspar. Neither hard basaltic stone that will take a smooth polish under traffic,

nor soft or weather worn stones nor syenite will be accepted.
The blocks must by rectangular in form, of not less than three
(3) nor more than four and one half (4 ½) inches in thickness, nor less than six (6) or more than seven (7) inches in
depth, nor less than eight (8) or more than twelve (12) inches in
length, and so split and dressed with fair and true surfaces on top,
bottom and ends so that when laid close together the end joints
will fit close together, and the side joints will not exceed three
fourths (¾) inch in width. The blocks will be imbedded in
the sand bed and laid at right angles to the line of the street,
except at street and alley intersections, where the same will be
laid at an angle of about 45 degrees with the line of the street.
The stone will be laid close together with the top surface
smoothly conforming to the crown of the street. Each course
is to be of uniform width, with each longitudinal joint broken
by a lap of not less than two inches. The blocks are to be
immediately covered with sufficient, clean, fine, hot, screened
gravel to fill the joints, to not more than 3 ½ inches from the
top after which the blocks will be tamped with a heavy paver's
ram to a firm, unyielding and uniform smooth surface. The
joints will then be filled flush with top of pavement with a hot
paving cement obtained by direct distillation of coal tar, immediately after which fine, dry, hot gravel will be run into the
joints. Not less than three (3) gallons of paving cement shall
be used to each square pard of pavement.*   M.

138. Specification for Granitoid Sidewalks. Sidewalks made after the following specifications are now exclusively
used in St. Louis, and have been in use in that city for many
years. Where granite can be obtained at a reasonable price, it
is thought this composition is more durable and satisfactory for
sidewalks than any other material or mixture which has ever
been used. The making of these sidewalks has grown to be a
very large industry in St. Louis and the price has been gradu-
ally reduced because of the great amount of this kind of work
done, until in 1895, the total cost of removing old pavement,
regrading, laying the foundation and pavement as here described
in the most approved manner, and strictly in accordance with
this specification is from eighteen to twenty cents per square
foot for the "ordinary single flagging."

* This treatment of the joints is especially satisfactory.—AUTHOR.

The sidewalks shall be of three separate and distinct thickness and kinds, and shall be classified as follows: "ORDINARY SINGLE FLAGGING," "EXTRA DOUBLE THICK FLAGGING," and "DRIVEWAY OR ENTRANCE FLAGGING," and shall be laid in the different localities within the above described limits at the discretion of the street commissioner, who shall determine which of the above named kinds shall be laid.

Preparation of Bed. The sidewalks shall be excavated and shaped to the proper depth and grade as directed by the street commissioner, and all the refuse material therefrom shall belong to the contractor and shall be promptly removed from the line of work.

Ordinary Single Flagging. After the shaping is done a foundation of cinders not less than eight (8) inches thick shall be placed upon the subgrade, which shall be well consolidated by ramming to an even surface, and which shall be moistened just before the concrete is placed thereon.

After the sub-foundation has been finished the artificial stone flagging shall be laid in a good workmanlike manner.

The same to consist of two parts: 1st. A bottom course to be three and one half (3½) inches in depth. 2d. A finishing or wearing course, to be one half (½) inch in depth.

The bottom course shall be composed of crushed granite and the best Portland cement, equal to the Dyckerhoff brand, and capable of withstanding a tensile strain of 400 pounds to the square inch after having been three hours in air and seven days in water, and shall be mixed in the proportion of one part cement to three parts of crushed granite.

The crushed granite shall consist of irregular, sharp-edged pieces, so broken that each piece will pass through a three fourths (¾) of an inch ring in all its diameters, and which shall be entirely free from dust or dirt.

The crushed granite and the cement in the above mentioned proportions shall first be mixed dry, then sufficient clean water shall be slowly added by sprinkling, while the material is constantly and carefully stirred and worked up, and said stirring and mixing shall be continued until the whole is thoroughly mixed.

This mass shall be spread upon the sub-foundation and shall be rammed until all the interstices are thoroughly filled with cement.

Particular care must be taken that the bottom course is well rammed and consolidated along the outer edges.

After the bottom course is completed, the finishing or wearing course shall be added. This course to consist of a stiff mortar composed of equal parts of Portland cement and the sharp screenings of the crushed granite, free from loamy or earthy substances, and to be laid to a depth of one half (½) of an inch and to be carefully smoothed to an even surface, which, after the first setting takes place, must not be disturbed by additional rubbing.

When the pavement is completed it must be covered for three days and be kept moist by sprinkling.

Extra Double Thick Flagging. After the grading and shaping is done, a foundation of cinders not less than six (6) inches thick shall be placed upon the subgrade, which shall be well consolidated by ramming to an even surface and which shall be moistened just before the concrete is placed thereon. After the sub-foundation has been finished the artificial stone flagging shall be laid in a good, workmanlike manner.

The same to consist of two parts: 1st. A bottom course to be five (5) inches in depth. 2d. A finishing or wearing course to be one (1) inch in depth.

The bottom course shall be composed of crushed granite and the best Portland cement equal to the Dyckerhoff brand, and capable of withstanding a tensile strain of 400 pounds to the square inch after having been three hours in air and seven days in water, and shall be mixed in the proportion of one part of cement to three parts of crushed granite.

The crushed granite shall consist of irregular, sharp-edged pieces, so broken that each piece will pass through a three fourths (¾) of an inch ring in all its diameters, and which shall be entirely free from dust or dirt.

The crushed granite and the cement in the above mentioned proportions shall first be mixed dry, then sufficient clean water shall be slowly added by sprinkling, while the material is constantly and carefully stirred and worked up, and said stirring and mixing shall be continued until the whole is thoroughly mixed.

This mass shall be spread upon the sub-foundation and shall be rammed until all the interstices are thoroughly filled with cement.

Particular care must be taken that the bottom course is well rammed and consolidated along the outer edges.

After the bottom course is completed the finishing or wearing course shall be added. This course to consist of a stiff mortar composed of equal parts of Portland cement and the sharp screenings of the crushed granite, free from loamy or earthy substances, and to be laid to a depth of one (1) inch and to be carefully smoothed to an even surface, which, after the first setting takes place, must not be disturbed by additional rubbing.

When the pavement is completed it must be covered for three days and be kept moist by sprinkling.

Driveway or Entrance Flagging.—After the grading and shaping is done, a foundation of crushed limestone and hydraulic cement mortar shall be laid to a depth of six (6) inches on the subgrade. The stone used in this concrete shall be broken so as to pass through a two (2) inch ring in its largest dimensions. The stone shall be cleaned from all dust and dirt and thoroughly wetted and then mixed with mortar, the general proportion being: One part of cement, two parts of sand, and five parts of stone. It shall be laid quickly and then rammed until the mortar flushes to the surface. No walking or driving over it shall be permitted when it is setting, and it shall be allowed to set for at least twelve hours, and such additional length of time as may be directed by the street commissioner or by his duly authorized agents, before the pavement is put down.

After the subfoundation has been finished, the artificial stone flagging shall be laid in a good, workmanlike manner. The same to consist of two parts: 1st. A bottom course to be five (5) inches in depth. 2d. A finishing or wearing course to be one (1) inch in depth.

The bottom course shall be composed of crushed granite and the best Portland cement, equal to the Dyckerhoff brand, and capable of withstanding a tensile strain of 400 pounds to the square inch after having been three hours in air and seven days in water, and shall be mixed in the proportion of one part . cement and three parts of crushed granite.

The crushed granite shall consist of irregular, sharp-edged pieces, so broken that each piece will pass through a three fourths (¾) of an inch ring in all its diameters, and which shall be entirely free from dust or dirt.

The crushed granite and the cement in the above mentioned proportions shall first be mixed dry, then sufficient clean water sha'l be slowly added by sprinkling, while the material is constantly are carefully stirred and worked up, and said stirring and mixing shall be continued until the whole is thoroughly mixed.

This mass shall be spread upon the subfoundation and shall be rammed until all the interstices are thoroughly filled with cement.

Particular care must be taken that the bottom course is well rammed and consolidated along the outer edges.

After the bottom course is completed, the finishing or wearing course shall be added. This course to consist of a stiff mortar composed of equal parts of Portland cement and the sharp screenings of the crushed granite, free from loamy or earthy substances, and to be laid to a depth of one (1) inch and to be carefully smoothed to an even surface, which, after the first setting takes place, must not be disturbed by additional rubbing.

When the pavement is completed, it must be covered for three days and be kept moist by sprinkling. St. L.

SPECIFICATIONS FOR SEWERS.

139. Specifications for Brick and Tile Sewers.

The following specifications for brick and tile sewers are those used in the city of St. Louis, so far as they relate to the construction proper, except that part relating to the use of cement, concrete, and rubble masonry. As specifications on these subjects are given elsewhere, they are not included here:

Excavation.—All excavation shall be done by open cut from the surface, except where tunneling is shown on the plans or is expressly permitted or directed by the sewer commissioner. Wherever the material is of such a nature ás to allow it, the bottom of the excavation up to the greatest horizontal diameter of the sewer shall be made with a template so as to conform to the exact shape of the brickwork. Above this line the cut may, in all ordinary cases, be carried to the surface at such a slope as the contractor may desire, but it will be calculated with a slope of one horizontal to——————— vertical, whatever may be the actual slope. Should the contractor think it best to keep the sides of the excavation vertical by bracing or otherwise, it is expressly·understood that it shall be done at his own cost and risk.

Rock shall be excavated so as to conform as nearly as possible to the lower half of the sewer, and all irregularities shall be filled with masonry or concrete so as to make a smooth bed for the brick work. The amount of the excavation in rock cuts will be calculated with a base at the bottom of the brickwork equal to the greatest horizontal diameter of the sewer, and with side slopes of the same inclinations as in other excavations. All the rock taken from the excavations shall belong to the contractor for his own use.

Wherever the excavation can not be adapted to the shape of the brickwork, it shall be done according to such directions as may be given in each case.

The sides of the excavation shall, whenever it may be necessary, be supported with suitable plank and shoring, but no allowance will be made therefor unless the same is left in by express orders of the sewer commissioner, when it will be paid for at ——————— dollars per thousand feet, board measure. In all other cases it will be drawn as the work progresses and not paid for by the city.

The contractor shall, at his own cost, keep the trenches free from water during the progress of the work. Excavated material must be so placed as not to interfere with travel on the street or to incommode occupants of adjoining property. Trenches shall not be opened more than 200 feet in advance of the laying of the sewer.

Back Filling.—Back filling shall follow close after the construction of the sewer, and in no case be more than 100 feet in the rear.

The filling of the earth around and on top of the sewers shall be done with the utmost care, and in a manner to obtain the greatest compactness and solidity possible. For that purpose the earth shall be laid and rammed in regular layers not more than nine inches thick up to the surface of the street, or thoroughly soaked with water, as may be directed by the sewer commissioner.* The macadamizing, if any has been removed, shall be carefully replaced on the top of the said filling; and when paving has been removed it shall be replaced in the same manner as when originally constructed, and the street or alley left in as good condition as it was before. If any new materrials are needed for such repairing, they shall be of the best quality, and shall be furnished and put down by the contractor at his own cost.

The gutter paving in front of and adjoining sewer inlets shall be taken up and replaced in proper shape, so as to conduct the storm water into the sewer inlets.

All work of restoring the surface of the streets and alleys shall be done to the satisfaction of the street commissioner, or his duly authorized agents, immediately after the sewer is laid. If not so done within five days after notice, the work may be done by the street commissioner, and the cost thereof shall be paid by the contractor; and in default of payment, the cost may be retained by the city of St. Louis, out of any money that may be due or become due to the contractor under this contract.

Surplus Earth.—All surplus earth shall be hauled away promptly to such places, within a distance of 3,000 feet, as the sewer commissioner shall designate, and be spread according to his directions; but if no such place is designated, the contractor shall dispose of such surplus at his own risk and expense. No surplus earth shall be deposited on private · property, if within the limit just named, it can be used on the streets or alleys or other public places. But if no such use can be found for it, it may, with the consent of the sewer commissioner, be deposited on private property; but all earth so deposited without the consent of said commissioner, shall be measured, and the amount thereof deducted from the measurement of the excavation.

The price paid for earth and rock excavation shall cover the whole cost of excavating the trenches and refilling the same

* A better plan is to thoroughly ram the layers in nine-inch courses and then to *also* thoroughly soak with water every four or five feet, whenever water is available.

with earth, restoring the street and hauling away the surplus materials, as well as the whole cost of pumping, bailing, planking, and shoring, excepting such planking as may be left in by express orders as hereinbefore specified.

Bricks.—All the bricks used shall be of uniform texture, hard-burned entirely through, free from lime or other impurities, that will affect them in water, and shall have straight edges and square angles. Broken bricks must not be brought on the ground, and such as are broken afterwards in handling shall be used only in making closures, or as shall be otherwise specially directed.

The bricks are to be culled as they are brought on the ground, and all bricks of improper quality thrown out and removed from the ground. The culling to be done at the expense of the contractor, who shall furnish the inspector with men for this and similar purposes, when required.

Brick Masonry.—In building brick masonry, none but careful and skillful bricklayers shall be employed.

The bricks shall be clean and thoroughly wet just before being laid, unless otherwise specially directed. Every brick shall be laid with a *push joint;* that is, by placing sufficient mortar on the bed and forcing the brick into it in such a manner as to thoroughly fill every joint, whether on the bottom, side or end of the brick with mortar. The joints shall be made as nearly as possible of uniform thickness, not exceeding three eighths of an inch, and in the inside of the invert or lower arch, they shall not exceed one eighth of an inch.

The bricks in each course shall be all stretchers, and to break joints with those in the adjoining courses. The bricks of the inside course shall be laid to a line and to the true cylindrical or other form given for each case. The inside course shall also be made of the smoothest and hardest bricks, carefully selected for this purpose.

The upper arch shall be built on strongly made centers, which shall be drawn with great care, so as not to disturb the brickwork. The crown of the arch shall be properly keyed with stretchers, and all the joints be well filled with the mortar. The exterior surface of the upper arch shall be covered with a coating of mortar, not less than three eighths ($\frac{3}{8}$) of an inch thick.

The mortar joints on the inside of the sewer below the center line shall be carefully struck when laid, and those above be scraped smooth with the brickwork immediately after the centers are drawn, and the mortar scraped off and entirely removed from the sewer, which is to be left perfectly clean throughout.

All unfinished brickwork must be racked back in courses, except when otherwise specially directed or permitted, and when new work is to be joined to'it, the surface of the bricks must be cleaned and moistened.

Openings for branch sewers shall be made and junction pieces inserted in the main sewers in such manner and at such places as may be directed. Every junction piece shall be closed with a cover of earthenware, or with bricks and cement.

All brickwork will be measured and paid for by the cubic yard of solid wall.

TUNNELING.

In tunneling, the excavation shall be made so as to conform neatly to the regular section of the sewer, and nothing will be allowed for any excavation beyond this. All holes or irregularities outside of the regular section must be filled up solid with bricks and mortar, but no extra allowance will be made therefor.

All timbers used in sustaining the excavation must be removed as the brickwork progresses.

Points, by which to get the proper line of the sewer, will be given from time to time as may be needed, and from these the contractor will be required to continue the line of the excavation at his own risk of its accuracy, and to correct at once any errors of alignment that may be discovered before the brickwork is finished.

In tunnels, the quantities paid for will be the earth or rock excavated in the regular section of the sewer, and the brick or stone masonry required for this section, together with any foundation work that may have been expressly ordered, and the amount paid for these items shall be in full for furnishing all materials, and finishing the sewer; the cost of sinking shafts, pumping water, shoring, restoring falls and all accessory works of every kind being borne wholly by the contractor. Those parts only of the sewer will be paid for as tunnels, which are so marked on the plans exhibited at the time of the letting; all the rest will be paid for as open cut, regardless of the manner in which the work is actually done.

PIPE SEWERS.

All pipe sewers shall be made of the best quality of vitrified clay pipe with smooth interior surface. Each piece shall be straight or evenly curved, as may be required, and in section shall not vary more than half an inch from a true circle. The thickness of six-inch pipes shall not be less than three quarters of an inch; of twelve-inch pipes, not less than one and one eighth inches; of fifteen-inch pipes, not less than one and one quarter inches; and of eighteen-inch pipes, not less than one and one half inches. Junction pieces, for use in brick

13

sewers, shall be smoothly beveled off to an angle of forty-five degrees, and be not less than two feet long, exclusive of the socket. For pipe sewers the junction piece shall be a part of the main pipe, and no right angle junction shall ever be used.

So far as the specifications for the excavation of trenches, shoring and pumping, preparation of foundations, backfilling and restoring the street surface, already given for brick sewers, can be made to apply to the construction of pipe sewers, they shall be followed.

Each pipe is to be laid on a firm bed and in perfect conformity with the lines and levels given. The bottom of the trench must be shaped so as to fit the lower half of the pipe as nearly as possible, with places cut at the joints for the sockets to rest in, so that the pipe shall have a uniform bearing on the ground from end to end.

The pipes shall be joined by filling the socket with a mortar of pure cement without sand, with only water enough to give it a proper consistency. Great care must be taken to make the joint throughout the lower three fourths of the pipe perfectly water tight. The upper one fourth of joint, when so directed, shall be left open.

The interior of the pipes shall be carefully cleaned from all dirt, cement and superfluous material of every description, and a wad made of a sack filled with hay, large enough to fill the pipe and attached to a rod or cord, shall, at all times be kept in the pipe and drawn forward as the swork proceed, care being taken not to loosen the joints.

After the pipes are properly laid and joined, any space between them and the sides of the excavation must be filled with sand, either washed in or well rammed, up to the middle of the pipe. From this point for at least twelve inches above the top of the pipe, the earth shall be filled in so as not to disturb the pipes, and thoroughly rammed; after which, up to the surface, it may be either rammed in layers or thoroughly soaked with water, as may be directed by the sewer commissioner, so that the least possible settling will take place after the work is completed.

Pipe sewers will be paid for by the linear foot of finished work, the price so paid to be in full payment for furnishing and laying the pipe, including the earth excavation, shoring and pumping, backfilling, restoring the street surface, hauling away surplus material, and all other work and material required by the specifications or necessary to give a finished result.

Where rock is encountered in pipe sewers, such rock excavation shall be paid for at the price named herein————————————the amount to be estimated with a base of six inches more than the inside diameter of the pipe and the side slope of one horizontal to eight vertical.

140. Specification for Sewer Pipe. The following specification for sewer pipe and specials is probably the most carefully worked out of any found in current American practice. While these specifications are very full and complete in many details which are usually overlooked, they are not unreasonably severe. They simply describe clearly what kinds of faults will serve as cause for rejection, and are as valuable to the manufacturer of the pipe in enabling him to select those specimens which he feels will be accepted, as to the inspector himself, who is called upon to accept or reject the material when supplied upon the ground. This specification, therefore, has the great merit of extreme definiteness of meaning, which is the most vital and necessary quality of all specifications. They were prepared by an engineer who knew from experience exactly what could be furnished by the best sewer pipe manufacturers without greatly increasing the cost.

Sewer Pipe and Specials—Pipe sewers are composed of straight sections which are herein termed *"pipe,"* and of branches, bends, reducers, etc., which will here be called *"specials"* or *"special pieces."*

The main sewer, as well as all surface and lot lateral sewers, shall be constructed of the best quality of salt-glazed, vitrified stoneware sewer pipe, and all special pieces that may be required in the work shall be of the same description and quality.

The pipes and specials must be carefully selected and examined by the contractor before or while being delivered upon the street, and all such material which may be used in the work must conform to the following requirements and conditions:

All hubs or sockets must be of sufficient diameter to receive their full depth the spigot end of the next following pipe or special without chipping whatever of either, and also to leave a space of not less than 1-8 inch in width all around for the cement mortar joint. Pipes and specials which can not be thus freely fitted into each other shall be rejected.

In the case of pipes and specials of 12 inches and upward in diameter, at least 40 per cent. of all such that will be used in the work must be truly circular or substantially circular in cross-section, and in the case of pipes and specials less than 12 inches in diameter, at least 60 per cent. of the whole number required must be truly circular or substantially circular in cross-section. Of the remainder, in each case, the allowable divergence from a truly circular cross-section shall never exceed the following limits: *a*. For an *elliptical* cross-section, the greatest internal diameter must not be more than from 6 to 7 per cent. longer than the least internal diameter in the same cross-section. *b*. For an *oval or egg-shaped* cross-section, the same rule as for eliptical cross-sections shall apply. *c*. Pipes and specials having cross-sections which exhibit *angles, sharp curves* or *flat places* of appreciable magnitude in the circumference, will be rejected.

A single fire-crack, which extends through the *entire thickness* of a pipe or special, must not be over two inches long at the spigot end, nor more than one inch long at the hub or socket end, measured in the latter case from the bottom, or shoulder, of said hub or socket. Two or more such fire-cracks, however, at either end of said pipe or special will cause the same to be rejected.

A single fire-crack, which extends through only *two thirds of the thickness* of a pipe or special, must not be over four inches long at either end thereof, measured in the direction of its length. Two or more such fire-cracks, however, at either end of said pipe or special will cause the same to be rejected.

A single fire-crack, which extends through only *one half of this thickness* of a pipe or special, must not be over six inches long at either end thereof, measured in the direction of its length. Two or more such fire-cracks, however, at either end of said pipes or special will cause the same to be rejected.

A single fire-crack, which extends through *less than one half of the thickness* of a pipe or special, must not be over eight inches long, measured in the direction of the length of such pipe. Two or more such fire-cracks, however, anywhere in the pipe will cause the same to be rejected.

A transverse fire-crack in a pipe or special must not be longer than one sixth of the circumference of such pipe, nor shall its depth be greater than one third of the thickness thereof. Two or more such fire-cracks will be cause for rejection.

No fire-cracks of any description shall, however, be more than one eighth inch wide at its widest point.

No combination of the foregoing six limitations will be allowed, except with the express consent of the executive board and the city surveyor, as the intent and meaning of these restrictions or limitations is to insure the furnishing of the best marketable quality of pipe and specials by the contractor. In general, any pipe or special which exhibits more than one fire-crack of the magnitudes above mentioned should be rejected at once by the inspector in charge of the work of laying the pipes, unless there be time to make a thorough aud minute examination of the other fire-cracks which it may display, and to become thereby convinced that they are of trifling significance.

Any pipe or special which is found to be cracked through its whole thickness from any other cause except the process of burning in the kiln, shall be rejected at once, regardless of the extent of such crack. This refers particularly to damage done by transportation, by cooling or by frost.

Irregular lumps or unbroken blisters on the interior surface of a pipe or special of sufficient size and number to form an appreciable obstruction to the free flow of the sewage, will be cause for rejection. A few small, unbroken blisters, not exceeding one fourth of an inch in height and one or two inches in diameter, upon the inner surface, need not reject a pipe or special. If there is a broken blister or a flake on the interior of a pipe or special which is thicker than one sixth of the normal thickness of said pipe or special, and whose largest diameter is greater than one twelfth of the inner circumferenc of said pipe or special, the latter shall be rejected. Furthermore, if such broken blister or flake is as large or smaller than just defined, then, unless said pipe or special can be properly fitted and laid so as to bring such broken blister or flake on the top or upper side of the sewer, the said pipe or special shall also be rejected.

Irregular lumps and small, unbroken blisters on the outside of a pipe or special need not reject it. A large and broken blister or a flake on the outside of a pipe or special, which is thicker than one sixth of the normal thickness of said pipe, and whose largest diameter is greater than from one ninth to one twelfth of the outer circumference of said pipe, will cause the same to be rejected. Should, however, the broken blister or flake be within the limits of size just defined, and should the pipe or special admit of being properly laid so as to bring said blister or flake on the upper part of the sewer, then said pipe or special may be accepted, if otherwise sound in all respects.

Any pipe or special which betrays in any manner a want of thorough vitrification or fusion, or the use of improper materials and methods in its manufacture, shall be rejected. Attention of inspectors is particularly called to the character of the material composing the interior of a pipe or special where the same is exposed by the breaking of a blister, the removal of a flake, or the face of the spigot end of such pipe.

All pipe and specials which are designed to be straight shall not exhibit any material deviation from a straight line.

Special curves or bends shall substantially conform to the degree of curvature and general dimensions that may be required.

If a piece be broken out of the rim forming the hub or socket of a pipe or special without injuring the body of such pipe, the latter shall be rejected if the length of said broken piece, or the gap left thereby, is greater than one tenth of the circumference of said hub. In case that a defect of this nature, and within the limits just defined, occurs in a pipe or special, the latter shall also be rejected unless it can be so fitted in the sewer as to bring said defect on the upper part thereof.

The attention of the inspector in charge of the work of laying the sewer pipe is herewith particularly directed to the foregoing requirements as to the quality of the pipe and specials that will be allowed in the sewer, and in all cases of doubtful interpretation of said requirements, the necessary definitions will be given by the city surveyor and the executive board. Said board also reserves the right to add to the foregoing requirements, at any time during the progress of the work, such further restrictions and conditions respecting the quality of the said pipe and specials as it may deem for the best interests of the tax-payers, in order to secure the best materials which can practically be obtained. All such explanations or definitions of said requirements, in cases of doubtful interpretation, together with all said further restrictions and conditions relating to the quality of said pipe and specials, shall have the same force as though a part of this specification, and the contractor shall be required to comply therewith without extra compensation beyond the prices bid by him for performing the work. E. K.

141. Specification for Laying Sewer Pipe.

The following specification for the laying of sewer pipe and specials has all the merits ascribed to the specification for sewer pipe as given in the previous article, and has been prepared by the same engineer. For the purpose of removing any cement mortar which may have been forced through the joints, and which may, when hardened, form serious obstructions in the sewer, probably no specification will insure such excellent results as that given in the St. Louis specifications for pipe sewers in Art. 139, where the contractor is required to provide "A wad made of a sack filled with hay, large enough to fill the pipe and attached to a rod or cord, which shall at all times be kept in the pipe, and which shall be drawn forward as the work proceeds, care being taken not to loosen the joints." It is an easy matter for the inspector to examine at any time to see whether or not this wad is being drawn forward, and when drawn forward it must of necessity remove any protruding fins of mortar, and leave the interior smooth and entirely free from such obstructions.

LAYING THE SEWER PIPE AND SPECIALS.—Previous to laying the pipe and specials which have been delivered upon the street, into the trench, they shall all be subjected to a rigid inspection by both contractor and inspector, and those which do not come up to the foregoing requirements shall be rejected.

Additional tests by sounding said pipe for cracks, and examining closely all blisters and flakes, shall also be applied. Before lowering the pipes and specials which have passed the inspections into the trench, they shall first be properly fitted together upon the surface of the street in the order in which they are to be used; and to facilitate the process of laying, the top of each pipe or special, after said fitting, shall be plainly marked with chalk or paint, so that the pipe previously laid in the bottom of the trench shall be disturbed as little as possible.

All pipes and specials in which the spigots and sockets can not be made to fit together, while on the surface, must be rejected, as no chipping of either socket, hub or spigot will be allowed.

The faces of all spigot ends and of all shoulders in the hubs or sockets must be true, and be brought into fair contact, and all lumps or excrescences on said faces shall be carefully cut away before the pipes are lowered into the trench.

In all cases where the rim of any hub or socket has been broken, as aforesaid, the pipe or special shall be rejected unless it can be so fitted as to bring the broken portion on the top, or upper portion of the sewer. The same condition shall also be applied to the case of broken blisters and flakes, as above mentioned, on either inside or outside of the pipes and specials. All special pieces required in the work, such as branches, bends, curves, reducers, etc., shall likewise be subject to the same conditions as the straight pipe.

The pipes and specials shall be so laid in the trench that after the sewer is completed the interior surface thereof shall conform on the bottom accurately to the grades and alignment fixed and given by the city surveyor. The main sewer will be divided by man-holes and lamp or hand-holes into a number of distinct divisions or working sections, in each of which the grade and alignment shall, under ordinary circumstances, be truly straight. Changes of grade or direction, or both, in said main sewer will generally be made at man-holes or lamp or hand-holes, although under special conditions, to be defined only by the executive board and city surveyor, such changes may be made at intermediate places.

While the pipe and specials are being laid in each of the aforesaid straight divisions or working sections of the main sewer, a light or a burning lamp must be maintained continually by the contractor at the beginning of such section, and each pipe and specials must be so laid that such light or lamp shall remain constantly in plain view throughout the entire length of such section or division. The same test shall also be applied during the work of refilling the trench, so that when the sewer is in all respects fully completed and accepted by the executive board a light which may be applied at one end of such a division of the main sewer shall be clearly and plainly seen by looking through said sewer from the other end of said division or working section. The length of any such division or the distance between a man-hole and the next following lamp or hand-hole, or between any two consecutive openings of such kind in the main sewer, will, in general, not exceed 300 feet, although in particular cases it may be somewhat greater.

The trenches must, in all cases, be wide enough to admit of the laying of the pipe and specials as above mentioned, and wherever they have not been thus excavated, all necessary widening thereof must be done before the pipe and specials are lowered therein. Ample room or space must likewise be left on each side of said pipe and specials, both to admit of proper refilling underneath and also to allow of free access to all parts of the hub or socket while making the cement joint. Wherever any additional excavation or enlargement in the sides of the trench is required for such purposes, it shall be satisfactorily performed before the pipe and specials are laid or put into place, as no cutting away of the banks will be permitted after any such pipe or special has been set.

Furthermore, before any pipe or special is put into place, a small excavation must be made in the bottom of the previously graded trench to receive the projecting part of the hub or socket, so that each pipe will have a firm and uniform bearing upon said graded bottom over virtually its entire length. All adjustment of the pipes to line and grade must be done by scraping away or filling in the earth under the body of the pipe, and not by blocking or wedging up the spigot or the hub or socket. Special attention must be paid to this part of the work, since the stability and permanence of the sewer depend largely upon the manner in which the pipes are bedded.

The joints between the individual pipes and specials shall, in all cases, be made water-tight by completely filling out the entire annular space between the exterior of the spigot end and the interior of the hub or socket with hydraulic cement mortar, of such composition as is hereinafter specified. To prevent the mortar from reaching the interior of said pipe, the contractor may if he desires, use a narrow gasket of oakum or hemp, which shall be properly caulked into each joint, after which the mortar shall be introduced therein; but no extra compensation for the use of such gaskets will be allowed. Special care must be taken to secure a perfect filling of the aforesaid annular space at the bottom sides of the pipes, as well as at the top; and previous to the introduction of the mortar, said space, together with the surfaces of the pipe bounding the same, shall be thoroughly free all around from dust, sand, earth, dirt, small stones and water. After said space has been filled as described, a neat and proper finish shall be given to the joint by the further application of similar mortar to the face of the hub or socket, so as to form a continuous and even beveled surface, from the exterior of said socket to the exterior of the connecting spigot all around. The pipes must also be thoroughly cleaned before being laid; and any mortar, earth or other material which may have found its way through a joint or otherwise, into any pipe or special must be carefully removed before the next succeeding pipe is laid, in order that the interior of the sewer shall be left smooth and clean.

As soon as the cementing of any joint, whether in a main sewer or in a lateral sewer, has been completed, the excavation previously made in the bottom of the trench for the reception of the hub or socket must be carefully and compactly filled with sand, loam or fine earth, so as to hold the external mortar finish of said joint securely in its place; and such refilling shall also be carried up around the sides or circumference of the socket, as far as may be necessary. Any water which may have accumulated in said excavations must first be removed, or else said excavations must be completely filled out with the cement mortar specified, in which event no extra compensation will be allowed.

When a pipe or special is used in any main or lateral sewer, which is affected by a broken hub or socket, or a boken blister or flake, or a fire-crack on its exterior surface, as limited and defined in the foregoing, such pipe or special must be set so as to bring said permissible defect on the top or upper part of the sewer; and said defect must thereupon be completely and liberally covered over with a thick layer of hydraulic cement mortar, of the quality specified for the joints, to the full satisfaction of the city surveyor, and the executive board.

As the work proceeds, all of the required specials that are indicated upon the plan of the street, or that may be required during the progress of the work, shall be introduced and set in their proper positions.

Any omissions of the required specials intended to be laid, and indicated upon the plan for the sewer, or that may especially be ordered beforehand by the surveyor, shall be corrected by the contractor without additional compensation; but in case that any special not indicated upon the said plan, or not distinctly required to be introduced beforehand by the surveyor is inserted into the sewer after the latter has been laid, the expense of such insertions will be paid by the executive board upon proper certificate from said surveyor.

Before leaving the work for the night, or during a storm, or for any other reason, care must be taken that the unfinished end of the main sewer, or of any lateral sewer is securely closed with a tightly fitting iron or wooden plug. Any earth, or other material that may find entrance into said main sewer, or into any lateral sewer, through any such open end or unplugged branch, must be removed at the contractor's expense. The cost of all such plugs, and the labor connected therewith, moreover, must be included in the regular prices bid for the sewers.   E. K.

142. Specifications for the Manufacture and Delivery of Cast Iron Water Pipe.

The following specifications for the manufacture of cast iron water mains are in use in the city of Rochester, N. Y. Although water pipe is now manufactured and sold as a standard article of commerce, and is often purchased without any test or inspection whatever, it must be admitted to be a poor practice, and if the contract is a large one, the material should be thoroughly inspected and tested in all the stages of manufacture. Special attention should be given to the tests of the strength and resilience of the material. When cast iron water mains burst, it is due to a water ram or shock, and the more elastic the material is of which the pipes are composed, the less will be the force of the ram the more able the pipes will be to withstand the shock. The resilience of the iron is measured by the product of the strength into the deflection, and in the following specifications both tensile and cross-breaking tests are required, and the requisite deflection in the cross-breaking test is also specified. The deflection here named will insure a very good quality of cast iron, so far as its resilience is concerned, although the strength requirement is not particularly high. The author has had a large experience in testing the strength of cast iron, and he can approve of the standards of strength and resilience here named for water pipe metal.

Specifications for Water Pipe.

Dimensions and Weight of Pipe.—The pipe shall be of the kind usually known as "Hub and Spigot," and in general each straight pipe shall be about twelve feet in length from the bottom of the hub to the end of the spigot. No straight pipes will be received that will lay less than 11 feet 8 inches; but it is understood that not more than two per cent. of the total number of pipes required in each class may be 10 feet or more in length, produced by properly cutting off in a lathe a defectively cast spigot end. The form and dimensions of the hub and spigot ends of all pipes and castings shall be subject to the approval of the Engineer, when specific drawings therefor are not furnished by him, and shall conform accurately in shape and dimensions to all drawings that may be furnished by him from time to time.

(See accompanying figure for these dimensions for the St. Louis standard water pipe.)

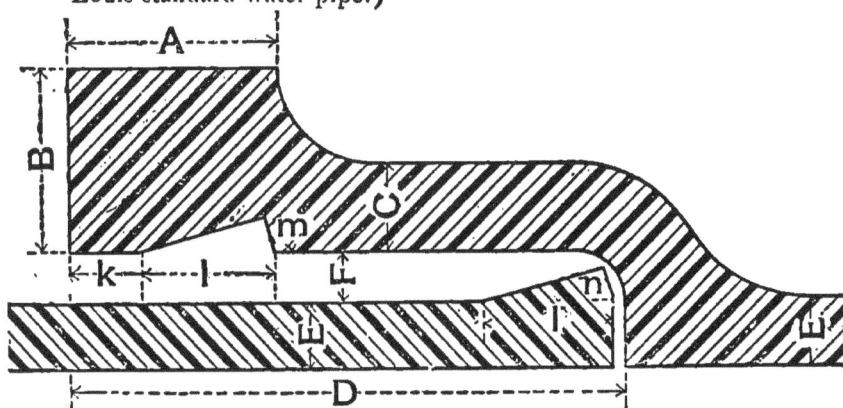

The weights and dimensions of the straight pipes shall conform to the figures in the following Table, it being stipulated that the same may be modified at any time hereafter by the Engineer:

TABLE OF WEIGHTS AND DIMENSIONS OF STRAIGHT PIPE.

Nominal internal diameter of pipe.	Class.	Thickness of barrel.	External diameter of Barrel.	Thickness of lead joint.	Depth of hub.	Standard weight of pipe laying 12 feet. Per lineal foot.	Standard weight of pipe laying 12 feet. Per pipe.	Permitted deviation in weight of pipe laying 12 ft.	Maximum weight of pipe laying 12 feet.	Minimum weight of pipe laying 12 feet.	Deduction from standard weight for each inch of less laying length than 12 feet.	Addition to standard weight for each inch of greater laying length than 12 feet.
in.		inches.	in.	inches.	in.	lbs.	lbs.	p. c.	lbs.	lbs.	pounds.	pounds.
36	A	1¼	38½	7-16 to ½	4½	492	5,904	3	6,081	5,727	51	41
36	B	1⅛	38½	7-16 to ½	4½	444	5,328	3	5,488	5,168	47	37
36	C	1	38½	7-16 to ½	4½	397	4,764	3	4,907	4,621	43	33
30	B	1	32¼	⅜ to 7-16	4½	330	3,960	3	4,079	3,841	35	26
20		¾		⅜ to 7-16	3¾	165	1,980	4	2,059	1,901	20	14
12		9-16		⅜ to 7-16	3½	75	900	4	936	864	8	6
10		½		5-16 to ⅜	3½	56	672	4	699	645	7	5
8		7-16		5-16 to ⅜	3¼	41	492	4	512	472	4	3
6		$\frac{27}{64}$		5-16 to ⅜	3¼	30	360	4	374	346	3	2

The specified internal diameter of the pipe is nominal, but no pipe or special casting of any class shall have a less internal diameter than the nominal diameter. The external diameters of all classes of said pipe shall be the same throughout, and all variations in thickness of metal of the shells or barrels shall be made by changing the internal diameter.

The thickness of the metal of the pipe and castings will be measured after they have been thoroughly cleaned, and before being coated. No pipe of any class will be received when the thickness of the metal is over one sixteenth ($\frac{1}{16}$) of an inch less in any part than the thickness above specified, or hereafter required by the engineer.

No pipe of full length will be received whose weight is less than the above specified minimum weight, and no excess of weight in any such pipe, beyond the specified maximum weight, will be paid for. It is also expressly understood that the average weights of the straight pipe of the several classes shall not exceed the said standard weights by more than two per cent. of the latter, and that no greater over-weight than this percentage will be paid for in the final settlement. The standard weight of the straight pipes will depend upon the laying length of the pipes actually furnished, and will be determined by the engineer.

Quality of Metal.—The materials, details of manufacture, and the testing of all pipe and special castings herein referred to, shall at all times be subject to the inspection and approval of the engineer. The metal, which must be remelted in the cupola or air furnace, shall be made without admixture of cinder-iron or other inferior metal, and shall be of such character as to make a pipe strong, tough, and of sound, even grain, free from uncombined carbon when examined under the microscope, and such as will satisfactorily bear drilling, chipping and cutting. Its tensile strength and resilience, when tested in proper samples, shall meet all the requirements hereinafter expressed.

Specimen rods of the metal used, of a size and form suitable for a testing machine, shall be made and carefully tested to ascertain its tensile strength. Another set of test bars, each being twenty-six (26) inches long, two (2) inches wide, and one (1) inch thick, shall also be made as often as the engineer shall direct, and shall be tested both for transverse strength and deflection by placing them horizontally and flatwise upon supports twenty-four (24) inches apart, and then applying a steadily increasing load at the middle of each bar.

The bars for testing the transverse strength or resilience of the metal shall be cast from regular patterns in dry or green sand, and as nearly as possible to the required dimensions without being finished up; proper corrections will, however, be made in the results for slight variations of width and thickness. The rods for testing the tensile strength of the iron, on the other hand, must be turned down on a lathe in order to remove the rough exterior and enable the diameter to be accurately measured.

At least one set of four test bars, of each kind above designated, shall be made and tested as described on each working day during the manufacture of the pipes and specials. These test bars must be poured from the ladle either before or after any particular pipes or special casting are poured, and must present true samples of the iron used in said pipes or castings. Records shall be kept of the tests of all bars made, and a duly certified copy of such records shall be forwarded weekly to the engineer.

The quality of the metal used for the pipe and specials must be such that said bars for testing resilience, as aforesaid, shall each carry a center load of not less than nineteen hundred (1,900) pounds before breaking, and exhibit a deflection of not less than five sixteenths ($\frac{5}{16}$) of an inch; also that the tensile strength of said metal shall be at least 17,000 pounds per square inch, as determined by the tests with the first named set of rods. In estimating the suitability of the metal from said tests, the average of the three highest results obtained from each set of four bars will be considered as representing the actual strength of the iron.

Manufacture of Pipe and Special Castings.—All the straight pipes shall be cast in dry sand moulds, vertically with the hub end down. Every pipe is to have the initials of the maker's name cast distinctly upon it, and also the year, the class letter, and a number signifying the order of its casting, in point of date; the several different classes of pipe each to have its own series of numbering; the figures and letters to be at least two inches in length, with a proportionate width; the weight of each pipe to be conspicuously painted on the outside, before delivery, with white lead paint at the contractor's expense.

The branches and all other special castings must conform in weight and thickness of iron to the drawings and directions to be furnished by the engineer, and no allowance will be made for making or altering patterns for the pipe or any special castings, or for any machine work in properly facing and drilling flanges, etc., where bolted joints are to be made. All required machine work on said castings shall be done in the best and most workmanlike manner, in accordance with said plans and directions of the engineer, and to his entire satisfaction. Said castings shall be subjected to the same examinations and tests at the foundry, except the water-pressure proof, as the straight pipe, and shall be marked in a similar manner. The engineer may reject, without proving, any pipe or casting which is not in conformity with the specifications or the drawings furnished.

Pipes and special castings shall not be taken from the pit and stripped while still showing any color of heat, but shall be left in the flasks for a sufficient length of time to prevent unequal cooling and contraction by subsequent exposure.

On being removed from the flasks, all pipes and special castings shall be subjected to a careful examination and hammer test for the purpose of detecting imperfections of any kind. They shall then be thoroughly dressed and made clear and free from earth, sand or dust, which adheres to the iron in the moulds; iron wire brushes must be used, as well as softer brushes to remove the loose dust. No acid shall be used in cleaning the castings. After having been properly dressed and cleaned, they shall again be subjected to a thorough inspection and hammer test. The contractor will be required at the foundry to place all castings in such positions as may be deemed necessary. by the Engineer for convenience of inspection.

The pipes and special castings shall be free from scoria, sand-holes, air-bubbles and other defects or imperfections; they shall be truly cylindrical in the bore, straight in the axis of the straight pipes, and true to the required curvature or form in the axis of the other pipes; they shall be internally of the full specified diameters, and shall have their inner and outer surfaces concentric. To insure proper diameters of sockets and spigots, a circular iron templet of the required dimensions shall be passed to the bottom of every socket, and a circular ring over every spigot. Care shall also be taken to avoid all excess in diameter of the sockets. No pipes or special castings will be accepted which are defective in joint room, whether in consequence of eccentricity of form or otherwise. No lump or rough places shall be left in the barrels or sockets, and no plugging or filling will be allowed. All pipes and special castings with defective hubs or flanges will be rejected.

When a defective spigot end is to be cut off from any straight pipe, such cutting must in all cases be done in a lathe, and a suitable bead or fillet of half-oval wrought iron, about three fourths ($\frac{3}{4}$) inch wide and five sixteenths ($\frac{5}{16}$) inch thick shall be shrunk upon the new end of the pipe; and there shall be deducted from the proper original weight of the pipe an amount as determined from the rate specified in the foregoing table.

Coating the Pipe and Special Castings.—After the above described cleaning and inspection, every pipe and special casting shall be heated in a suitable oven to a temperature of about 320° F. and, while at this temperature, be immersed in a bath of hot coal tar pitch varnish, prepared in general according to Dr. R. Angus Smith's process. Special care shall be taken to have the surfaces of all pipes and castings entirely clean and free from rust immediately before putting them into said bath. If any pipe or casting cannot be dipped in said bath soon after its removal from the mould, it shall at once be thoroughly coated with pure linseed oil in order to prevent the formation of any rust before applying said varnish.

The varnish above mentioned shall be made from coal tar, distilled until the naptha is entirely removed and the material deodorized, also until it attains the consistency of wax when cold. Pitch which becomes hard and brittle when cold will be rejected. To this material from five to six per cent. of its weight of pure boiled linseed oil shall be added and thoroughly boiled therewith. The relative portions of pitch and oil, as well as the details of mixture and boiling, are to be carefully determined by experiment.

The coating must be durable, smooth, glossy, hard, tough, perfectly water-proof, not affected by any salts or acids found in the soil, free from bubbles or blisters, strongly adhesive to the iron under all circumstances, and with no tendency to become soft enough to flow when exposed to the sun in summer, or to become so brittle as to sca'e off in winter. As one test of the quality of the coating, a properly coated specimen casting will be plunged into a freezing mixture, and kept therein until the metal has acquired the temperature of said mixture, after which the casting shall be well hammered. If the coating remains tough and adhering closely to the metal, it will be considered proper, provided that it be satisfactory in all other respects.

After a varnish of the proper quality has been obtained, it shall be heated in a suitable dipping tank to a temperature of about 300° F., or such other temperature as may be found expedient, and shall be maintained thereat uniformly during the time of dipping. Fresh materials must be added from time to time in the right proportions to keep the mixture of the proper consistency. The exact proportions will be determined by the Engineer, and will be varied also according to the season of the year, as may be directed by the said Engineer, or found necessary to produce a coating of the required quality. The tank shall also be occasionally emptied of its contents and refilled with fresh material, the frequency of such operation depending both on the character of the mixture and the manner of conducting the coating process.

Every pipe and special casting, after having been inspected, cleaned and dressed as above described, shall be heated in a suitable oven to a temperature about 20° F. higher than that which was found most expedient for the bath of coating material aforesaid, and while at such temperature, shall be immersed or dipped in said bath. All pipes or castings shall remain in the tank at least twenty (20) minutes, or as much longer as may be necessary to insure the soundness of the coating.

Whilst any pipe or casting remains in said bath, the hot mixture must be kept thoroughly stirred by a frequent rolling, turning or churning motion of such casting, and upon its removal from the tank, the coating shall fume freely for a short time, and set perfectly hard within one hour thereafter. Proper facilities for handling the castings and allowing all surplus material to drip off, shall be provided by the contractor. The cost of all labor and material involved in the coating of the pipes and castings must be included in the prices bid for furnishing said pipes and castings.

Testing.—After the said coating has become thoroughly set and hard, every pipe shall be subjected to a proof by water-pressure of from 200 to 300 pounds per square inch, according to its class and diameter, and as will be determined by the Engineer. Each pipe while under the required pressure, shall be sharply rapped from end to end with a hand hammer, to ascertain whether any defects have been overlooked; and any pipes which may exhibit any defects by leaking, sweating or otherwise, shall be rejected.

All the above inspections, manipulation and tests of the pipe and test bars shall be made at the expense of the contractor for the pipe, said expense, however, not to include salary of any inspector who may be appointed by the Executive Board. If required by the said Board, the affidavit of the superintendent of the foundry, or that of the foreman employed by him to perform the above described testing, shall also be furnished to the Engineer from time to time ; said affidavits to be recorded upon the pipe inspector's sheets, and stating in detail that the pipes or castings therein described have been carefully tested at the foundry in accordance with these specifications, and that no defects were discovered or discoverable.

Weighing for Payment.—The pipes and castings will be weighed for payment after all cleaning, dressing and machine work has been done and the coating has been applied, and the contractor must furnish, at his own expense, accurate and properly sealed scales, together with the necessary labor for the purpose. The Executive Board also reserves the right to reweigh on similar scales, any pipe or casting upon or after its arrival at the designated point of delivery; and if any discrepancy be discovered between the weight marked upon said pipe or casting and that which was found on such re-weighing, the latter weight will be adopted in the final settlement. Payment for all material furnished in accordance with these specifications will be made at the prices bid per net ton (2,000 lbs.) for straight pipe and special castings.

Transportation of Pipes and Castings.—All pipes and castings must be delivered in all respects sound and in conformity with these specifications. Upon their delivery at the point designated, the Executive Board reserves the right to subject the said pipe and castings to the same water-pressure proof and hammer tests as are above specified to be applied at the foundry; and all defective pipes or castings which may have passed the inspector at the foundry, or which may have been broken in transportation from the foundry to said point of delivery, will be rejected when there discovered, unless the same may be cut as hereinafter provided. Care must also be taken

14

in handling the pipes and castings not to injure the coating, and no material of any kind shall be placed in said pipes and castings during transportation, or any time after being coated.

If, upon its arrival at the designated point of delivery, the spigot end of any straight pipe should be found cracked or broken, during transportation from the foundry to the said point or otherwise, such defective portion will be cut off at the contractor's expense, provided that the same does not exceed a length of four (4) feet, and a suitable fillet or bead shall then be shrunk on the new spigot end, as above specified. A deduction from the proper original weight of such pipe shall also be made in each such case at the rate specified in the above table for every inch of length so cut off. No pipe or special casting in which the hub is found to be cracked or defective in any respect, will be accepted at said point of delivery or elsewhere; nor will any special casting with a defective spigot end be received, or permitted to be cut off, without the written order of the Engineer.　　　　　　　　　　　　　　　E. K.

143. Specifications for Laying Water Pipe.

The following clauses referring to the methods of laying water pipe, and making the joints, are taken from the complete specifications on this subject used by the water commissioner of St. Louis. All that portion of the specification referring to the trenching, protection, tools, alignment, grades, connections, back-filling, etc., together with the general clauses are here omitted.

The reducers, bends, caps and such other parts as are liable to draw, shall be firmly secured by straps and bolts, and in addition to this a firm blocking shall be set behind all caps, curves, fire hydrants and three way branches, said blocking to have a large surface bearing against the undisturbed earth, and to be wedged up tight. All applications necessary to the perfect working of the distribution, when the water is let on, shall be made and completed.

The straps and bolts used shall be made from the best American refined iron, and the size and workmanship, as well as the material, shall be in all respects satisfactory to the water commissioner.

Any omission of branches, stop-cocks, or other appurtenances intended to be laid, shall be corrected when required, by re-opening the trench, if it has been filled up, and introducing what may have been omitted.

At the time when laid, the spigots of the pipe shall be so adjusted in the sockets as to give a uniform space all around, and if any pipe does not allow sufficient space, it shall be replaced by one of proper dimensions. The joint shall, at all points, be at least five sixteenths of an inch in thickness. In the lead and gasket joints, the depth of lead shall not be less than three and one quarter inches for the fifteen inch pipes and over, nor less than two and three quarter inches for smaller pipes. Gaskets of clean, sound hemp yarn, braided or twisted, and tightly driven, shall be used to pack these joints, when required, a space of one quarter inch shall be left between the contiguous pipes.

The lead used shall be of the best quality of pure and soft lead, and suitable for caulking and securing a tight and permanent joint.

Before running the lead, the joints shall be carefully wiped out to make them clean and dry; the joint shall be run full at one pouring, and the melting pot shall always be kept within fifty feet of the joint about to be poured.

The joint shall be caulked by competent mechanics. The caulking to be faithfully executed, and in such a manner as to secure a tight joint without overstraining the iron of the bell. In all cases the caulking shall be done *towards* the place of the gate and other points where the lead is likely to be porous, so as to drive it together at these points. The lead, after being driven, shall be flush with the face of the socket.

The pipes and all other castings shall be carefully swept and cleaned, as they are laid, of any earth or rubbish which may have found place inside, during or before the operation of laying. Every open end of a pipe shall be plugged or otherwise closed before leaving the work for the night.

In refilling the trenches, the earth filled into the bottom of the trench, under and to the top of the pipes and other castings, shall be carefully packed and well rammed with proper tools for the purpose.

Whenever written directions so to do are given, the contractor shall fill the trench with river sand, said filling to be done in exact accordance with the orders and directions of the water commissioner. For all sand filling done as above, the sum of $ per cubic yard will be paid, which sum shall include all expense of materials, tools and labor for the sand filling, and removing the surplus earth from the work.

Care shall be taken to give the pipe a solid bearing throughout its entire length. The earth filling above the pipes shall

also be sufficiently packed and rammed to prevent after settlement, and the material used shall be free from stones or rock
fragments. The trenches shall, in all cases; be refilled with
the material furnished by their excavation, provided that it be
of a proper quality, and the necessary haul be not more than
500 feet. Earth borrowed or hauled over 500 feet, to refill the
trenches (excepting trenches where rock has been excavated),
will be paid for as embankment, at the price given under item
of section seven.

In streets and roads, the class of surface before existing,
shall be replaced, so as to be in every way equal to that surface in
materials and workmanship, and satisfactory to the water
commissioner.

Whenever trenches are excavated in or across streets
paved with granite or wood blocks, or with asphalt, the contractor will be required to have the back-fill of trench thoroughly
rammed (not less than three men ramming to each man filling
the trench), and to replace the paving temporarily, so as to
make the street passible for traffic; the permanent laying of the
pavement in these cases, will be assumed by the city.

A wooden box or vault shall be furnished and set over each
of the stop cocks, air cocks, and fire hydrants, and the iron
frames and covers shall be properly fastened to them. These
boxes are to be made of the form and dimensions shown by
samples furnished and approved by the water commissioner;
they shall be made from sound, well seasoned oak lumber;
the corner posts shall be of four-inch scantling, and the sides
shall be formed from two-inch plank, set close, and securely
nailed.   M. L. H.

144. Specifications for Stop Valves.

The follow-
ing specifications for stop valves for water mains are thought to
be particularly strong in the requirements governing the strength
of the material used in the different parts. These requirements are followed up very carefully by numerous tests of the
strength of the material, and in this way the character of the
composition metal used has come to be very superior to that
formerly employed, and much superior to that which would be
obtained without such rigid specifications and tests. They are
the standard specifications used in the St. Louis water department.

All the iron castings shall be made from a superior quality
of iron, remelted in the cupola or air furnace, tough and of even.

grain, and shall possess a tensile strength of not less than 18,000 pounds per square inch.

Test bars of the metal 3 inches by ½ inch when broken transversely, 18 inches between supports and loaded in the center shall have a breaking load of not less than 1,000 pounds, and shall have a total deflection of not less than $\frac{3}{16}$ of an inch before breaking. Said bars to be cast as near as possible to the above dimensions without finishing, but correction will be made by the water commissioner for variations in thickness and width, and the corrected result must conform to above requirements.

Specimen bars of the metal used, of a size and form suitable for testing, shall be prepared when required.

These specimen bars shall be poured from the ladle at any time, either before or after the casting has been poured, as may be required, and shall present a true specimen of the iron used for making the castings.

If any two test bars cast the same day do not show the required cross breaking load and deflection, all the castings made from the same mixture to be rejected.

Each valve shall have the maker's initials, the numbers showing point in time of casting, and the year cast upon it. The year above and the number below, thus: $\frac{1890}{1}$, $\frac{1890}{2}$, etc.

The figures and letters will be from 2 to 2 ½ inches long, and shall have at least ⅛ inch relief.

All the wrought iron used shall be of the first quality of American refined iron.

All the composition metal used, except the valve stem, shall be composed of the following proportions, viz: 85 per cent. copper, 10 per cent. tin, and 5 per cent. spelter; and shall have a tensile strength of not less than 22,000 pounds per square inch, with 5 per cent. elongation in 8 diameters, and 5 per cent. reduction of area at breaking point.

All castings must conform in shape and dimensions to the drawings. The castings must be clean and perfect, without blow or sand holes, or defects of any kind. No plugging or other stopping of holes will be allowed.

The valve guides must be straight and smooth. Irregularities, if any, must be planed or chipped off smooth. All face joints must be planed true and smooth, in the most workmanlike manner, so as to make a perfectly water-tight joint, with a *very thin* layer of strictly pure lead cement.

All bolt holes must be accurately drilled from templates. The upper part of valve to be finished to receive the valve stem, collar and stuffing box, and the fitting at this point must be such as to secure a perfect working joint.

The valve to be a two-faced wedge valve; the castings for same to be as shown on drawing. The raised rims to be turned true with dovetailed channel to hold the composition rings. The faces must be brought to the exact angle before the rings are put on. The face rings are to be of composition metal, of quality hereinbefore specified, and are to be turned to fit the dovetail in the iron wedge. The composition rings of valves must be shrunk on, and also fastened by copper studs, placed not over three inches apart—the whole to be then brought to a true plane surface.

The upper portion of the wedge to be arranged to receive the composition nut as shown. Care shall be taken to give the composition nut a perfect bearing surface—both top and bottom.

On the 36 inch and 30 inch valves, the brass bearings of side guides shall be of the full dimensions, and have the exact clearance shown on drawings, and be secured in place by countersunk copper studs, placed not over three inches apart, after which the guides shall be brought to a true and smooth surface.

The seats for rings in body of valve shall be turned true and smooth, and to the required angle as shown on drawings.

The seat rings shall be of form and dimensions as shown on drawings, and faced true and smooth. Seat rings to be forced into position and thoroughly and securely fastened in place, and a perfectly water-tight joint secured.

All valves of 10 inch diameter and upwards to be provided with indicator as shown on drawings.

All wrought iron bolts and nuts to be made from the best quality of American refined iron. The nuts to be hexagonal and the heads square. Heads, nuts and threads to be standard size.

Valve stem shall be made of phosphor bronze, quality B; or Crescent bronze, quality No. 2; or of first quality of "Stuckstede" bronze, and shall be free from flaws or defects of any kind, and have a tensile strength of not less than 30,000 pounds per square inch. Screw threads on the stems and nuts to be cut in most perfect manner, and of the exact pitch shown on the drawings, and so as to work true and smooth, and in perfect line throughout entire lift of valve.

There shall be two dowel pins, made of composition, set in the flanges connecting the dome and main casting, as shown on drawings, for the purpose of centering and bringing into perfect alignment these castings. Holes for dowel pins to be drilled and reamed tapering, and pins turned to perfect fit. Pins for the 36 inch and 30 inch to be 1 inch in diameter: for the 20 and 15 inch, ¾ inch diameter; for the 12 and 10 inch, ⅝ inch diameter; and for the 8 and 6 inch, ½ inch diameter.

Gearing to be extra strong, and of the form and dimensions shown. Pinion post to be of a good quality of steel; key seats shall be truly cut, and keys made of steel, and of the full dimensions.

Cap nuts for valve wrench to be of the following outside dimensions: for all 6 to 15 inch valves (inclusive), to be 2 inches square; for the 20 inch, to be 2¾ inches square; and for the 30 and 36 inch, to be 3¼ inches square.

All iron work, after being thoroughly cleaned, to be painted with three good coats of paraffine varnish, applied hot. The valves shall be tested by hydraulic pressure, as follows:

First. Heads shall be secured at each end of casting, the valve opened, and a pressure of 200 pounds per square inch applied.

Second. Each face joint of valve shall be tested by closing the valve, leaving one end of the casting open, and applying a pressure of 100 pounds per square inch to the other—this operation to be reversed to test the other face.

Any and all defects developed in testing shall be thoroughly corrected to the satisfaction of the water commissioner. After testing all valves to be thoroughly drained.

All parts of valves of the same size to be perfectly interchangeable.

The water commissioner may take at random any wrought iron bolt or nut, and have it broken in a testing machine. If bolt shall not fulfill the requirements of table below, the whole lot of that size and make to be rejected:

SIZE OF BOLT.	TENSILE BREAKING STRENGTH.	REDUCTION OF AREA AT BREAKING POINT.
5-8 inch.	9,000 lbs.	20 per cent.
3-4 "	13,000 "	20 per cent.
7-8 "	19,000 "	20 per cent.
1 "	25,000 "	20 per cent.
1 1-8 "	31,000 "	20 per cent.
1 1-4 "	40,000 "	20 per cent.
1 1-2 "	58,000 "	20 per cent.

The Water Commissioner may take at random any valve stem with nut, either finished or unfinished, for 6, 8, 10 or 12 inch valves, and have it broken in a testing machine.

If any stem or nut shall not fulfill the requirements of the table below, the whole lot of that make and size to be rejected.

SIZE OF VALVE.	TENSILE BREAKING STRENGTH OF STEM. (Including Nut and Collar.)	DUCTILITY IN 8 Diameter.
6 inch.	34,000 lbs.	8 per cent.
8 "	34,000 "	8 "
10 "	34,000 "	8 "
12 "	42,000 "	8 "

All valve stems for 15 inch and larger valves to be cast with a coupon on one end, 15 inches long by 1½ inches diameter. Any one or all of these coupons may be taken by the Water Commissioner and broken in a testing machine. If any coupon shall show a breaking strength of less than 30,000 lbs. per square inch, or shall have a ductility of less than 8 per cent. in 8 diameters, the stem from which it was cut shall be rejected.

For all materials taken by the Water Commissioner for testing which are found to conform to the above requrements, there shall be added to the final estimate :

For all wrought iron.......................... 7 cents per pound.
For all Phosphor bronze...... . ...25 cents per pound.
For all Croscent bronze...... ....25 cents per pound.
For all Stuckstede bronze.... . . .. 25 cents per pound.

The broken material to belong to the party of the second part. For all materials taken for testing which do not come up to requirements there shall be no allowance, and the broken material shall be returned to party of the first part.*

The whole to be put together in a thorough and workmanlike manner, and delivered, packed, ready for use. The working parts to be perfectly fitted together and working true in line. The joint between the face rings, when the valve is closed, must be absolutely water-tight. The whole to be in material, workmanship and finish, to the satisfaction and acceptance of the water commissioner.   M. L. H.

LUMBER GRADING AND CLASSIFICATION.

145. Rules of the Southern Lumber Manufacturers' Association. The rules given in the following articles were adopted by the Southern Lumber Manufacturers' Association at Memphis, Tennessee, February 21, 1895. They are given here entire to assist the engineer to use descriptive terms in the same sense in which they are used by the lumber manufacturers and dealers. While they are intended to apply only to southern yellow pine, they can be understood to apply in a general way to all merchantable lumber. Since lumber is always sold under certain grade names, and since in the large

* In the St. Louis specifications the contractor is the party of the first part.

markets the lumber is officially graded, it is sufficient for the
engineer and architect to use these technical terms in his specifications, provided he knows that he is using it in the same
sense in which it is used by lumber dealers in that market. If
he does not feel safe in limiting his description to the use of
such technical class terms, he will still find considerable information in the following official rules, which will enable him
better to describe the kind of lumber which he wishes to have
supplied.

146. General Rules for Classifying Lumber.

The following general rules are intended to serve as a guide to
lumber inspectors in enabling them to classify the lumber in
accordance with the grades named below in subsequent
articles.

1. Yellow pine lumber shall be graded and classified
according to the following rules and specifications as to quality;
and dressed stock shall conform to the subjoined table of standard sizes, except where otherwise expressly stipulated between
buyer and seller.

2. * Recognized defects in yellow pine are knots (pin,
round, spike, black, encased, loose or rotten), knot holes,
splits (either from seasoning, ring-heart or rough handling),
shake, wane, crooks, warp, rotten streaks, dote, rot, worm
holes, pitch pockets, seasoning or kiln checks, blue sap and
pitch streaks.

* Some of the following terms may need defining: Ring-heart is a "shake" or
cleavage along the plane of an annual ring, usually about half way between the pith
and the circumference. "Shake" or "wind shake" is a cleavage of the trunk of a tree
while yet standing, due to the action of the wind in bending the trunk. It is usually
along the plane of an annual ring, that is to say, concentric with the center or pith of
the tree.

"Heart-shake" is a diametral or radial cleavage through the tree or log. If it
occurs after the logs are cut, or in large timbers after they are sawed, it is due to shrinkage in drying. This is the common defect of all oak logs or large timbers.

"Wane" is a deficiency in width, either over the entire edge or on one corner,
caused by a crook in the log.

"Crooks" are permanent distortions of the board, due to defective piling or from
other causes.

"Warp" is a twisting of the board into a warped surface.

"Seasoning or kiln checks" are either very small or large cracks, caused by drying the surface of the board with its accompanying shrinkage, while the interior is still
wet.

"Blue sap" is a discoloration, which green yellow pine is subject to, especially the
sap portion, if not at once piled for drying or placed in a dry kiln.

"Pitch streaks" are longitudinal openings, sometimes of considerable size, as ½
inch to ¼ inch wide and several inches (or even feet) long, filled with rosin.

3. Bright sap shall not be considered a defect in any of the grades provided for and described in these rules. The restriction or exclusion of bright sap constitutes a special class of material which can be secured only by specific contract.

4. Firm red heart shall not be considered a defect in common grades.

5. Defects in rough stock, caused by improper manufacture or drying, will reduce grade, unless they can be removed in working such stock to standard sizes.

6. Imperfect manufacture in dressed stock, such as chipped, grain splintered or torn places, broken knots on edge of shiplap, insufficient tongue on flooring, etc., shall be considered defects, and reduce grade accordingly.

7. A standard knot is sound, and not over $1\frac{1}{4}$ inches in diameter. A pin knot is sound, and not over half an inch in diameter.

8. Any piece that will not work one half its size shall be classed as a dead cull.

9. The grade of all regular stock shall be determined by the number and position of the defects visible in any piece. The enumerated defects admissible in any given grade are intended to be descriptive of the coarsest pieces such grade may contain. The average quality of the grade should be midway between such pieces and the defects allowed in the next higher grade.

10. Lumber or timber sawed for specific purposes, such as wagon tongues, bridge timbers, car sills, etc., must be inspected with a view to the adaptability of the piece for the use intended.

11. In finishing, flooring, etc., the enumerated defects admissible in a given grade apply only to the face side of the piece, but reverse face should not admit defects that would render the piece unsuitable for the purpose intended.

12. Standard lengths are multiples of 2 feet from 10 to 20 feet, inclusive, for boards and strips, and from 10 to 24 feet, inclusive, for dimension, joists and timbers. Longer or shorter lengths than those herein specified are special. Odd lengths, if below 24 feet, shall be counted as of the next higher even length.

13. On stock width shipments of 8-inch and under no board shall be admissible that is more than $\frac{1}{4}$ inch scant; on 10-inch not more than $\frac{3}{8}$ inch, and on 12-inch not more than $\frac{1}{2}$ inch scant of specified width.

14. Yellow pine of better grade than No. 1 common up to 4 inches in width is classified according to grain as edge grain and flat grain. Edge grain yellow pine has been variously designated as rift-sawn, straight grain, vertical grain and quarter-sawed, all being commercially synonymous terms.

Edge grain stock is specially desirable for flooring, and admits no piece in which the angle of the grain exceeds forty-five degrees from vertical, thus excluding all pieces that will sliver or shell from wear.   Such stock as will not meet these require-ments is known as flat grain.

15.   All dressed and matched stock shall be measured and sold "strip count," *i. e.*, full size of rough strip from which such stock is made—3, 4, 5 and 6 inches wide.

16.   The foregoing general observations shall apply to and govern the following detailed descriptive enumeration of recognized grades.

147.   Rules for Grading Finishing Lumber.   The

following rules for grading apply to all kinds of finishing stock, whether for interior or out-door work.   In these rules such expressions as "S. 1 S." or "S. 2 S." mean "surfaced one side," or "surfaced two sides," respectively. Also "S. 1 S. 1 E." will be understood to mean "surfaced one side and one edge."   By surfacing is meant planing or running it through a planing machine.   It may still require hand dressing for the best work. Nearly all saw mills now dry their lumber and run it through the planer, in order to save the extra freight on the rough and green lumber.

(Grades: First and second clear; third clear; barn and roofing stocks).

17.   *First and Second Clear Finish*, 1 inch, S. 1 or 2 S., up to and including 10 inches wide, must show one face clear from all defects; 33⅓ per cent. of any shipment of 12 or 14 inches wide will admit two pin knots or one standard knot, slight pitch streak, or small pitch pocket, or sap stain not over 1½ inches wide running across the face, or small kiln or seasoning checks, but no two of these defects shall appear in a single piece; 16 inches wide will admit of two defects allowed in 12-inch or their equivalent.   Wider than 16-inch will admit proportionately more defects.   Pieces otherwise admissible in which the point of the grain has been loosened or slivered in dressing on the face side should be put in lower grade.   Defec-tive dressing or reverse face of finishing is admissible.   In case both faces are desired clear special contract must be made.

18.   *Third Clear Finish*, 1 inch, S. 1 S. or 2 S., up to and including 10 inches in width, may have not more than two, of the following defects on best or face side: Three pin knots one standard knot; three sap stains 2 inches wide running across

the face or their equivalent; two pitch pockets; slight pitch streaks, kiln or seasoning checks; torn places, and wane which does not enter more than 1 inch, nor extend more than 2 feet; 12-inch will admit three of the above defects, or their equivalent. This grade is suitable for paint finish.

19. 1¼, 1½ and 2 inch, S. 1 or 2 S., shall take 1 inch inspection, and unless otherwise agreed between buyer and seller, shall be subject to inspection on face or best side only.

20. Barn and novelty siding, shiplap and grooved roofing shall be 8, 10 and 12 inches wide, and consist of boards falling below third clear which are sound and water-tight, free from coarse knots and wane over 1 inch wide extending more than 3 feet in any piece. Pitch, except in narrow streaks, should be excluded.

21. *Edge-Grain Flooring.* (Grades: First clear, second clear). First clear edge-grain flooring must be well manufactured, and free from all defects on face side of strip.

22. Second clear edge-grain flooring will admit of three pin knots or one standard knot, or small pitch pocket, or blue sap stain not to exceed 10 per cent. of the face.

23. *Flat-Grain Flooring.* (Grades: A flat, B flat). A flat flooring may contain two pin knots or one small pitch pocket, but shall be free from other defects, and must be well manufactured. Pieces in which the point of the grain has been loosened in dressing should be put in lower grade.

24. B flat flooring may have any two of the following defects: Three pin knots or one standard knot, slight sap stains, small pitch pockets, slight torn places and defects in manufacture, narrow pitch streaks and seasoning checks. When all other defects are absent, blue sap stain in any quantity shall be admitted.

25. *Common Flooring.* (Grades: No. 1 common, No. 2 common). No. 1 common flooring must be manufactured from sound stock. In addition to the defects described in B flat, also admits of sound knots, blue sap and firm red heart in any quantity, pitch and slight shake, but must lay without waste. No division as to grain is made in this grade.

26. No. 2 common flooring includes all pieces that will not grade No. 1 common which can be laid without wasting more than one fourth the length of any piece. This grade will admit imperfections which do not render the piece unfit for use in cheap floors and roof sheathing.

27. Center-matched flooring shall be required to come up to grade on one face only.

28. *Ceiling.* (Grades: A, B, C). A ceiling shall be free from all defects on face side and well manufactured.

29. B ceiling will admit slight imperfections in dressing—three pin knots or one standard knot, pitch streaks or small pitch pockets, or blue sap stain not to exceed 10 per cent of the face; but not more than two of these defects to be admitted in any piece.

30. C ceiling conforms to grade of No. 1 common flooring and is suitable for paint finish. Will admit imperfections that do not prevent its use without waste.

31. *Wagon Bottoms.* (Grades: A, B). Wagon bottoms shall be graded the same as flat grain flooring.

32. *Bevel and Drop Siding.* (Grades: A, B and C). Shall be graded according to ceiling rules, but will admit more blue stain, and, except in C grade, should exclude pitch. Slight additional imperfections on the thin edge of bevel siding which will be covered by the lap are admissible.

33. *Partition.* (Grades: A, B and C). Partition shall conform to ceiling grades, but must meet the requirements of the specified grade only on one face. The reverse face shall not be more than one grade lower.

34. *Molded Casings and Base.* (Grades: First clear, second clear). First clear shall be free of all defects on face and perfect in manufacture.

35. Second clear is suitable for work that is to receive a paint finish, and usually consists of rejections, made after dressing, from stock inspected in the rough as first clear. The defects admitted in B ceiling would be allowed.

148. Rules for Grading Common Boards and Rough Lumber.

COMMON BOARDS AND SHIPLAP.

36. No. 1 common boards, S. 1 S., and No. 1 common shiplap shall be manufactured from sound stock, of even thickness the entire length. Will admit of any two of the following defects: Wane one half inch deep on edge and one sixth the length of any piece; tight sound knots, none of which shall be larger than three inches in diameter, or equivalent spike knots; one split not more than sixteen inches long, and blue sap. These boards should be firm and strong, suitable for use in all ordinary construction and serviceable without waste.

37. No. 2 common boards and No. 2 common shiplap admit pieces that fall below No. 1 common which are free from the following defects: Rotten streaks that go through the piece, through heart shakes which extend more than one half the length of the piece, and wane over two inches wide exceeding one third of the length of the piece. A knot hole 1½ inches in diameter or its equivalent will be allowed, provided the piece would otherwise grade No. 1 common. Worm

holes and straight splits one fourth of the length of the piece are admissible.

FENCING S. 1 S.

38. No. 1 common fencing must be manufactured from sound stock. May contain sound knots equal in diameter to not over one third the width of piece at any given point throughout its length, but must be free from spike knots the length of which is over one half the width of piece. Also, free from wane over one half inch deep on edge and one half the length of any piece measured on one side. This grade must work its full length without waste.

39. No. 2 common fencing shall admit of pieces that fall below No. 1 common which are free from through rotten streaks.

40. Miscut 1 inch stock in boards and fencing which does not fall below ¾ inch thick shall be admitted in No. 2 common, provided that the grade of such thin stock is in all other respects as good as No. 1 common.

DIMENSION S. 1 S. 1 E.

41. *No. 1 Common Dimension* shall be manufactured from sound stock, and be free from loose and unsound knots, and large knots so located as to materially impair the strength of the piece ; will admit of seasoning checks and heart shakes that do not go through, of slight wane and such other defects as do not prevent its use as substantial structural material.

42. *No. 2 Common Dimension* admits all pieces falling below No. 1 common which are free from through rotten streaks, and sound enough to be used without waste.

43. Miscut 2 inch stock which does not fall below 1½ inch shall be admitted in No. 2 common, provided that the grade of such thin stock is in all other respects as good as No. 1 common.

44. In boards, fencing and dimension, stock falling below No. 2 grade and excluding dead culls shall be classed as No. 3.

45. Dressed timbers shall conform in grade to the specifications applying to rough timbers of similar size.

ROUGH YELLOW PINE—FLOORING STRIPS AND FINISHING.

46. Flooring strips are 3 inches, 4 inches, 5 inches and 6 inches wide when green ; square-edged and evenly manufactured.

47. Finish must be evenly manufactured, and shall embrace all sizes from 1 inch to 2 inches thick by six inches and over in width.

48. No finishing lumber, unless otherwise ordered, should measure when dry and rough less than $\frac{1}{16}$ inch scant in thick-

ness. No piece in any shipment of boards and strips shall be more than ¼ inch scant on 6 and 8 inch stock, ⅜ inch scant on 10 and ½ inches scant on 12 inch and wider stock.

49. Wane and seasoning checks that will dress out in working to standard thicknesses and widths are admissible.

50. Subject to the foregoing provisions rough finishing shall be graded according to the specifications applying to dress finishing. When like grade of both faces is required special contract should be made.

COMMON BOARDS, FENCING AND DIMENSION.

51. *Rough Common Boards and Fencing* must be evenly manufactured, and should not be less than ⅞ inch thick when dry, nor more than ½ inch scant of specified width.

52. *Rough 2 inch Common* shall be evenly manufactured and not less than 1⅞ inches thick when green, or 1¾ inches thick when dry. The several widths must not be less than ⅛ inch over the standard dressing width for such stock.

53. The defects admissible in rough rock shall be the same as those applying to dressed stock of like kind and grade, but such further defects as would disappear in dressing to standard size of such material shall be allowed.

54. Rough timbers 6x6 and larger shall not be more than ¼ inch scant when green and be evenly manufactured from sound stock with not less than three square edges, and must be free from knots that will materially weaken the piece.

55. Timbers 10x10 in size may have a 2 inch wane on one corner, or its equivalent on two or more corners, one fourth the length of the piece. Other sizes may have proportionate defects.

56. Seasoning checks, and shakes extending not over one eighth the length of the piece, are admissible.

149. Standard Dimensions of the Southern Lumber Manufacturers' Association.*

Flooring. The standard of 1x4 and 6 inch shall be $2\frac{7}{32}$x3¼ and 5¼ inches; 1¼ inch flooring, $1\frac{3}{32}$ inches.

Ceiling. ⅜ inch ceiling, $\frac{5}{16}$ inch; ½ inch ceiling, $\frac{7}{16}$ inch; ⅝ inch ceiling, $\frac{9}{16}$ inch; ¾ inch ceiling, $\frac{11}{16}$ inch; same width as flooring.

Finishing. 1 inch, S 1 S or S 2 S, to $\frac{27}{32}$; 1¼ inch, S 1 S or S 2 S, to $1\frac{3}{32}$ inch; 1½ inch, S 1 S or S 2 S, to $1\frac{11}{32}$ inches; 2 inch, S 1 S or S 2 S, to 1¾ inches.

Boards and Fencing. 1 inch, S 1 S or S 2 S, to 13-16.

*These particular dimensions can not be assumed to hold for all parts of the country.

Dimension. 2x4, S 1 S 1 E, to 1⅝x3⅝ inches; 2x6, S 1 S 1 E, to 1⅝x5⅝ inches; 2x8, S 1 S 1 E, 1⅝x7½ inches; 2x10, S 1 S 1 E, to 1⅝x9½ inches; 2x12, S 1 S 1 E, to 1⅝x11½ inches; 4x4, ⅜ inch off side and edge; 4x4, S 4 S, ¼ inch off each side.

150. Specification for Thoroughly Seasoned Lumber.

There is no difference between "seasoned" lumber and "dried" lumber. "Thoroughly seasoned" or "thoroughly dried" lumber is lumber which has been dried, either in the open air or in a dry kiln, until it has reached that state of dryness which is relatively permanent. It then contains water equal to about ten per cent. of its weight. This is what might be called the atmospheric moisture. This will remain in the wood unless driven off by evaporation at a temperature of 212 degrees Fahrenheit or more. The word "thoroughly" when used in this connection, means "uniformly" as well as "effectually." That is, "thoroughly dried" lumber is dried uniformly throughout its entire cross-section and throughout its entire length.

To determine the percentage of moisture of lumber it is only necessary to cut a section from a board or stick and weigh it; then dry in an ordinary stove oven with a slow fire for an hour or two and then weigh again; the difference in weight divided by the dry weight is the percentage of moisture. As determined by this test, "thoroughly dry lumber" should not contain more than ten or twelve per cent. of water, and the interior should be as dry as the exterior.

The necessity for using thoroughly dried lumber where shrinkage is to be avoided, arises from the fact that *below about 30 per cent. moisture lumber shrinks nearly as much as it dries.* That is to say, when lumber dries down from 30 per cent. moisture to 10 per cent. moisture it dries out, or loses in weight, 20 per cent. of its dry weight. It also loses about 20 per cent. of its dry volume, or say 15 per cent. of its volume at 30 per cent. moisture. The shrinkage lengthwise is very slight, hence it has lost about 15 per cent. of its cross-section, or say six or seven per cent. of each of its lateral dimensions. That is to say a board one foot wide at 30 per cent. moisture is only about 11⅜ inches wide at 10 per cent. moisture; or a flooring board 4 inches wide at 20 per cent. moisture is only about 3¾ inches wide at 10 per cent. moisture. On account of the very large radial fibres (medullary rays) in oak wood, this kind of lumber shrinks mostly in a circumferential direction, and all timber shrinks more circumferentially than radially since all woods have these medullary rays to a greater or less extent. It is for this reason that "quarter sawed" (radial sawed) lumber is more satisfactory than "flat sawed" for all kinds of furniture and house trimmings. For flooring quarter sawed, or "rift sawed" boards, presenting an "edge-grain" surface, is far preferable to "flat-grain" because it wears evenly and does not sliver on the surface.

The specification may read as follows:

All the lumber delivered under this contract, to be used for purposes of ————, shall be thoroughly seasoned or dried, either in the open air or in a kiln or both. By "thoroughly seasoned" as here used is meant a seasoning or drying uniformly throughout the entire sections of the various sizes delivered, and the average percentage of moisture contained in the lumber when delivered shall not be more than ten per cent. of its weight, as determined by actual experiment.

———————

SPECIFICATIONS FOR IRON AND STEEL.

151. Specification for Cast Iron. There is probably no material in engineering structures which can more profitably be governed by specifications involving tests than cast iron. Since cast iron usually breaks under some kind of shock or blow, it is more necessary to test the iron for resilience than for strength. The most convenient test for resilience is

15

the cross-bending test, in which deflection is measured. The half product of the deflection multiplied by the breaking load is the mathematical measure of the resilience in inch pounds. This can be reduced to an absolute unit by dividing by either the weight or the volume of the bar, and if all the bars tested in this way are rectangular in cross-section and of uniform size from end to end, the unit obtained in the above manner will be comparable, notwithstanding great variations in the dimensions. It is best, however, to have the test specimens always made from the same pattern, using the thickness of metal which corresponds closely to the average thickness of web of the castings required. If uniform test specimens be employed, there is no necessity of dividing the half product of deflection and breaking load by the volume or by the weight, since this volume or weight remains a constant. In this case the relative resilience of the material will be indicated by the product of the breaking load into the maximum deflection. The *strength* of the material will be indicated by the breaking load alone.

The following specification is the one commonly employed for all castings made for the water department of St. Louis, and is designed to answer the above requirements.

Cast Iron.

All of the iron castings shall be made from a superior quality of iron, remelted in the cupola or air furnace, tough and of even grain, and shall possess a tensile strength of not less than 18,000 pounds per square inch.

Test bars of the metal 3 inches by ½ inch, when broken transversely, 18 inches between supports, and loaded in the center, shall have a breaking load of not less than 1,000 pounds and shall have a total deflection of not less than 3-10 of an inch before breaking.* Said bars to be cast as near as possible to the above dimensions without finishing; but correction will be made by the water commissioner for variations in thickness and width, and the corrected result must conform to above requirements.

*The tensile strength may be raised to 20,000 or even to 25,000 pounds per square inch, while the deflection may be made ⅜ inch for ordinary good cast iron and ½ inch for a better quality. For a superior quality it may be made ⅝ inch, with a breaking load of 1250 pounds.

Specimen bars of the metal used, of a size and form suitable for testing, shall be prepared when required.

These specimen bars shall be poured from the ladle at any time, either before or after the casting has been poured, as may be required, and shall present a true specimen of the iron used for making the castings.

If any two test bars cast the same day show a breaking strength of less than 18,000 pounds per square inch, or do not show the required cross-breaking load and deflection, all the castings made from the same mixture to be rejected.

All castings shall conform to the shape and dimensions required by the drawings, and shall be clean and perfect, without blow or sand holes, or defects of any kind. No plugging or other stopping of holes will be allowed.

Particular care shall be taken to secure perfect lugs, where such are required by the drawings. Whenever any doubt exists of the exact interpretation as to the shape or dimensions shown on the drawings, the contractor must consult with the water commissioner, or his duly authorized agent, in regard thereto. M. L. H.

152. Specification for Wrought Iron.

Since the cost of the manufacture of soft and mild steel has been so greatly reduced as to enable this material to compete in price with that of wrought iron, the wrought iron mills have been driven to cheapen their product, and they do this by hurrying the metal through the puddling process too rapidly. The result is that whereas steel has been constantly improved for structural purposes wrought iron has constantly degenerated. It is necessary, therefore, now to examine and test the wrought iron very rigidly to insure against obtaining a comparatively worthless product. It is thought the following specification is sufficient for this purpose, provided it is followed up by suitable tests.

Wrought Iron.

All wrought iron used must be tough, ductile and fibrous, of a uniform quality, free from crystalline structure, cinders, flaws or cracks. In bars, it must have an ultimate strength of 50.000 pounds per square inch, with 26,000 pounds elastic limit and an elongation of 25 per cent. in eight inches. Angle iron must have an ultimate strength of 48,000 pounds per square inch, 24,000 pounds elastic limit, with 20 per cent. elongation in eight inches. A. P. B.

153. Specification for Structural Steel. While
ordinarily it is not wise for a civil engineer to specify methods
of manufacture, the author of this work believes that it is advisable in the case of structural steel to limit the manufacture to
the open-hearth process, and also to limit the product to that
supplied by the manufacturers of established reputation. It is
now an easy matter to obtain any desired grade of structural
steel from that having a strength of 55,000 pounds to that having a strength of 70,000 pounds, wherein the ultimate elongation of a test specimen eight inches in length will be from 33
per cent. to 35 per cent. for the softer grades, to 20 per cent.
or 25 per cent. for the higher grades. It is important also to
limit the phosphorus, as this produces brittleness. The best of
this material can be bent cold upon itself and mashed flat without showing signs of failure up to a thickness of plate of ⅝ of
an inch. Steel manufactured by the open-hearth process is
usually more uniform in character, and if care is taken in its
manufacture as is necessary under rigid specifications, it is
always possible to obtain the desired results. The following
specification is probably fully up to the present practice in this
direction.

All steel used shall be open-hearth, made at works of
established reputation, and which have been successfully manufacturing steel for at least one year. All melts must be made
from uniform stock in which phosphorus shall never exceed
eight hundredths of one per cent. A sample bar must be
rolled from each melt, of three quarters of an inch an diameter,
the method of obtaining the same being uniform for all melts.
Tests upon such samples to be made without annealing. Tests
shall also be made upon specimens cut from the finished
product. The three-quarter-round tests must conform to the
requirements hereinafter given, and the finished product tests
cut from shapes and plates must conform to the same within
four per cent. Every piece of steel shall be stamped with a
number identifying the melt, and a full record of all laboratory
tests kept. Three qualities of steel will be required.

First—For the main trusses, floor beams and stringers
and cross girders, exhibiting an ultimate strength of sixty-three
to seventy thousand pounds per square inch, a minimum

elastic limit of thirty-seven thousand pounds per square inch, with an elongation in eight inches of not less than twenty-two per cent., and a reduction of area of not less than forty-four per cent.

Second—For buckle plates, caisson plates, rivets, or where wrought iron is permitted (and in lieu thereof), a steel having an ultimate strength not exceeding sixty-two thousand pounds per square inch, or less than fifty-five thousand pounds, with fifty-eight per cent. elastic limit, a reduction in area of fifty per cent. and an elongation in eight inches of twenty-eight per cent.*

Third—For wheel treads of track circle in turn-table, a steel having an ultimate strength of from seventy thousand to eighty thousand pounds per square inch, with sixty per cent. elastic limit.

The first two steels to be subject to bending test, before and after quenching, and the metal when cold must bend one hundred and eighty degrees upon itself without sign of fracture in convex side. Specimens must withstand such punch, drifting, and forge tests as may be required to test soundness, temper and ductility.

All metal, whether steel or iron must be clean rolled, and any imperfect work, such as ragged or cracked edges, surface imperfections, or imperfectly rolled shapes, will be sufficient cause for rejection. Materials will be also rejected varying more than two and one half per cent. from weights or sizes.

A. P. B.

154. Tests, Inspection, and Acceptance of Medium Structural Steel. The following specifications were prepared in 1895 for the Northwestern Elevated Railroad of Chicago, and represents the latest and best American practice:

All steel shall be manufactured by either the acid or the basic open hearth process, preference being given to the former, and must be uniform in character for each specified kind. Any attempt to substitute Bessemer or any other steel for the open hearth product, will be considered as a violation of the contract, and a good and sufficient reason for canceling the same.

It is understood also that, if the contract be let on the basis of using acid open hearth steel, no basic open hearth steel will be permitted to be employed in any part of the work; but, if the contract be let on the basis of using basic open hearth steel, the employment of the acid open hearth product

*This is the common specification for steel plates used for boilers, stand-pipes, plate-girders, etc.

will be permitted whenever the contractor may desire to use it, provided, of course, that it comply with these specifications.

The maximum limits for phosphorus shall be as follows:

For acid open hearth steel, eight hundredths of one per cent. (o.oS per cent.).

For basic open hearth steel, five hundredths of one per cent. (o.o5 per cent.).

Each ingot which is cast shall be stamped or marked plainly with its proper melt number, and this melt number must br stamped or painted plainly on all blooms, billets or slabs made from such ingots, in order to identify the material throughout its various processes of manufacture; and the proper melt number together with the furnace heat number must be stamped plainly on each piece of finished material.

All finished material must be free from injurious seams, flaws or cracks, and must have a clean, smooth finish.

All slabs for rolling p'ates must be hammered or rolled from ingots of at least twice their cross section.

Test Bars.

The tests shall be made in the following manner:

The tensile strength, limit of elasticity and ductility shall be determined from standard test pieces, not less than three eighths ($\frac{3}{8}$) of an inch in thickness, and from full sized pieces, tooled to parallel sides. If the cross section be reduced, the tangent between shoulders shall be not less than eight (S) inches, and the area of the minimum cross section in either case shall not be less than one half ($\frac{1}{2}$) of a square inch, and preferably not more than one and a half ($1\frac{1}{2}$) square inches. When of a rectangular cross section, two opposite sides of the piece are to be left, if practicable, as they come from the rolls, but the finish of opposite sides must be the same in this respect.

A full sized piece, when not exceeding the above limitations, may be used as its own test piece.

Tensile Tests.

The ultimate tensile strength of all steel, except that used for rivets and adjustable members, shall be sixty four thousand (64,000) pounds per square inch.

Steel for rivets and adjustable members shall have an ultimate tensile strength of fifty seven thousand (57,000) pounds per square inch.

All test bars must have a tensile strength within four thousand (4,000) pounds per square inch of that specified ; and for medium steel an elastic limit not less than one half of the tensile strength of the test bar, a percentage of elongation not less than 1,500,000 divided by the tensile strength in pounds per square inch, and a percentage of reduction of area not less than 3,000,000 divided by the tensile strength in pounds per square inch. In determining the ductility, the elongation shall be measured after breaking on an original length of eight (8) inches, in which length must occur the curve of reduction from stretch on both sides of the point of fracture.

For rivet steel the elastic limit must not fall below thirty thousand (30,000) pounds, the elongation in eight (8) inches shall not be less than twenty-five (25) per cent., and the reduction of area shall be at least fifty (50) per cent.

All broken samples must show a silky fracture of uniform color.

Bending and Drifting Tests.

Specimens of medium steel, when heated to a cherry red and cooled in water at seventy (70) degrees Fahrenheit, shall be capable of bending one hundred and eighty (180) degrees around a circle whose diameter is equal to the thickness of the test piece, without showing signs of cracking on the convex side of the bend.

Punched rivet holes in medium steel, pitched two (2) diameters from a sheared edge, must stand drifting until their diameters are fifty (50) per cent. greater than those of the original holes, and must show no signs of cracking the metal.

Specimens of rivet or soft steel shall be capable of bending cold to one hundred and eighty (180) degrees and closing down, flat, upon themselves without cracking.

Number of Test Pieces.

At least three tests for tensile strength, etc., and three for bending shall be made on specimens from different ingots of each melt. These tests shall be made by the contractor without charge, and if the consulting engineer or his authorized inspector be not satisfied that the tensile test correctly indicates the effect of the heating and rolling, such additional tests for tensile strength, limit of elasticity and ductility, as he may desire, shall be made for him on test pieces conforming to the provisions named above, at the rate of one dollar ($1.00) each; or, if the contractor desire additional tests, he may make them at his own expense under the supervision of the consulting engineer or his authorized inspector, the quality of the material to be determined by the result of all the tests in the manner set forth in the following paragraph. If material of various shapes is to be made from the same melt, the specimens for testing are to be so selected as to represent the different shapes rolled from such melt.

The contractor will be required to manufacture and test on his own or some other testing machine, free of all extra charge for either material or labor, a reasonable number of full-sized

details used in the design of the structure. It is understood
that no allowance will be made for the weight of any test
pieces, when the total weight of metal, for which the contractor
is to be paid, is computed by the consulting engineer.

Acceptance and Rejection.

Except for tensile strength, the respective requirements
stated are for an average of the tests for each; and the lot of
finished pieces from which samples were selected shall be
accepted, if the tests give such average results; but, if any test
piece give results more than four (4) per cent. below said
requirements, the particular piece from which it was taken may
be rejected, but such tests shall be included in making the
average. If any piece have a manifest flaw, its test shall not
be considered. For each piece thus giving results more than
four (4) per cent. below requirements, tests from two additional pieces shall be furnished by the contractor without
charge; and, if in a total of not more than ten (10) tests, two
(2) pieces (or for a larger number of tests a proportionately
greater number of pieces), show results more than four (4)
per cent. below the requirements, it shall be cause for rejecting
the lot from which the samples were taken. Such lots shall not
exceed twenty (20) tons in weight; and plates rolled in
universal mill or in grooves, or sheared plates, shall each constitute a separate lot, as shall also angles, channels or beams.

Variation in Weight.

A variation in cross section or weight of rolled material of
more than two (2) per cent. from that specified shall be cause
for rejection, and no excess of metal above that computed from
the drawings exceeding one (1) per cent. will be paid for.

J. A. L. W.

155. Specification for Riveted Work. The following specification for riveted work in structural designing
has been carefully drawn to give the highest possible efficiency
and strength without causing extravagant expense. If the
highest perfection in this kind of work is not required, some
reduction in cost can be effected, without materially lowering
the standard, by punching to full size of hole all plates having
a thickness of $\frac{5}{8}$ of an inch or less, and for such plates resort-
ing to reaming only after the plates are assembled to secure the
necessary coincidence of parts. For all plates greater than $\frac{5}{8}$
of an inch in thickness the holes should be punched $\frac{1}{8}$ of an

inch less in diameter than required, the parts then assembled and the holes reamed out to the required diameter.*

After plates or bars are carefully straightened the rivet holes will be accurately spaced, so that when members are brought into position, the holes shall be truly opposite before rivets are driven. Drifting under no circumstances will be allowed. In steel work, for all thicknesses up to three quarters of an inch, the rivet holes shall be punched one eighth of an inch smaller than the rivet required, the parts assembled, and the holes reamed out to just pass the rivet before driving. For all thicknesses over three quarters of an inch, punching will not be permitted, and the holes must be drilled, parts assembled, and reamed as above. Reamed work is not required for lacing bars, transverse, diagonal, or lateral bracing, or for caissons, excepting the cross girders belonging thereto, and their connection to caisson sides. In the lattice girders for the web system of fixed spans, and plate girders of draw span, the sharp edges left by drilling or reaming to be eased off before driving rivets. Surfaces in contact shall be thoroughly coated with boiled linseed oil and pure red lead before being assembled for riveting. Power riveting to be alone used wherever possible. All rivets to have neatly capped hemisphered heads. Tightening loose rivets by recupping or "setting up" will not be allowed; they must be cut out and redriven, whether in shop or field. Every built member or girder must be true and out of wind, neatly finished to length, and field driven rivets of all main girder connections shall be laid out with iron templates and accurately drilled so as to just pass the rivet cold.

A. P. B.

156. Specification for Wrought Iron and Steel Railroad Bridge Superstructure. The specifications for iron and steel superstructures have been undergoing gradual changes at the hands of the leading American civil engineers for many years. The changes have all been based on numerous experiments on full-sized bridge members and they have always leaned to the safe and conservative side. The following specification is fully up to the best practice in the year 1895 and probably will not be varied much in the near future, as to either materials or workmanship:

The superstructure will consist of one pivot draw 440 feet long, divided into fourteen panels of 30 feet each, and one cen-

* See Modern Framed Structures, Appendix A, for a full discussion of this subject.

tral panel of 20 feet, carried on a rim bearing turntable, and of
two spans each 330 feet long, divided into eleven panels of 30
feet each.   It will include the iron fence which sustains the
ends of the ties.

The trusses will in all cases be spaced 20 feet between
centers.

Full detail plans showing all dimensions shall be prepared
at the expense of the contractors under the direction of the
chief engineer, these plans to be made in accordance with the
standard practice of the chief engineer.

The superstructure shall be proportioned to carry the following loads:

The actual weight of the structure including floor, rails and
all other parts.

A moving load of 3,000 pounds per lineal foot on all
lengths exceeding 120 feet.

A moving load of 6,000 pounds per foot on a length not
exceeding 20 feet.

A graduated moving load decreasing by 30 pounds for
each foot from 6,000 pounds per foot on 20 feet, to 3,000
pounds per foot on 120 feet.

A concentrated load of 60,000 pounds upon a single axle.

Provision shall be made for impact by adding fifty per
cent. to the portion of the strain of the web members which is
due solely to variable load.

The tension members of the trusses shall be proportioned
on the basis of a strain of 20,000 per square inch for dead load
and 10,000 pounds per square inch for live load.

Stringers, floor beams and other riveted girders shall be
proportioned on the basis of a strain not exceeding 8,000 pounds
per square inch of net section.

The maximum compression strain allowed on mild steel
shall not exceed 14,000 pounds per square inch and the maximum compression strain allowed on soft steel shall not exceed
10,000 pounds per square inch when the length of the member
does not exceed sixteen times the least transverse dimension,
these strains to be reduced in proportions which will limit the
strain to 6,000 pounds per square inch when the length of the
member is thirty times its least transverse dimension.

In the draw, the turntable, eleven panels of each bottom
chord, the four central posts, the eye-bars, the pins, and some
details of the end lifting apparatus shall be of M steel.   The
coned wheels and the two principal pinions shall be of cast
steel.   The lower casting of the track and the center casting
shall be of cast iron.   All other parts shall be of soft steel or
of wrought iron at the option of the contractor.

In the fixed spans, the chords, end posts, eye-bars and pins
shall be of mild steel.   The expansion bearings shall be accord-

ing to the special provisions of the specifications. All other parts shall be of soft steel or wrought iron at the option of the contractors.

STEEL.

Classes.—Steel will be divided into four classes: HM, M, MS, and S, of which M and S will be standards, and HM and MS intermediates.

Class M will be known as medium steel, and will be used in those portions of every member which constitute the calculated section.

Class S will be known as soft steel, and will be used for rivets, fences and the lateral rods of the floor system.

HM and MS steel will be accepted for details and parts which do not form portions of the calculated sections.

Manufacture.—Steel shall be made by the open hearth process and no steel shall be made at works which have not been in successful operation for at least one year; but this provision shall not be held to exclude new furnaces erected in connection with old works.

If made in an acid furnace, the amount of phosphorus in the finished product shall never exceed eight one hundredths of one per cent., this being a maximum and not an average requirement.

If made in a basic furnace, the amount of phosphorus shall never exceed four one hundredths of one per cent., this being a maximum and not an average requirement and being considered necessary to show a proper amount of work in the furnace.

The finished product shall be perfect in all parts and free from irregularities and surface imperfections of all kinds. All steel must be free from piping.

The cross sections shall never differ more than $2\frac{1}{2}$ per cent. from the ordered cross sections as shown by the dimensions on the plans.

Steel for pins more than four inches in diameter shall be hammered.

Every finished plate, bar or angle shall be stamped on one side, near the middle, with a number identifying the melt and this stamp shall be surrounded with a heavy circle of white paint. Steel for pins shall have the melt numbers stamped on the ends. Rivet steel and small pieces which do not form part of the calculated section of members may be shipped in bundles, wired together, with the melt number on a metal tag attached.

Tests.—A sample bar not more than two inches wide, and having a cross section of one square inch when the material is not less than one half inch thick, shall be cut from the finished product of every melt. When taken from metal more than two inches thick this sample may be a turned, round bar. The laboratory tests shall be made on this sample bar in its natural state without annealing.

Where a melt is rolled into several varieties of material, each variety shall be separately tested. A variety shall consist entirely of one of the following shapes: Sheared Plates, Universal Mill Plates, Angles, Z's, Channels, I beams, Flats, Rounds, Squares, Pin Steel and Eyebar Steel. Flats will include all flats not intended to be forged into eyebars. Where several sizes of the same variety are rolled, the cross section of the largest size shall not be more than twice that of the smallest size, and the sample shall be taken from the size which comes nearest to a mean.

In the laboratory tests, measurements to determine elongation shall be made on a length of eight inches.

A piece of each sample bar shall be bent 180 degrees and closed up against itself. In no case shall any crack appear until the circle around which the bar is bent becomes less than the thickness of the bar. Except when the sample is taken from a pin, the sample bar shall close up against itself without showing any crack or flaw on the outside of the bent portion.

The sample bar shall be tested in a lever machine and the following requirements fulfilled:

	CLASS OF STEEL.			
	HM.	M.	MS.	S.
Ultimate strength, lbs., per sq. in....	70,000	66,000	62,000	58,000
Elastic limit, " " " "	35,000	33,000	31,000	29,000
Percentage of elongation in 8 in......	18	22	24	26
Percentage of reduction at fracture...	36	44	48	52

Where the sample is taken from a pin, the elongation and reduction will be reduced to 15 and 30 per cent. for the HM steel and to 18 and 36 per cent. for the M steel.

The entire fracture shall be silky.

The requirements for ultimate strength are means, and steel will be accepted when the ultimate strength does not differ more than 4,000 pounds from the requirements of the table.

The requirements for elastic limit, elongation and reduction are minimum requirements, and no steel will be accepted which falls below these conditions.

The elastic limit will be observed by the falling of the beam of the testing machine.

Duplicate tests may be made when the first sample tested fulfills four of the five requirements. If the second test and also the average of the two tests meet all the requirements, the melt may be accepted. Cases in which the tests are thought not to give fair indications of the character of the material, shall be referred to the engineer.

Analyses shall be made, showing the amount of phosphorus and carbon in every melt, the drillings for these analyses being taken directly from one of the ingots. Besides this a set of analyses of phosphorus, carbon, silicon and manganese shall be made from every ten melts, the drillings to be taken from a sample test bar.

SHOP REQUIREMENTS.

The work shall be done in all respects according to the detail plans furnished by the engineer.

Where there is room for doubt as to the quality of work required by the plans or specifications, the doubt shall be decided by using the best class of work which any interpretation would admit of.

All workmanship, whether particularly specified or not, must be of the best kind known in use. Past work done for the same engineer will never be recognized as a precedent for the use of other than the best kind of work.

All material shall be cleansed, and, if necessary, scraped and given one heavy coat of Cleveland Iron Clad Paint, Purple Brand, put on with boiled linseed oil before shipment. This applies to everything except machine finished surfaces.

The same paint shall be used wherever painting is required.

All machine surfaces shall be cleansed, oiled and given a heavy coat of white lead and tallow before shipment. The inspector must see that this is a substantial coat, such as is used on machinery and not a merely nominal covering.

All small bolts, all pins less than six inches in diameter, the expansion rollers, and everything with special work on it, shall be carefully boxed before shipment.

Riveted Work.

Riveted work will be of two classes: 1. Reamed work. 2. Punched work. The provisions immediately following will apply to both classes.

All plates, angles and shapes shall be carefully straightened at the shop before they are put together; mill straightening will not be considered as meeting this requirement.

If the rivet holes are marked with templets, these templets shall lie flat without distortion when the marking is made.

The size of rivets shown on the plans is the size of the cold rivet before heating.

The diameter of the finished rivet hole shall not be more than one sixteenth inch greater than the diameter of the cold rivet. The heated rivet shall not drop into the hole but require a slight pressure to force it in; the relative size of the rivet and rivet hole must be such as to meet this requirement.

In all cases where riveting is to be done in the field the parts so to be riveted shall be fitted together in the shops and the rivet holes reamed out while they are so assembled, or an iron templet shall be made and both parts reamed to fit this templet.

All surfaces in contact shall be cleaned and painted before they are put together.

The rivets shall be driven by power wherever this is possible. The manufacturer will be required to procure special riveting machines to meet special positions.

All rivets shall be regular in shape, with hemispherical heads concentric with the axis, absolutely tight and shall completely fill the hole. Tightening by calking or recupping will not be allowed. This applies to both power driven and hand driven rivets.

The angles of stringers must be square and straight. The web plate must not project above the angles, and the outside edges of the two angles must never be above a true plane and never more than one sixteenth inch below the true plane coincident with the roots of the angles.

The outside angle at the root of the angles connecting stringers with floor beams, floor beams with posts, or in other like details, shall never be less than a right angle, and the excess over a right angle shall never be greater than one eighth inch in the longer leg of the angle; the angle shall be perfectly straight.

Reamed Work.

All work which is of mild steel shall be reamed.

All sheared or rough edges shall be carefully planed off.

The material may be punched with holes one eighth inch smaller than the size of the rivets shown on the plans, except as provided below.

When the thickness of the metal is greater than a thickness one eighth inch less than the diameter of the rivet, the punched hole shall be one quarter inch smaller than the diameter of the rivet.

When the thickness of the metal is greater than a thickness one eighth inch more than the diameter of the rivet, no punching will be allowed but the holes must be drilled.

After the several pieces have been punched (or drilled) they shall be assemb'ed. The holes shall then be reamed to the diameter required by the size of the rivets, while the pieces are together.

After reaming, every hole shall be entirely smooth, showing that the reaming tool has everywhere touched the metal. In special cases where this fails, the engineer may authorize the hole to be reamed to a larger size and larger rivets used.

A reamer shall be run over the outer edges of every hole so as to remove the sharp edges and make a fillet of at least one sixteenth inch under each rivet head.

After the reaming is completed the several pieces shall be taken apart and cleaned.

Punched Work.

All work which is of wrought iron or soft steel may be punched without reaming.

All pieces whose thickness does not exceed the diameter of the rivet shall be punched with holes not more than one sixteenth inch larger than the diameter of the rivet.

All pieces whose thickness exceeds the diameter of the rivet shall be punched with holes one sixteenth inch smaller than the diameter of the rivet, and subsequently reamed to one sixteenth inch larger than the diameter of the rivet; this reaming may be done before the several parts are assembled.

Where the thickness of the metal is more than one quarter inch greater than the diameter of the rivet, the punching shall be one eighth inch smaller than the diameter of the rivet.

The several parts shall be cleaned before they are assembled.

When the several pieces are put together, the work shall be sufficiently true for a cold rivet to pass through every hole without reaming or drifting. In special cases the engineer may authorize the holes to be reamed and larger rivets used.

Forged Work.

The heads of eyebars shall be formed by upsetting and forging into shape by a process acceptable to the engineer. No welds will be allowed.

After the working is completed the bars shall be annealed in a suitable annealing furnace by heating them to a uniform dark red heat and allowing them to cool slowly.

The form of the heads of steel eyebars may be modified by the contractors to suit the process in use at their works, but the thickness of the head shall not be more than one sixteenth inch greater than that of the body of the bar, and the heads shall be of sufficient strength to break the body of the bar.

The heads and enlarged ends for screws in laterals, suspenders and counters shall be formed by upsetting and shall be of sufficient strength to break the body of the bar.

Nuts, swivels and clevises, if made of steel, shall be forged without welds; whether made of steel or wrought iron, one of each size shall be tested, and shall develop a strength ten per cent. greater than the strength of the bars to which they are attached as determined by required strength of material in such bars.

Eyebars shall be bored truly and at exact distances, the pin holes to be exactly on the axis of the bar and at exactly right angles to the planes of the flat surfaces.

When six bars of the same billed strength are piled together the two pins shall pass through both pin holes at the same time without driving. Every bar shall be tested for this requirement.

Pin holes shall, be bored with a sharp tool that will make a clean, smooth cut. Two cuts shall always be taken, the finishing cut never to be more than one eighth inch. Roughness in pin holes will be sufficient reason for rejecting bars.

One full sized eyebar for every fifty bars made shall be selected from time to time from bars made for the bridge for testing.

No bars known to be defective in any way shall be taken for test bars, but the bars shall be selected as fair average specimens of the good bars which would be accepted for the work.

The tests of full sized eyebars shall be made in the large testing machine at Athens, Pa., unless some other machine is especially accepted by the engineer.

These bars shall be required to develop an average stretch of twelve per cent. and a minimum stretch of ten per cent. before breaking. The elongation shall be measured on a length of not less than twenty feet, including the fracture.

The bars will be required to break in the body.

They shall also show an elastic limit of not less than 32,000 pounds and an ultimate strength of not less than 60,000 pounds, as indicated by the registering gauges of the testing machine at Athens.

In case of bars too long for the machine, the bars shall be cut in two, each half reheaded, and both halves tested in the machine, the two tests, however, to count as a single test bar.

In these tests a failure to meet the required elongation will be considered fatal, and be a sufficient cause for condemning the bars represented by the bars so tested, but the engineer shall examine carefully into the cause of the breakage of any bar which does not meet the requirements, and may order additional tests if he sees fit.

The failure of the bar to break in the body shall not be considered sufficient reason for rejection, provided the required elongation is obtained and not more than one quarter of the bars break in the head.

Machine Work.

The planing, drilling, and reaming required under the provisions for riveted work shall always be performed.

The ends of the chord sections shall be faced so as to be perfectly true, after they are riveted up complete, excepting only the projecting splice p ates.

When four chord pieces are fitted together complete in the shop, there shall be no perceptible wind in the length of the four sections.

All chord sections shall be stamped at each end on the outside with letters and numbers designating the joints in accordance with the diagram furnished by the engineer.

All pin holes and holes for turned bolts passing through the whole width of a riveted member shall be bored or drilled after all other work is completed.

Pin holes shall be bored truly and at exact distances parallel with one another and at exactly right angles to the axis of the member.

Pin holes shall be bored with a sharp tool which will make a clean smooth cut. Two cuts shall always be taken, the finishing cut never to be more than one eighth inch. Roughness in pin holes will be sufficient reason for rejecting a whole member.

Pin holes shall be bored to fit the pins with a play not exceeding one fifteenth of an inch. These requirements apply to lateral connections as well as to other pins.

The plans show the distance between the centers of pin holes. Shop requirements shall be between the bearing edges of tension or compression members, with a proper allowance for the diameter of the pin. An iron standard of the same temperature as the piece measured shall always be used.

All pins shall be accurately turned to a gauge and shall be of full size throughout.

The ends of stringers and of floor beams shall be squared in a facer, as shall also all other similar connections.

All bearing surfaces shall be truly faced.

All surfaces so designated on the plans shall be planed.

All screws on iron shall have a truncated V thread, United States standard sized.

All screws cut on steel shall have a truncated V thread, United States standard, with eight threads to the inch when the diameter is two and one half inches or less and six threads to the inch for all larger sizes.   G. S. M.

157. Specification for the Material and Workmanship of a Steel Stand-Pipe.

The following specification for the material and workmanship suitable for a steel water tower or stand-pipe have been prepared by Mr. Wm. D. Pence, after a very long and careful investigation of the numerous failures which have occurred in such structures and also of the materials and workmanship suitable and necessary

16

for this kind of work. The specification includes a phosphorus limit of 0.06 of one per cent., whereas those given in articles 154 and 156 allowed an upper limit of 0.08 of one per cent. There is no question but that a limit of from 0.06 to 0.08 of one per cent., is necessary in order to exclude high phosphorus steels which are of necessity brittle. The author heartily commends these specifications, not only for the purpose named, but for all similar kinds of work.

Material. The metal composing the stand-pipe shall be soft, open-hearth steel, containing not more than 0.06 per cent. phosphorus, and having an ultimate tensile strength of not less than 54,000 nor more than 62,000 pounds per square inch, an elastic limit not less than one half the ultimate strength, an elongation of not less than 26 per cent. in eight inches and a reduction of area of not less than 50 per cent. at fracture, which shall be silky in character. Before or after being heated to a cherry red and quenched in water at 80° F., the steel shall admit of bending while cold, flat upon itself, without sign of fracture on the outside of the bent portion.

Test Pieces. All test samples shall be cut from finished material. Tensile test pieces to be at least 16 inches long, and to have for a length of 8 inches a uniform planed-edged sectional area of at least ½ square inch, the width in no case to be less than the thickness of the piece. Bending test pieces to be 12 inches long, and to have a width of not less than four times the thickness, with edges filed smooth.

Number of Tests. For the purpose of identification the number of the melt or heat of steel shall be stamped on each plate produced therefrom. At least one full series of tests, both chemical and physical, as above specified, shall be made of each melt, and such additional tests may be made as, in the judgment of the inspector, seem essential for corroborative purposes under varying conditions or methods of treatment of the metal.

Finish of Material. All plates must be free from laminations and surface defects, and shall be rolled truly to the specified thicknesses.

Facilities for Testing. Complete facilities for the tests and inspections shall be provided by the contractor, as required.

Inspector. Material will be inspected at the mill by (name of a trustworthy testing concern equipped to make both chemical and physical tests) or such other party as may be approved by the engineer.

Additional Test Pieces. If required by the engineer, the contractor will provide four certified samples of each thickness of plate used in the work, these samples to be 2 inches wide and 16 inches long.

Workmanship. All workmanship must be first-class in every particular.

Working Steel. The plates and angles shall be shaped to the proper curvature by cold rolling. No heating and hammering shall be allowed for straightening or curving, or for other purposes.*

Punching. The work shall be carefully and accurately laid out in the shop, and the rivet holes punched with a center punch, sharp and in perfect order, from the surface to be in contact. The diameter of the punch shall not exceed that of the rivet by more than 1-16 inch, and the diameter of the die shall in no case exceed that of the punch by more than 1-16 inch. Rivet holes in plates having a thickness of ¾ inch, and over shall either be drilled or if punched, shall be reamed not less than ⅛ inch larger than the die sides of the holes, and sharp edges shall be trimmed.

Beveling, etc. All calking edges shall be planed to a proper bevel. All parts must be adjusted to a perfect fit, and properly marked before leaving the shop.

Erection. In assembling the work, the rivet holes shall match so that hot rivets may be inserted without the use of a hammer. Drifting is prohibited. Eccentric holes, if any, must be reamed, and if required, larger-sized rivets shall be used in such holes.

Rivets and Riveting. The best grade of soft charcoal iron rivets to be had in the market shall be used. Sufficient stock must be provided in the rivets to completely fill the holes and make a full head. (The rivets shall be driven at such a heat as will admit of their being finished in good form with a button set before the rivet has cooled to a critical point.) As often as may be deemed advisable for the purpose of testing the work, rivets shall be cut out at the direction of the inspector. The quality of the rivet metal and of the workmanship shall be such that the fracture of the rivet so removed at random shall show a good, tough, fibrous structure without any crystalline appearance, and there shall be no evidence of brittleness. Loose rivets must be promptly replaced, no rivet calking being permitted.

Calking. All seams must be calked thoroughly tight with a round-nosed calking tool by workmen of acceptable skill. Great care must be taken not to injure the under plate.

Rejections. Defective material and workmanship may be rejected at any stage of the work, and must be properly replaced by the contractor as directed.

*If lap riveting is used, omit the expression "or for other purposes," and insert the following sentence: "No scarfing shall be done at a temperature below that of igni- tion of a hard-wood hammer handle, and no work shall be done upon the steel between such temperature and that of boiling water."

Final Tests. After completion the work shall be tested by filling the stand-pipe with water, and the leaks, if any, shall be promptly and thoroughly calked. The stand-pipe must be water-tight before acceptance.

Superintendence. All inspections shall be made under the direction of the engineer who shall have general supervision of the work.   W. D. P.

MISCELLANEOUS SPECIFICATIONS.

158. Specifications for Pile and Trestle Bridging.

The following specifications for pile foundations and timber trestles are those used by the Union Pacific Railway Co. These specifications may, however, serve as a standard for all kinds of pile foundations, and for the selection of large timbers for engineering structures. The formula for obtaining the safe bearing resistance of pile foundations is that generally known as the "Engineering News formula."

All piles to be made from straight, sound, live timber, free from cracks, shakes and rotten knots, cut from the following kinds of timber: White Oak, Burr Oak, Red or Yellow Oregon Fir. They must be so straight that a straight line taken in any direction from the center of each end of the pile, and run the length of it, shall show that the pile is at no point over one-eighth of its diameter at such point out of a straight line. They must show an even, gradual taper from end to end. Ends must be cut square, all bark taken off, branches and knots trimmed off smooth, finishing the pile in a workmanlike manner. They must not be less than fourteen (14) inches in diameter at the narrowest point of measurement of butt or large end, nor less than ten (10) inches in diameter at narrowest point of measurement of point or small end, and at no part more than seventeen (17) inches in diameter.

All piles must be properly sharpened before driving. They must be driven until they will carry a safe working load of—— pounds, computed by the following formula:

$$L = \frac{2wh}{s+1}$$

In which L = Safe load in pounds.
$\qquad$ w = Weight of hammer in pounds.
$\qquad$ h = Fall of hammer in feet.
$\qquad$ s = Last penetration in inches.

They will be estimated and paid for by the lineal foot. 1. As delivered at the site of the structure, according to bills furnished by the engineer. 2. For driving, straightening and cutting off ready for the caps, and only the length actually left standing in the structure to be paid for.

All timbers must be of the exact dimensions given and figured on the plans, to be cut from sound, live timber, free from loose or rotten knots, worm holes, wind shakes or splits; reasonably well seasoned, straight grained, square edged, and free from any and every defect calculated to impair its strength and durability. It will be estimated and paid for in the work by the thousand feet, board measure. The following kinds of timber will be accepted:

All bridge ties will be White or Burr Oak, Oregon Red or Yellow Fir, Tamarack, or Yellow Pine.

All track stringers and guard timbers will be Oregon Fir or Yellow Pine, of the long leaved, southern hard pine variety.

All posts, caps, sills, bracing and end plank will be White or Burr Oak, Red or Yellow Oregon Fir, White or Yellow Pine, or Tamarack.

All wrought iron must be of the best quality of refined iron, tough, ductile, and capable of standing a tensile strain of fifty thousand (50,000) pounds per square inch of sectional area. The manufacture of the bolts must be perfect in every respect, and have nuts and screws of the United States standard dimensions, length of thread to be not less than three inches.

All washers and spacing blocks, etc., must be well manufactured of good gray iron and to the exact dimensions shown on the drawing. The cost of placing all bolts, spikes, and washers in the structure will be included in the price paid for framing and erecting the timber.

All bridge ties will be furnished and placed in the bridges by the contractor.

The surface of the ties must be brought to a true plane under the rail, so that the rail will get a full bearing on every tie.

All of the track stringers shall be brought to a true plane, so that ties will get an even bearing on all the stringers.

Where any timber or pile trestle bridge is built on a curve, the blocking for elevating the outer rail, or other means for elevating it, will be as per drawings for the same, a copy of which will be furnished from the office of the chief engineer.

The culverts will be put in place and finished ahead of the grading, so that it will not interfere with or detain the grading, in any way.

Bridging shall begin when directed by the engineer, and progress at a rate sufficiently rapid to keep out of the way of the tracklayers.

When directed by the engineer drain pipes will be used instead of culverts; they will be of cast iron or vitrified terra cotta; this will be carefully bedded and jointed and of such size as may be directed by the engineer.

All framing shall be accurately fitted; no blocking or shimming will be allowed in making joints; the holes for the bolts shall be bored with an auger of the exact size of the bolts. The nuts on all bolts shall be screwed so the washers shall pinch hard upon the wood and bring all the parts of the structure close together.

On completion pick up and remove all rubbish from the premises.

All material will be inspected on the above specifications, at points of shipment or destination as agreed, and the owners required to remove all rejected material from the company's premises within thirty (30) days from the date of notice to do so. The company after that time will not be responsible for the return or safe keeping of the same.

When from any cause bridge materials are unloaded from cars at material yards or end of track, it shall be reloaded by the contractor at his own expense.   U. P. R'y.

159. Specifications for the Steam Plant of a Small Electric Light Station.

The following specification for the steam plant of a small electric light station includes specifications for the engine and its attachments, feed water heater, boiler feed-pumps, boilers, furnace, stack and pipe connections. While not especially elaborate, they have been prepared by a mechanical engineer of large experience in this field of practice. They are given here, however, not for the purpose of being copied, but simply as an illustration of such a specification. The reader will note that three kinds of engines have been provided for, and that large liberty is retained by the engineer in the selection of the engine from those submitted for competition in the bids. A particular feed water heater was here specified, because it was thought to be best suited for the kind of water which was to be used, the advantage of this heater being that it largely removes the scale from hard water before it enters the boiler:

ENGINE.

Type: There will be one engine, of the High Speed Automatic pattern; cylinders 10½ inches or 11 inches diameter; capable of operating continuously at 600 feet piston speed per minute, without undue heating.

Regulation: The automatic governor must permit a cut-off as late as ½; and must be so adjusted, that the difference in speed, when running with 100 pounds initial pressure and no load, as compared with 75 pounds initial pressure and cut-off ½, shall not exceed a guaranteed amount to be stated by bidder; with a correspondingly less variation inside of the limits named. The regulator must be so constructed, as to permit this guaranteed regulation to be easily maintained, without racing.

Fixtures and Fittings: Standard cast iron sub-base, and two heavy driving pulleys, of such diameter and face as may be required to suit dynamo pulleys; a full set of foundation bolts, nuts, and plates; template for foundation; throttle and drain valves; cylinder lubricators, automatic oil cups, wrenches, indicator motion, etc.; and two one-inch relief valves set at 110 pounds pressure.

Dimensions: State diameter and material of shaft and crank pin, and submit drawing or blue print indicating clearly size of cylinders, speed, diameter of pipe openings, space occupied, and dimensions of foundations.

Corliss Engine: Bids will also be considered under the same conditions on a Corliss engine; 11 inches diameter of cylinder, shaft 6 inches; flywheel and frame extra heavy; speed 100. Regulation to be guaranteed.

Direct Connected Engines: This type will also be considered, together with suitable dynamo. High speed. Vertical or horizontal. Compound or single expansion. Full details must accompany proposals.

Belting: Double thickness, even and pliable, equivalent in strength and adhesiveness to the Shultz Leather Belting. Of selected stock, stretched twice before being made into belts. No shoulders or flank leather to be used. Its tensile strength must not be less than 3200 lbs. per square inch of section, and must not be worked beyond 65 lbs. per inch of width. The belt will be thoroughly stretched again after making, and before shipment.

FEED WATER HEATER.

One No. 5 Hoppes exhaust steam feed water heater and purifier capable of heating 3,000 pounds of water per hour to the highest point attainable, without back pressure on the engine. To have steel shell, oil extractor and trap; crane for removing head; automatic water regulator and openings for water and steam as required.

BOILER FEED PUMPS.

One Worthington Duplex pump, $5\frac{1}{4}$x$3\frac{1}{2}$x5, water ends to have packed pistons. Piston rods, water cylinder linings and water pistons to be of gun metal. Valves suitable for hot water; complete with a full set of oil cups and wrenches.

BOILERS.

To be two in number as follows:

Dimensions: Fifty-four inches diameter, eighteen feet long; thickness of shells, five sixteenths; heads, seven sixteenths, to have half smoke-box extension, bolted on, sixteen inches in length.

Material: Park Bros.' Open Hearth Homogenous flange steel of 60,000 pounds tensile strength. All plates to be stamped with name of maker, quality and tensile strength.

Construction: The heads are to be machine flanged, to have an easy radius; and amply braced, with braces of best refined iron, uniformly distributed, so that each brace will carry its full share of strain. Before beginning construction a plan of the bracing proposed must be submitted to the engineer for approval.

Tubes: Thirty-eight—4 inches in diameter, 18 feet long, of lap welded, charcoal iron; carefully and properly expanded and beaded over.

Dome: Thirty inches diameter, thirty inches high. To be of same material as shell; well braced and double riveted.

Seams: There will be one longitudinal, double riveted seam, in each sheet, well removed from the fire. Other riveting single. The make, size and spacing of rivets shall be in accordance with the best modern engineering practice.

Supports: There will be two extra heavy cast iron lugs for each side; each $4\frac{1}{2}$ feet from end of the boiler. The forward lugs to rest directly on cast iron plates 12 inches square, supported by the masonry. The rear lugs will rest upon 9 one-inch rollers, which in turn will rest upon 12x12 plates.

Stack: Of sheet steel, No. 12 gauge in thickness, diameter 32 inches, height 50 feet. Lower end Y shaped to fit stack plates. Furnish sufficient $\frac{3}{8}$ inch galvanized iron guy wire to make two complete sets of guys. Support stack underneath Y to brick work or floor.

Fittings: One 5-inch chime whistle; one soot sucker, complete, with hose and handle; one flue scraper; one steel barrow; complete set of firing tools, consisting of shovel hoe, slice-bar, and poker; 2 eight-inch steam gauges; 2 one and onequarter inch combination water columns, with gauge cocks, and water glasses; two 4-inch safety valves, with levers marked to 150 pounds pressure; 2-inch check, stop and blow-off valves.

Castings: Two square top, full flush fronts of approved ornamental design, with tight fitting doors, and anchor rods extending the entire length of brick work; six 9-foot binding bars with cross and archor rods; soot door and frame; stack plate and damper. Cast iron skeleton frames suitable for standard sizes of fire brick, to be used in place of back plates. Rocking grates will be furnished and erected by the city.

Openings. The man hole in front head under tubes, and one in shell back of dome, both properly reinforced, and provided with heads, arches and bolts complete; two 1¼ inch openings for water column; one 2-inch for feed and blow off pipe; one 4-inch main steam outlet, and one 4-inch for safety valve; all to be properly reinforced and located as directed by the engineer.

Inspection and Insurance. Before shipment the boilers will be tested and made tight under a water pressure of 150 pounds. Certificate of inspection and insurance policy in the Hartford Steam Boiler Inspection and Insurance Company, for the sum of $500, for one year must be furnished, for each boiler.

FOUNDATIONS AND BRICK WORK.

(See Drawing.)

The dimensions of foundations for engine, boilers, heaters, pumps, and brick work for furnaces, will be clearly shown in drawings, which must be accurately followed.

Foundations: All are to be of concrete composed of one part best domestic cement, three parts of clean, sharp sand, and five parts clean, broken stone of sizes that will pass through a 2½ inch ling; all to be thoroughly mixed, laid quickly, and rammed down solid.

Excavations: As per drawing. Remove promptly all earth and other debris. Bottom to be level, and rammed if necessary.

Iron Work: All bolts and plates for engines and dynamos must be put thoroughly and permanently into position by the contractor. ' Outside of each bolt place a piece of 2-inch iron pipe, so as to permit some adjustment of the bolts.

Cap Stones: Foundations for heater, and boiler feed pumps will project somewhat above floor line. Each of these will be surmounted by a neatly cut cap stone, 8 inches thick and of proper dimensions.

Boiler Furnaces: To be of well burned red brick, thoroughly wetted before laying; all joints flushed solid; all courses level and straight. Every sixth course both inside and outside to be a header. Brick to be laid in mortar composed of one part lime to five parts of sharp sand. Build into side and rear walls a 1-inch air space, which shall be air tight; except immediately under the supporting lugs of boilers, where the walls shall be carried up solid.

Fire Brick Lining: The entire inside of the furnace where exposed to flame, will be lined with A No. 1 hard burned fire brick, laid in dry milled fire clay, with very thin joints, flushed full; headers every sixth course. Use the following special fire brick "Angle B," to form the top and front corner of same, use the "4½ to give the batter on front of bridge wall: at top and bottom, Jamb," also for the inner corners of cleaning out doors openings. Front of bridge all headers.

Closing In Tile: For the sides of furnace, use fire brick tile 6x12x2¼; and for the rear above tubes lay ordinary fire brick special skeleton arch frame.

Iron Work: Place in position all cleaning out doors, cast iron plates and anchor rods.

PIPE CONNECTIONS.

To be as per drawing, which will be furnished.

Steam: Four inches from boiler to 6-inch header leading to engine room, where it will reduce to size required by engine, thence to engine proper size, through a Hine separator suitably drained. Leave Tee having plugged outlets for additional engine and water works pump.

Exhaust: Four inches from engine to main line; thence 6 inches through heater to 5 feet above roof. Leave plugged inlet Tee for additional engine and pump.

Drains: Both the steam and exhaust pipes are to have suitable drains of ample size wherever there is any possibility of water accumulating. Run these drains outside of building.

Small Piping: Feed, blow-off and steam and exhaust pipes for boiler feed pumps to be as per details shown in drawing.

Fittings: Of the best construction, with threads true and clean. Use in all cases what is known as "water" or "sweep" ells and fittings, having extra long radii for curves.

Valves: Of the Jenkin Brothers, or asbestos disk pattern. Use gate and angle valves in preference to globe. When globe valves are used they must be so placed as not to form water pockets.

Supports: All pipe work to be well supported in such a way as to bring no unusual strain on the pipe or fittings; either from their weight, or from expansion or contraction.

Covering: All live steam pipes, domes and top of boilers to be covered with a high grade non-conducting material, such as magnesia ectional.

In General: The arrangement of the pipe work must be such as to provide for all differential strains arising from expansion and contraction. The work to be of the best and most thorough possible. The steam pipe will be tested to 150 pounds.   W. H. B.

160. Specifications for Leather Driving Belts.

The following specifications for large leather belts were prepared for the large water power electric plant at Austin, Texas, in 1894. It is thought they conform to the latest and best practice in the manufacture of leather belting.

There are to be six main driving belts and seven belts to drive dynamos, of dimensions as hereinafter scheduled.

These belts are to be of leather made from the best selected, large steer hides, of pure oak-bark tannage. The cuts are to be taken from the centre solid portions of the hides, and are not to include shoulders, flank or soft parts of the hides. Each piece is to be of fine, close fibres and all pieces are to be scarfed to a uniform thickness. No piece taken from one hide is to exceed a net length of fifty inches.

The individual pieces of the leather are to be thoroughly stretched after currying and again machine tested and the utmost stretch, within elastic limits, given to the belts when they have been made up complete.

The transverse lap joints are not to exceed four inches in longitudinal length, are to be scarfed in the best manner, thoroughly cemented and are to be made fast and durable without the use of pegs or rivets. All belt edges are to be properly rounded.

All belts are to be thoroughly water-proofed.

The complete belts are to be soft, pliable, and finished with smooth polished surfaces.

The belts of thirty-eight inch width are to be of double thickness. The outer face pieces are each to be in a single width, with centre conforming to the back-bone centre of the hide. The inner, or running face pieces of the thirty-eight inch belts are to be n'neteen inches in net width and to have one and one-half inch scarfed and lapped longitudinal joints. One edge of each half-width will be cut along the back-bone centre of the hide and in the makeup of the belt, these inside half widths are to be placed with the back-bone edges at the outer edges of the belt. These seam sides of the thirty-eight inch belts are to be run next the pulleys.

All the remaining belts are to be of double thickness in single width pieces, with centres of each piece conforming to the back-bone centres of the hide. The belts are to be finished

with uniform thicknesses respectively not less than as follows
for each stated width:

38 inches width of not less than $\frac{28}{64}$ inches thickness.

25   "   "   "   "   " $\frac{24}{64}$ "   "

24   "   "   "   "   " $\frac{26}{64}$ "   "

14   "   "   "   "   " $\frac{24}{64}$ "   "

13   "   "   "   "   " $\frac{24}{64}$ "   "

11   "   "   "   "   " $\frac{22}{64}$ "   "

The speeds of the belts will be at rates of about 5,000
lineal feet per minute.

All these belts are to be transported to the power house
now being constructed by the Board of Public Works of Austin, Texas, in Austin, and are to be placed upon the pulleys in
the power house and spliced and cemented in place.

The hides and manufacture, finish and fitting of the belts
are to be first-class in every respect and the belts are to be
guaranteed to run smoothly and straight upon the pulleys and
to work successfully for the space of one year from the time of
the starting up of the power house for regular work.

If any defect tending to impair the usefulness or life of any
belt supplied under this specification, shall develop within one
year that belt shall at once be made good by the manufacturer
or replaced by a belt conforming with this specification.

Proposals for these belts, as fitted in place ready for the
starting of the machines, are to be delivered to the Hon. John
McDonald, Mayor, and President of the Board of Public Works,
Austin, Texas, on or before the 8th day of December, 1894,
and all belts are to be delivered and fitted in place ready for
use within six weeks of the date of the order for their manufacture and delivery.

The Board of Public Works reserves the right to reject any
and all proposals as may be for the best interest of the City of
Austin.

Blue-prints showing relative positions of the pulleys and
inclinations of the belts are submitted herewith.

SCHEDULE OF BELTS.

	Width of Belt.	Thickness in 64ths of an inch.	Diameter of Driving Pulley.	Diameter of Driven Pulley.	Distance between Centres of Shafts.	Total Net Length Belt in Place.
	H. P.	In.	64ths.	In.	In.	Feet.	Feet.
Main Driving Belt...............	520	3S	2S	54	54	43.10S	100.333
" " " ....:....	520	3S	2S	54	54	42.763	99.663
" " "	201	25	24	54	26	44.970	100.412
" " "	201	25	24	54	26	48.075	106.622
" " "	201	25	24	54	26	45.379	101.230
" " "	201	25	24	54	26	48.687	107.S46
Dynamo Belt...............	241	24	26	56	32	17.851	47.352
" "	134	14	24	50	2S.5	18.625	47.222
" "	134	14	24	50	26.5	17.6SS	45.34S
" "	100	13	24	55	1S	15.293	39.70S
" "	100	13	24	55	1S	15.293	39.70S
" "	80	11	21	46	1S	15.211	3S.S00
" "	So	11	22	46	1S	19.912	4S.102

J. T. F.

161. Specifications for Pumps to be Operated by Water Power. The following specification was prepared for the city of Austin, Texas, in 1892, for the construction of two pumps for a city water supply to be driven by water power machinery.

PUMPS.—There are to be two pairs of horizontal, double-acting, plunger pumps, having two pump cylinders to each pair. Each pair of pumps is to have capacity to deliver four million gallons of water per 24 hours.

The dimensions of plunger will be approximately as follows: 1½ foot diameter, 2¾ foot stroke, with 25 revolutions per minute; the plunger speed not to exceed a mean rate of 137½ feet per minute when pumping water at the rate of four million gallons per twenty-four hours.

These pumps are to be adapted for pumping to a reservoir and also for pumping directly into the city distribution pipes with direct pressure, the pump house being located between the reservoir and the city. The static head of the reservoir is 245 feet and the dynamic head approximately 265 feet and the force main to the reservoir is 7,600 feet in length, of 24 inch pipe.

These pumps are to receive motion from vertical turbine shafts having beveled gears which drive a jack shaft. On the jack shaft is to be a spur pinion, which will drive a spur mortise gear on the main pump shaft. On each end of the main shafts are to be balanced crank disks, which will drive the pumps.

The receiving and delivery chambers are to be ample in dimensions, and are to have nests of valves of the best bronze composition of approximately 3 inches diameter, and sufficient in number so that the valves shall not lift more than $\frac{3}{8}$ inches each when the rate of delivery of the pumps is at four million gallons per day. The pump chambers will be well provided with hand-holes that will give easy access to each of the valves.

The water supply for each pair of pumps is to be taken from a 30 inch branch in the horizontal penstock in the basement beneath the pump room floor. The necessary admission and discharge pipes, of ample size and easy curves, and a tall air vessel for each pair of pumps are to be provided. The force mains will be connected with the force mains leading to the reservoir, on the outside of the pump house wall, and not exceeding 10 feet distance from the face of the wall.

The pump cylinders will be connected with the main and jack shaft pillow blocks by continuous, heavy cast-iron girders, adapted to carry the bearings and the strains of the connecting rods without tremble or elasticity.

The main and jack shafts are to be of the best wrought iron forgings turned to the diameters indicated upon the drawings, and bossed up to receive the gears. The large gears, of both the spur and bevel pairs, will be mortise gears with their mortises planed, and having thoroughly seasoned, machine cut, smooth maple cogs, fitted and keyed in the most rigid manner. The cogs are to be thoroughly boiled in oil.

Each spur pinion is to be machine moulded, with teeth planed on both sides to match and run with mortise gears. Both gears and pinions are to be bored to fit their respective shafts and to be keyed in place. The pinions are to be feather keyed on the jack shaft and fitted with approved screw motion to move them out or into gear and to hold them where placed.

The jack shafts are to be not less than $6\frac{1}{2}$ inches and the main shafts not less than nine inches diameter in the bearings.

The two jack shafts are to have their axes precisely in line and. are to have a connecting shaft fitted with couplings, bearings and distance plates as directed.

All bearings are to be babbeted and bored out, of ample lenght to insure moderate wear and easy lubrication, and are to be fitted with oil cups.

The cranks will be in balanced disk forms and fitted with mild steel pins. The connecting rods will be of the best wrought iron, and fitted with brasses, steel straps and adjusting keys and babbeted friction bearings. The cross heads will be of forged iron approved model with steel wrist pins. The guides will be rigidly bolted to the girders and adjustable to wear. The plunger rod and plunger stuffing box glands, the valves and all interior bolts and nuts will be of the best solid standard bronze metal. The plunger stuffing box will be packed with the best quality of "Seldon" or other approved packing. The crank disks, connecting rods, straps and cross-heads are to have polished surfaces.

All bearings and wrist pin brasses are to be fitted with the best oil cups and with proper drip pockets, and drip pipes are to be provided where required.

There is to be furnished and connected with each pair of pumps one 8 inch water pressure gauge, indicating the column of water in both feet head and pounds pressure. There will be a good approved revolution counter fitted to each pair of pumps as directed by the engineer. There will be a long Scotch tube water glass with proper cocks on each air vessel.

The general plan of the power house shows the position of these pumps, the method of connecting their power, and the positions of their suction and delivery pipes.

Full detail drawings of the pumps and their appendages are to be delivered to the consulting engineer and are to be subject, in all respects to his approval.　　　　J. T. F.

162. Specification for a Pump Well. The following specification describes a good method of sinking a well from 20 to 50 feet in diameter and to a depth not usually exceeding 50 to 60 feet. Such wells are usually sunk through water-bearing strata, where an open cut would have to be supported by some kind of curbing and this curbing may as well be the masonry well curb itself. In this case it is necessary to rest this masonry upon a shoe which may be made either of wood or iron. In the former case an iron cutting edge should be provided. It is also best to reduce the external diameter of the masonry curb a short distance above the shoe in order to prevent the whole mass from sticking in the process of sinking by its becoming tightly bound by the surrounding earth. It is necessary to anchor the lower portion of the masonry curb to the shoe by means of iron bolts, as indicated in these specifications.

Pump Well.—The pump well shall be constructed within lines to be given by the Engineer, and will consist of a brick curb, laid in cement mortar, on an oak shoe provided with an angle iron ring bolted to the under side of shoe flush with its outer periphery.

Shoe.—The shoe shall be in three rings of unseasoned oak, three inches thick, made up in segments, the segments to be a true arc upon the outer periphery and bolted together with seven eigths (⅞) inch bolts and wrought iron washers to break joint as shown in drawing marked "Section and Details Pump Well," on file in the office of the Water Works Committee. The width of the shoe shall be twenty-two (22) inches, and when bolted up shall measure nine (9) inches in depth or thickness. The ring shall be of 4x4 inch angle iron five eighths (⅝) inch thick weighing 16.2 pounds per foot, twenty-one (21) feet external diameter, with two fishplates eighteen (18) inches long at each joint, riveted on hot, with four (4) three quarter (¾) rivets staggered, each side of joint; the heads of rivets to be countersunk and finished flush on outside of angle iron ring and shall be attached to the oak shoe with sixty (60) seven eighths (⅞) bolts, placed about one (1) foot and one (1) inch, center to center, on a circle twenty (20) feet, seven and one half (7½) inches diameter. These bolts shall have forged square heads and nuts and wrought iron washers. The oak shoe shall have a diameter of twenty-one (21) feet.

Curb.—The curb shall be of select hard burned front or paving brick, straight, sound and solid, when broken. No over burned or salmon brick will be accepted, and the brick shall be laid on the oak shoe in courses; in full beds of cement mortar, consisting of one part Louisville [fresh burned] cement to 2 parts of clean sharp sand which shall be mixed only as fast as used by the masons. No cement mortar which has had an over dose of water or sand, or which has begun to set in the mortar box shall be used, but all such must be thrown aside. The bricks shall be free from dust and wet with a sprinkling hose or can, or be dipped by the mason in clean water before being placed in the curb and all joints shall be slushed with mortar pressed in with the trowel, as rapidly as the courses of brick work are laid up. No GROUTING WILL BE ALLOWED.*

The outer and inner faces of the curb shall be laid in true circles of the dimensions shown by the drawing above mentioned, and shall have at the bottom (or on the shoe) an external diameter of twenty (20) feet ten (10) inches, and an internal diameter of seventeen (17) feet four (4) inches, these dimensions for a height of two (2) feet six (6) inches, when the internal diameter shall be contracted by a gradual racking inward of the courses to a height of four (4) feet above the shoe, where the internal diameter shall be sixteen (16) feet eight (8) inches.

From a depth or height of four (4) feet above the shoe to the top of well the internal diameter shall be eighteen (18) feet, and the external diameter shall be twenty (20) feet ten (10) inches from the shoe to the top of well, the thickness of wall at the bottom or on the shoe, and for two and one half

---

* The author would prefer the grouting to the slushing with mortar, as better calculated to obtain perfectly solid joints.

(2½) feet above, shall be twenty-one (21) inches, four (4) feet above the shoe the thickness shall be twenty-five (25) inches, and for the remainder of heighth or depth, the thickness shall be seventeen (17) inches.

The bricks shall be laid stretchers and headers or with hoop iron bond, as may be directed by the engineer. The contractor to state in his proposal the difference in price, (if any), should all bricks be laid as stretchers and the curb bonded from inside to outside with hoop iron.

Should hoop iron bond be used this will be of charcoal iron, No. 16, B. G., one and one half (1 1-2) inches wide cut four (4) inches longer than thickness of wall, with each end turned at right angles for a length of two (2) inches, and shall be placed eighteen (18) inches apart around the wall, and shall occur at every fifth horrizontal joint.

The curb of the well shall be truly cylindrical for all depths, and shall be carried down plumb. The outer surface of the brick work must be true to the arcs and smooth, to prevent sticking as the material is undermined from the shoe, and the inside joints shall be neatly struck as the courses of brick are laid.

Anchor Bolts. The lower or first eight (8) feet of the curb shall be attached to the shoe by ten (10) anchor bolts, eight (8) feet long, of seven eighths (7-8) inch round iron, with square nuts and wrought iron washers at the bottom and square nuts and plate iron washers of one quarter (1-4) inch boiler iron eight (8) inches square at the top, the bolts shall be placed about six (6) feet apart, and built in as shown by the drawing herein mentioned, and after the course of brick next under the washers (at top) has been laid, the nuts shall be all screwed down on the washers, and the excess of thread on the bolts carefully cut off with a hack saw, to avoid disturbance of the brick work just set.

Coping. The well shall be finished with a coping of sand stone ten (10) inches thick, twenty-one (21) inches wide, with an internal diameter of seventeen (17) feet eight (8) inches, to project two (2) inches inside and outside brick work at top of curb. The inner and outer edges of coping on top shall have a chiseled wash one and one half (1½) inches "in" on the bed and one (1) inch "down" on the face. The coping shall be made up in ten (10) segments of uniform length of arc, and shall be dressed to lay with less than one quarter (¼) inch joint. The joints shall be filled with mortar worked in with the trowel and the several segments shall be cramped together with iron cramps of best "½x2½ f. b. iron with legs 2¾ inches long, and width of cramps between legs fifteen (15) inches. These cramps shall be "let" into the beds of stones at the mid-width, flush with the surface of coping, and the leg pockets

17

shall be cut slightly dovetail with a flare downwards, and after the cramps are set shall be run with hot lead caulked in place. The lower bed shall be chisel dressed to make a joint on a full bed of mortar with the last course of brick, and the faces and upper beds shall be finely chiseled to a smooth even surface.

Excavation. The well will be constructed by the under-mining method, a circular hole, twenty-one (21) feet diameter will be sunk in the clay eight (8) or ten (10) or more feet, depending upon the capacity of the material to stand vertical, and at the option of the contractor and upon approval of the engineer. The shoe will then be placed in the hole and care-fully leveled, the anchor bolts being in place, the brick curb will be laid as herein provided until the brick work reaches a heighth of three (3) feet above the level of ground, when fur-ther excavation will be had by removing the material within the curb and under the shoe and allowing the shoe to settle from the superimposed weight of the curb. The excavation under the shoe to be carried down uniformly all around to maintain a true level of the last course of brick on the curb. The level shall be taken for each course of brick laid, and when found "out" the curb shall be truly leveled by additional excavation under the higher side.   J. W. H.

163. Specification for Turbine Water Wheels.

The following specifications for both horizontal and vertical turbine water wheels was used in the construction of the large water power system of Austin, Texas, of 1892.

Horizontal Turbines. There are to be four pairs of hori-zontal turbines of 506 horse power each, under 54 feet head. Each pair of these turbines will discharge into one common draft tube. On the shaft of each pair of turbines there will be two pulleys, each to be adapted to transmit the full power of the pair of turbines, and on the same shaft there will be a heavy balance wheel. The pulleys are to be for belt or rope driving as directed. The turbine shafts are to have ample bearing surfaces, and each exterior bearing is to be fitted with oil cups and proper drip pockets.

Each turbine case is to have a man-hole of 10x15 inches clear opening. Each of the turbine quarter-turns is to be of cast-iron and is to be fitted with a good stuffing box and is to be flanged, fitted and bolted to its 5½ feet diameter stop valve. The draft tube is to be flared at its mouth. A cast-iron plate is to be fitted on the floor of each tail race under each draft tube, which plate is to be 6½ feet diameter and raised coni-cally in the center to a point. The floor plates are each to be secured with twelve ¾ inch lag bolts.

The turbine cases, quarter-turns and feeder pipes are to have proper lugs upon them to rest upon the iron beams and masonry, which are to be their supports.

A strong and rigid frame of iron beams is to be furnished with each pair of turbines. Each frame is to be supplied with proper strong hold-down bolts and anchor plates. All anchor rods, plates, lugs and braces are to be supplied that may be necessary to secure the turbine cases, frames, draft tubes and feeder pipes securely in place, so they will be free from movements or vibration.

Vertical Turbines. There are to be also two turbines with vertical shafts, each of 185 horse power under 54 feet head. These vertical turbines will have draft tubes similar to those above described for the horizontal turbines. The draft tubes are to be quarter inch plate iron with seams riveted so as to be air tight and with seams caulked in a workmanship manner. The shaft of each vertical turbine is to extend up to a level proper to receive the beveled pinion at the pump room floor. A pair of bevel wheels is to be furnished with each of the vertical turbines of ratios as directed, approximately 2 and 75-100 to 1. The larger bevel will be a mortise gear wheel with planed mortises and fitted with the best maple cogs which have been thoroughly seasoned and boiled in oil and substantially keyed in place. The small bevel gear will be machine molded and have planed teeth.

The vertical feeder pipe with each vertical turbine and its quarter-turn will be supplied with each wheel case. Within this vertical feeder pipe are to be a sufficient number of bearings to keep the shaft truly in line. These feeder pipes are to be made of quarter inch plate iron of good ductile stock, of not less than 40,000 pounds tensile resistance per square inch, and are to be well riveted with hot rivets and are to be calked water tight in a workmanlike manner.

The quarter-turns will be flanged and fitted with bolts to connect them to the vertical and horizontal feeder penstock pipes or valves. Each vertical turbine is to have ample capacity when working under 54 feet head to start its pair of pumps when pumping at a rate of four million gallons in 24 hours against a pressure of 265 feet of water and to bring the speed of the pumps quickly and easily up to a rate of 25 revolutions per minute.

The turbines will have bronze buckets, approved gates and gate gears, and composition stuffing box glands. Approved cast-iron, bright standard and hand wheels will be set on the main floors of the house, where directed, and connected with the gate gears. Approved, sensitive regulators will be connected with the horizontal turbines. Substantial pedestals will be provided for the bearings of the horizontal shafts.

The turbines are each to be guaranteed to give a duty of not less than 8o per cent. by dynamometrical test, in a testing flume, or by similar test when driving their pumps at a rate to deliver four million gallons of water per 24 hours into the reservoir. *

The turbine cases, turbine and draft tubes, also the vertical feeder pipes of the vertical wheels and all the quarter-turns, are to be set in place in the power house being constructed in Austin, Texas, by the Board of Public Works of the city, and their materials and workmanship, and their trimmings and anchorages are to be of the best of their respective classes, to the full approval of the Engineer, and are to be guaranteed and maintained in perfect condition for the term of one year after their test and acceptance.

A general plan accompanying this specification shows the wheel pits, penstocks, feeder pipes and draft tubes, the floors of the power house and method of using the power. Full detail drawings of the turbines and their appendages are to be delivered to the Consulting Engineer and are to be subject to his approval in all respects.   J. T. F.

164. Specification for the Installation of an Electric Lighting Station in a Small City. The following specification for an electric lighting station were prepared in the year 1894, for a small city which required an economical installation. They are given here, not for the purpose of being copied, but as an illustration of what was considered good practice at the time they were drawn. The gentleman who prepared these specifications has had a large experience in Electric Light installation, having been at one time manager of an Electric Works, while at the same time being a mechanical engineer of thorough training and wide experience. The work was to be erected under his own supervision and inspection, so that it was only necessary to make such a specification as would warrant him in demanding first-class materials and workmanship in the execution.

STATION PLANT.

Dynamo: The dynamo shall be of constant potential alternating incandescent type, and to have a nominal or rated capacity of thirty to thirty-five kilo-watts, at one thousand to

---

*The author recommends that a bonus and forfeiture condition should accompany such a duty clause as this.

one thousand and one hundred volt at station. State number of alterations per minute. Exciter to be belt driven from alternator shaft. By "nominal" or "rated" capacity is meant that load at which the dynamo will run continuously, without undue heating. Proposals will state the capacity, and the amount of overload the machine will safely stand for three hours run in hot weather. Bids are also desired on direct connected generators, with high speed engine of approved design. Bids must give full details, and be accompanied by drawings.

Attachments: Machine to have insulated base frame, belt tightener, self-oiling bearings, automatic regulator, and all necessary station and switch-board apparatus, including lightning arresters. Submit a list of station equipment intended to be furnished.

Regulators: Must automatically control the current over the entire range of the capacity of the machine, without undue heating, or sparking; so that the power required is at all times proportionate to the number of lamps burning. It must provide a steady and uniform light, with variations in the engine speed not exceeding three per cent. The regulators must protect the dynamos in case of short circuiting on the line.

Erection: Contractors will deliver machine, and all station apparatus, and erect same in position, including substantial foundations of concrete. They will run wires in station between outlet where external construction begins, and switch-board, machines, arresters, exciter, grounds, etc. Wire to Okonite. Furnish and place switch-board, and erect all apparatus thereon. Furnish expert to erect, adjust, and run apparatus ten days, instructing the city's attendants in its care and operation. Contractor must keep informed as to the progress of the work, and arrange the time of his experts accordingly, and there will be no allowance for extra time or traveling expenses, not specially ordered. State charge per day for the time of expert longer than ten days.

Acceptance: Before leaving, the expert must satisfactorily make the capacity test, and such other tests as the city may require to satisfy itself that the provisions of the contract have been fully carried out. At the end of the ten days' run, the apparatus will be accepted, providing the requirements of the contract have been fully met.

Switch Board: Furnish and erect a switch-board, complete, of good, hard, well seasoned wood, providing for one dynamo and two mains for commercial, and for street lights as hereinafter provided. Arrange for easy access to rear of board. Submit list of apparatus to be placed on switchboard.

Lightning Arresters: Include six double pole lightning arresters of approved form for use on the circuits throughout the city.

MISCELLANEOUS APPARATUS AND SUPPLIES.

Converters: From one thousand to one hundred volts. Number and capacity to be as per the accompanying list. Each must be provided with fuse box and eye bolts, or wrought iron straps; with hooks on upper ends to hang directly from cross arm or cleats. The regulation must be within two and one half per cent. for the smallest size, and two per cent. for the largest, besides which, the leakage losses must not exceed five per cent. on the small, and one per cent. on the large, and the regulation and leakage must be uniform for all converters of the same size. Each converter must be ample to carry, in emergencies for three hours continuously, without dangerous overheating, twice its rated capacity, but, of course, with reduced efficiency.

Shunt Coils: Fifty in number; one to be used with each street lamp, of which there are five groups, of ten each. The shunt coils to take care of the current in the event of a lamp burning out. Furnish two extra coils for reserve.

Meters: Will read ampere or watts hours, and must be carefully adjusted and tested before shipment. See list appended. Furnish one extra meter of each of the three smallest sizes.

Lamps: To be of approved make, and furnished with such base as may be selected later. Efficiency fifty-five watts per sixteen candle power lamp. Furnish, now, 1,000 sixteen candle power, and 100 thirty-two candle power. All for one hundred volt current.

Sockets: One thousand of first-class construction, with porcelain base to fit such lamp as may be selected later.

Delivery and Erection: The converters, meters, lamps and sockets shown on accompanying lists and maps, are to be erected in position. The rest of the quantities above named are to be delivered to the city for future use.

Future Orders: The quantities hereinbefore mentioned are to be included in original proposal price, but a price must also be named at which additional orders may be placed within one year from signing contract.

EXTERNAL CONSTRUCTION.

Pole Line.—Furnish and erect in position all poles and cross-arms for the complete distribution system shown on blue print. All poles to be live peeled white cedar, 30 feet long, 6 inches diameter at top, housed and gained for two cross-arms. Set 4½ feet in ground and tamp well. Poles must be straight and sound. Any poles crooking more than one inch in five feet or having more than ten per cent. rot in butt, will be rejected. Location of poles will vary, between 100 and 150 feet apart, averaging probably 125. All locations to be approved by city. Furnish 12 extra poles for future use. Furnish all material and do all work, connected with the primary system, starting from station outlet, and including secondaries to house inlets. Corner pins to be 1½ inch extra quality locust, except in cases of unusually severe strain, where they must be of iron; all others 1½ inch painted oak; all wires to be carried on insulators of the deep groove double petticoat pattern.

Wire.—To have triple braided weather-proof insulation of superior finish and smoothness, tough and not easily abraded, and which will not disintegrate or deteriorate by exposure to the elements, and equal in all respects to "K K." The wires will be of sizes as shown on blue print. The drop from converter to house inlet when all lamps shown on map are burning at once must not exceed one per cent.

Street Lights.—Will be 47 of 32 candle power each, run in groups of 10-100 volt lamps, with shunt coils, each group in series. Location of street lamps and wires as per accompanying blue print. All wire No. 10 B. & S. Furnish and erect on switchboard at station, switches controlling all street lights. There will be two groups of 10 street lights each. The other three groups will each have 9 lights on street, and one in station.

· *Hanging Lamps.*—Contractor is to furnish all fixtures, material and labor; to hang in position the 47 incandescent street lamps shown on map, as high above grade line as possible; with cutters suspension street hoods, with cross-arms, insulators, nozzles, and petite pulleys and ¼ inch galvanized iron flexible lamp cord, with hemp core; also galvanized steel wire strand ¼ inch diameter, for suspending lamps in the centre of streets, by the cross suspension method. Use eye bolts with washers for suspension wires, projecting sufficiently to permit slack being taken up by tightening nut. Iron break arms are to be used where lines leave poles, or wherever a loop is made.

· *Returns from Street Circuits.*—Shown in broken lines on blue print may be cut into commercial circuits, instead of returning to station independently.

Incandescent Distribution.—Will be shown on accompanying blue print.

In General.—All joints are to be well soldered and taped. No wire must be lower than 20 feet above grade line. All streets, alleys, and other public places where work is done, must be left in as good condition as before starting. Use special iron brackets wherever necessary, always placing some soft moisture proof material between the iron screw and the insulator. Erect on incandescent mains where directed, the six lightning arresters.

SECONDARY INDOOR WIRING.

Capacity.—All secondary wiring must be sufficiently large to carry at one time 25 per cent. more lamps than the number shown on the accompanying map, without undue heating, and at 100 volts. The drop from house inlet to the most distant lamp with the above maximum load must not exceed 2½ per cent.

Erection.—All inlets to be in front of houses, except where some other place may be designated, as more convenient. Converters, meters, sockets, and lamps are to be furnished by this contractor, placed by him and connected permanently in position, complete. All other necessary material, such as fuse boxes, switches, cut-outs, etc., to be furnished and erected by this contractor.

Plans.—Name a lump sum for the complete installation of the lamps located on the blue print in accordance with these specifications. State also:

1st. Price per lamp at which this schedule may be added to or deducted from.

2d. Price per lamp which will be charged additional for concealed work.

Character of Work.—Except where otherwise arranged, all interior wiring will be open cleat work, using white double braided painted fire-proof wire. The details of all indoor wiring will be in accordance with the rules of the St. Louis Board of Fire Underwriters. The city will have the work inspected from time to time at its own expense, and any work which may be found, at any time previous to the acceptance of the plants not in accordance with those rules, must be put into satisfactory shape by this contractor at once. Drops to be No. 16 cotton flexible cable, with adjusting ball and fibre socket bushing.

Special Work.—The city grants the contractor the right to sell shades, fixtures, etc., and to do concealed and fixture wiring, for which extra work the customers will pay him direct, such work to be done under the supervision of the engineer, and to his satisfaction.   W. H. B.   .

165. Specifications for Electrical Distribution Circuits for Light and Power. The following is the descriptive portion of a set of specifications for electrical distribution for light and power prepared for the city of Austin, Texas, in 1894. They are thought to be a good example of such specifications and are here inserted exactly as used in the letting of the contract.

Power Station: The power station is located at the new dam in the Colorado river and is about three miles northerly of the corner of Congress avenue and Pecan streets in said city of Austin. There will be in the power station one 180 kilo-watt tri-phase generator adapted to generate alternating currents of 2,700 volts potential, two 100 kilo-watt generators adapted to generate direct currents of 550 volts potential, four alternators of 3,000 light capacity, and two alternators of 1,500 light capacity adapted to generate alternating currents of 2,200 volts potential and two arc machines each of capacity to supply currents for 100 arc lamps of 450 watts each.

The wires for power currents will be led out of the station for grouping on one set of poles and the wires for lighting currents for grouping on another set of poles. The currents of the three power generators will be transmitted by three-wire complete circuits.

All the wire circuits are to be connected with the switch-board and station apparatus so as to give the most complete switching, testing and regulating facilities with the least drop of potential consistent therewith. The leading wires are to be strung from the switch-board to the cupola of the power house and out through the panels of the cupola and are to be insulated from the building and panels in the best manner.

All the wires within the buildings will be covered with a firm water proof insulating material, such as shall be approved by the engineer, and to the safe insulation of currents with standard potentials of 2,500 volts.

Pole Lines: The line poles will be of peeled, white northern cedar. The poles are to be straight, sound, smooth and free from large or loose knots that might weaken them.

The dimensions of poles shall not be less than those stated in the following schedule and poles of each representative class are to be set at depths not less than those stated in the schedule, if set in earth, and six inches less if set in solid rock.

Poles will not be less than the schedule thickness at one half foot below their tops, and will be made roofed at their tops and their roofs will be painted with the best quality of mineral paint. Their lowest cross arms shall be at least 18 feet and 3 inches at center above the center of the street opposite the pole.

On the main two-feeder lines, between the power house and Congress avenue, the poles are to be spaced not exceeding 100 feet between centers, and on the sub-feeder and distribution lines poles are to be spaced not exceeding 132 feet between centers, and if lengths of blocks are such that three poles per block exceed this limit, four poles per block are to be used. Poles will be placed in the curb line or in a line parallel with the curb line if on streets and, if in the alleys, as directed by the engineer.

Corner and terminal poles and all other poles subject to extra unbalanced strains shall be securely guyed with No. 6 galvanized steel wire. Guys shall be so placed and secured as not to be obstructions or nuisances.

SCHEDULE OF MINIMUM DIMENSIONS OF POLES.

CLASS.	Number of cross arms.	Minimum length of poles.	Depth set in earth.	Height of lowest arms.	Dia. at top of poles.
A..................	One.	25 ft. 0 in.	4 ft. 0 in.	18 ft. 3 in.	$4\frac{1}{2}$
B..............	Two.	25 ft. 0 in.	4 ft. 8 in.	18 ft. 4 in.	5
C.............	Three.	27 ft. 0 in.	4 ft. 6 in.	18 ft. 5 in.	$5\frac{1}{2}$
D.............	Four.	29 ft. 6 in.	5 ft. 0 in.	18 ft. 9 in.	6
E.............	Five	31 ft. 6 in.	5 ft. 6 in.	18 ft. 7 in.	$6\frac{1}{2}$
F.............	Six.	33 ft. 6 in.	5 ft. 9 in.	18 ft. 8 in.	7
G..............	Seven.	35 ft. 0 in.	6 ft. 0 in.	18 ft. 3 in.	$7\frac{1}{2}$

Gains shall be cut in the poles so that the cross arms will fit snugly and rest at right angles to the axis of the poles. Proper gains are to be made to receive lightning arresters, transformers and other apparatus to be attached to the poles.

The contractor shall secure all necessary permissions for the trimming of private trees and shall do all trimming, and he shall secure the necessary permission for attaching any guy wires to private property.

Cross Arms.—The six-pin cross arms will be four and one quarter by five and one fourth inches section, and other cross arms of three and one quarter by four and one quarter inch section, and all are to be of sound, clear and smooth seasoned white oak. The two-pin cross arms will be of clear, hard Michigan white pine.

Each will be rounded on its top and each will be fastened with three and one half by seven inch lag screws with washers. The six-pin arms will be not less than five feet and ten inches long and four-pin cross arms not less than four feet ten inches long. All cross arms will have one good coat of the best "P. and B." paint compound for the purpose before being fastened to the poles.

The vertical distances between centers of cross arms shall not be less than twenty inches.

All four and six-pin cross arms will be stayed with one quarter by one and one half inch rolled iron japanned braces, not less than twenty-six inches long on the six-pin arms and twenty inches long on the shorter arms.

Each pair of braces will be secured with two lag screws, two by five-sixteenth inches, and one lag screw three by five-sixteenth inches dimensions each, with washers complete.

Pins.—All cross arms carrying No. 1 or larger wire will be furnished with the best quality of locust pins and for smaller wires with the best quality of white oak pins all with one and and one half inch diameter tenons. The interior pins shall be eight inches from centers of cross arms and other spacings of pins 12 inches between centers.

The pins shall be covered with "P. & B." paint compound, shall fit closely in the cross arm mortises, and shall be secured with steel nails.

Insulators.—Each pin shall have one of the best deep groove glass insulators of double petticoat pattern.

Pole Steps.—Screw pole steps of five eighths by eight inch wrought iron, galvanized, shall be placed on each pole on which there is a lightning arrester, transformer or cut-out. The lowest step shall be at eight feet from the ground and other steps at eighteen inches between centers vertically, but alternately on the opposite side of the poles.

Wire Circuits.—All of the circuits are to be of pure copper, of at least 95 per cent. conductivity, drawn true to gauge and of the best quality in every respect as electrical conductor wires.

The diameters of the circuit wires as herein described are stated in the dimensions of the Brown & Sharp gauge. The wires in the power house will have the best water proof insulation. The wires in all alternating current feeders and circuits are to have the best weather proof insulation of standard double braided and compounded coverings.

The arrangements of the power circuits on the poles from the power station to West Avenue are shown on an accompanying plan and the arrangements of the main alternating circuits and arc circuits are similarly shown on another plan.

On the top of the two main pole lines above described there will be one guard or protection galvanized iron standard, barbed double fence wire to be strung on pony insulators, and effectively grounded at distances not exceeding 500 feet.

All joints in wires must have full and durable contact and be soldered in the best manner so that the joints shall hold and maintain a degree of conductivity at least equal to that of the wires connected. All joints so made shall be thoroughly washed in an acid neutralizing solution and well wrapped with insulating tape, and the finishing end of the tape shall be wrapped with copper wire. The insulation resistance of the joint is to be equal to the insulation resistance on other parts of the line.

The power, arc and alternating circuits within the city will be arranged, as nearly as possible, as shown on the accompanying maps of the distribution system.

The commercial and domestic lighting by alternating currents will be divided into eight districts, as shown on the wiring map, and the wires will be proportioned for 16 candle power, alternating lamp transformers in each district as follows:

District No. 1, 1500 Lamps and 1200 Lamp capacity of transformers.
"   "   2, 3000   "   "   2400   "   "   "   "
"   "   3, 3000   "   "   2400   "   "   "   "
"   "   4, 1500   "   "   1200   "   "   "   "
"   "   5, 1500   "   "   1200   "   "   "   "
"   "   6, 1500   "   "   1200   "   "   "   "
"   "   7, 1500   "   "   1200   "   "   "   "
"   "   8, 1500   "   "   1200   "   "   "   "

The power generators will have their currents wired from the power house into the city by the Boulevard and Pecan street, to Red River street, and a branch current wire will extend along the alley between Congress avenue and Colorado street from 3rd to 10th streets. The transmission will be by three-wire circuits with complete returns, and the drop in potential in full power of the generators shall not exceed ten per cent.

Towers.—Thirty iron "Star" lighting towers of the Detroit pattern, 150 feet each in height to top of mast, are to be located in various parts of the city as shown in the accompanying maps of lighting towers. These towers are to be of the most substantial construction, substantially guyed, and equipped with six 450 watt arc lamps each.

Each of the two circuit systems of wires for lighting these tower arc lamps is to be of No. 6, weather proof, insulated copper wire, connected with the switch board in the power house.

Potentials.—In the wires for commercial and domestic lighting by alternating currents, the loss by drop in potential in the mains between the power house and West avenue shall not exceed twelve and one half per cent., and in the sub-feeders and branches shall not exceed an additional five per cent.

Transformers.—The schedule of transformers or converters as herein contemplated is as follows:

Twenty-one of ....12 Lamp Capacity, 50 watts per lamp.
Ninety-nine of....................25 Lamp Capacity, 50 watts per lamp.
Forty of ....50 Lamp Capacity, 50 watts per lamp.
Fifteen of....... ....70 Lamp Capacity, 50 watts per lamp.
Fifteen of..... . ...90 Lamp Capacity, 50 watts per lamp.
Eleven of.... ....125 Lamp Capacity, 50 watts per lamp.
Twelve of.....................250 Lamp Capacity, 50 watts per lamp.
One of.......... ....500 Lamp Capacity, 50 watts per lamp.

The said party of the first part hereby reserves the right to exchange converters by sizes, taking an equal capacity in smaller converters as the interest of its patrons shall require. The converters, as located by the engineer, are to be fully connected in the wiring circuits ready for attaching the domestic and commercial wires.

Grounds.—Effective grounds are to be prepared for each of the lightning arresters and for the ground connections of the guard wires. When no good ground connection is available one is to be prepared by placing two bushels of good coke or charcoal near the base of a pole and placing therein a copper plate, one eighth by four inches in section and three feet in length, and the ground wires are to be soldered thereto.

Apparatus.—All the circuits will be fully equipped with the requisite installation apparatus required for the safe and easy operation of the lines and for their testing, inspection and maintenance, such as feeder boxes, primary switch and fuse boxes, cut-outs, transformers, etc., each marked with their safe ampere carrying capacity, and all lines will be fully equipped with lightning arresters.

Each piece of this apparatus is to be located as directed, is to be of the best material and workmanship for the purpose and is to be set and secured in the best manner, and each is to be subject to the rigid inspection and test, aud to the approval and rejection of the engineer.

Guarantees.—All apparatus, materials, and workmanship herein specified and contracted for are, by the said party of the second part, hereby guaranteed against all electrical and mechanical defects, and defective workmanship for the space of one year from and after their completion and acceptance. The party of the second part also hereby guarantees that any of the lighting towers herein contracted for, when provided with six direct current arc lamps of 450 watts capacity each (2,000 nominal candle power) will illuminate any portion of a circle 3,000 feet in diameter, of which the tower is the center, sufficiently so that any ordinary watch may be read on the darkest night when the said towers are illuminated.   J. T. F.

PART IV

Illustrative Examples of Complete Contracts and Specifications.

EXAMPLES OF COMPLETE ENGINEERING SPECIFICATIONS, SO
FRAMED AS TO INCLUDE THE CONTRACT AND BOND,
TOGETHER WITH ALL THE GENERAL CLAUSES,
SO DRAWN AS TO BE DISTINCT AND
SEPARATE FROM THE
SPECIFICATIONS.

166. Contract and Bond Combined in One Document with the Specifications. It is often customary for corporations doing a great deal of work by contract to have a standard form of combined contract, specification, and bond, in which the contracting and surety clauses remain the same, and in which a large proportion of the general clauses remain unchanged, while the specifications proper vary in accordance with the different classes of work to be done. Of such an example is that given in the following article, this being the standard form used by the city of St. Louis. It will be noted that in this contract, the contractor is represented as the party of the first part, and the city of St. Louis as the party of the second part. In Part II of this work, wherein the general clauses of specifications were discussed, the party of the first part was supposed to indicate the employer, and the party of the second part, the contractor. It is, of course, a matter of indifference as to which custom is followed, so long as the document clearly defines the meaning of these terms.

270

In all the examples given in this portion of the work, the subjects of the clauses will be indicated by marginal titles. This is the common practice in all specifications, but it has not been followed in the previous portions of the work, since the examples chosen were fragmentary in their character, and did not seem to require this kind of indexing.   In actual practice, however, it is advisable to use these marginal titles for convenience of reference.   So also should the clauses be all numbered, as is done in the examples which follow, these numbers also having been omitted in the previous portions of this work, because of their fragmentary character.

167. Contract and General Specifications for Large Pumping Engines. The following complete contract and specifications was used in 1894 by the Water Commissioner of the city of St. Louis, in the letting of contracts for two large high service pumping engines.   They are what is known as general specifications, since they do not indicate any particular style of engine, and since no plans were drawn for the work.   It should be understood also that the city of St. Louis is obliged to let all public work by contract and always to accept the lowest bid or to reject all bids.   It has hitherto been customary for this city to prepare detail plans for all public work because of this provision requiring them to accept the lowest bid.   These specifications have therefore been drawn with the greatest care, and in such a way that the city may be able to accept the lowest bid without danger of obtaining an inferior product.   The gentleman who prepared these specifications is a thorough civil and mechanical engineer of about twenty years experience in the designing and operation of pumping engines, and therefore the requirements here embodied are likely to represent the latest and best practice.   They are given here, however, not for the purpose of being copied, but for the purpose of illustrating the care and foresight required in the letting of contracts under general specifications, in order that all the bidders may be placed on an even footing, and that even the lowest bid shall of necessity correspond to a first-class and in every way satisfactory result. In general, where it is obligatory to accept the lowest bid, it is advisable to have detail plans prepared. The privilege had been specially reserved, however, in the advertisement of this work, to reject all the bids, if none of them proved satisfactory, but the city is not allowed, under its charter, to reject a lower bid, and accept a higher.

Referring to clause D in these specifications and to the last portion of that clause, the wording here is evidently too inclusive. That is to say, the Water Commissioner would not be allowed by law to "decide all questions which may arise relative to the execution of this contract on the part of the contractor," with the condition that "his estimates and decisions shall be final and conclusive." See articles 12 and 13, part I, and article 109, part II, for a discussion of this question.

168. Contract and Specifications *for designing, furnishing and erecting at High Service Pumping Station No. 3, St. Louis, Mo., Pumping Engines Nos. 7 and 8, with Fixtures and all Appurtenances Complete.*

A   AGREEMENT made and entered into this ———— day of ———— ————, 18—, by and between ——————————————————————————, part   of the first part, and the City of St. Louis, party of the second part, *witnesseth :*

WHEREAS, The Board of Public Improvements of the said City of St. Louis, under the provisions of Ordinance No. 17006, approved December 30, 1892, and by virtue of the authority vested in the said Board by the Charter and general ordinances of the city, did let out unto the said —————— ————— the work of designing, furnishing and erecting, at High Service Pumping Station No. 3, St. Louis, Mo., Pumping Engines Nos. 7 and 8.

B   *Now, therefore,* in consideration of the payments and covenants hereinafter mentioned to be made and performed by said second party, the said

——————————————————— hereby covenant and
agree   to furnish and erect in the pump pits at High
Service Pumping Station No. 3, two pumping
engines, each of a capacity of ten million U. S.
gallons of water in twenty-four consecutive hours,
with all fixtures and appurtenances complete, and in
conformity to the requirements and conditions hereinafter specified.

Wherever the words "Water Commissioner" C
are used herein, they shall be understood to refer to
the Water Commissioner of the City of St. Louis,
and to his properly authorized agents, limited by the
particular duties entrusted to them.

Wherever the word "Contractor" is used herein,
it shall be understood to refer to the part   who
ha   entered into the contract to perform   the
work to be done under this contract and these
specifications, or the legal representative of such
part   .

To prevent all disputes and litigation, it is   D
agreed by and between the parties to this contract
that the Water Commissioner shall, in all cases,
determine the quantity and quality of the several
kinds of material to be furnished and work to be
done, the duty and capacity of the engines, and the
amount to be paid under this contract; and he shall
decide all questions which may arise relative to the
execution of this contract on the part of the Contractor, and his estimates and decisions shall be final and
conclusive.

The said part   of the first part hereby agree   E
that all materials and workmanship, of whatever
description, shall be subject to inspection and rejection by the Water Commissioner, and that the entire
work shall be done to his satisfaction. The said
part   of the first part further agree   that the Water
Commissioner may appoint such assistants as he
may deem necessary to inspect the materials to be
be furnished and the work to be done under this
agreement, and see that the same strictly correspond
with the specifications hereinafter set forth; and
that said Water Commissioner shall at all times have
the right to enter the works, shops, etc., where the
machinery is being constructed, for the purpose of
inspection and examination of the materials furnished
and work being done, and shall be afforded such
assistance as may be required to determine whether
the quality of the materials and the character of the

18

work are in accordance with the requirements and intentions of this contract.

F The part of the first part further agree that the materials used throughout the engines and ap- purtenances shall be of the qualities specified, and new and unused when put into the work, and that the engines and appurtenances shall be constructed and erected in the most workmanlike and substantial manner, and everything done and furnished neces- sary to complete and perfect the engines and appur- tenances according to the designs and intentions of this contract, whether particularly specified or not, but which may be inferred from the drawings and from this contract and the following specifications:

SPECIFICATIONS.

Work to be done. 1. The work to be done consists in making the design, furnishing general and detail drawings, constructing and erecting complete in place ready for service at High Service Pumping Station No. 3, St. Louis, Mo., two vertical triple expansion con- densing pumping engines. Each engine shall pump ten millions U. S. gallons of water in twenty-four hours.

GENERAL DATA.

Water Pressure................ ... . 125 pounds.
Steam Pressure... .... .125 pounds.
Elevation Bottom Pump Pit (City Datum 100). 90 feet.
Elevation Engine Room Floor .. . .118 feet.
Elevation Water in Wet Well (Approximate).110 feet.
Dimensions of Pump Pit......... ....56x57 feet.

PLANS.

Working de- tail plans. 2. A complete set of accurate and distinct detail working tracings, made in accordance with the general plans submitted by the Contractor with his proposal and approved by the Board of Public Improvements, shall be furnished by the Contractor and submitted to the Water Commissioner within four months after the award of the contract.

3. The tracings shall be of uniform size—25½ x39 inches—and shall have a clear margin of at least ⅜ of an inch.

4. The kind of material to be used in each and every part of the construction shall be clearly denoted in the tracings by different section lining or by distinct lettering.

5. The tracings shall show complete sectional outline and plan views, giving all necessary dimen-

sions and thickness of metal, radii of fillets and roundings in the various parts of the construction in plain and intelligible figures, and shall definitely state in printed letters, at all surfaces and details, the name of the parts and the kind of machine work and finish to be put upon them, thus enabling the machinery to be built and completed exclusively from blue prints taken from the tracings.

6. There shall be separate tracings showing the valve motion, as put together in working condition.

7. The tracings will be examined by the water To be Approved. commissioner, and if found in accord with this contract and specifications, will be approved; any change found necessary shall be at once made by the contractor to the satisfaction of the Water Commissioner.

8. The contractor shall also, within two months General Working Plans. after the award of the contract, furnish accurate and workmanlike general tracings, made in accordance with the drawings submitted by the contractor with his proposal, and filed in the office of the board of public improvements, and with the detail drawings approved by the Water Commissioner.

9. These general tracings shall show the position of the engines in the pits, with all required foundation piers and bolts, and all floors, girders, platforms, stairs, galleries, railing, pipes, stop valves and all appliances complete, giving all general dimensions required in the erection of the machinery.

10. If, during the construction, it be found Change of Design. expedient or necessary to change or modify the design of any of the details of the engines, working drawings showing the proposed changes shall be submitted to and approved by the Water Commissioner before any change is made.

11. All drawings rendered in any way incorrect through changes or modifications, must be completely replaced by new tracings.

12. Before the final payment for the engines, Book of Finished Drawings. the contractor must furnish and deliver to the water ings. commissioner a book of complete general and detail drawings of all parts of the engines and appurtenances, as built and erected.

The detail drawings shall show all details entering into the construction in sectional, outline and plan views, with all dimensions plainly written in neat and intelligible figures and names printed at

every detail, the kind of material used and the finish
of the various parts and surfaces.

The general drawings shall show the engines in
position in the pump pits in at least four different
views, viz.: Sectional side elevation, sectional end
elevation, contour or outline end elevation and plan,
and shall give necessary main dimensions, thickness
and kind of metals, location of foundation bolts, and
all important sizes of the machinery as erected.

These general and detailed drawings shall be
made on mounted double Elephant paper of a size of
25½x39 inches inside the margin lines, strongly
and substantially bound in book form, with the name
and date of the engines printed in gilt letters on the
covers of the book.

All drawings shall be accurately and neatly
executed in ink in a workmanlike manner and to an
appropriate scale. All sheets shall be uniformly
lettered and consecutively numbered and provided
with proper titles and headings.

DESIGN.

General Features.

Pit.   13. The two engines shall be designed to be
erected and operated independently in the south pit
of the engine house, which will be built by the city
of St. Louis, substantially as shown by the plans on
file in the office of the water commissioner.

Especial attention must be paid to the fact that
the engines will be used for direct pressure service.

14. Engines shall have ample space around all
their various parts for access and maintenance.

Suction.   15. The height of the water in the wet well
will depend upon height of water in conduit, which
will be approximately constant.

Steam.   16. The engines shall be designed for an
initial steam pressure of 125 pounds per square inch
and a water pressure of 125 pounds per square
inch.

Plunger.   17. The pumps shall be designed and constructed to deliver the stipulated quantity of water
at a plunger speed which will insure a smooth and
effective action of the pump valves, and all working
parts of the machinery, but in no case shall the
diameter of any pump plunger exceed 40 per cent.
of its stroke, or the plunger speed exceed 180 feet
per minute.

18. The arrangement and construction of the Balanced. engines shall be such that they will give equal steam cards on the up and down strokes.

19. The engines shall be designed and pro-Reliability,etc. portioned to have great working strength, stability and stiffness, and ample space around all parts for erection, repairs, lubrication, inspection and adjustment.

20. The steam cylinders and the plungers of Vertical. the engines shall be vertical.

21. The steam cylinders and the regulating Height. mechanism of the cut-off and valve motion shall be placed entirely above an elevation of 120 feet above datum.

22. The pump chambers and steam cylinders Frame. shall be rigidly connected and supported through the intervening frames and columns to make the whole construction of ample stability, strength and stiffness.

23. Each engine shall have vertical, single Plungers. acting outside packed plungers, and no construction will be allowed requiring internal stuffing boxes, glands or water packings in the pumps. All stuffing boxes shall be readily accessible for inspection and tightening up, while the engine is running.

24. The machinery shall be so constructed, Removal of Parts. supported and arranged that the pump chambers or any important part or piece of the substructure can be easily removed to such position that it can be hoisted out of the pump pit without necessitating the frame and fixed parts of the superstructure of the machinery being taken apart, disturbed or removed.

25. The two engines shall each be provided Condenser. with a surface condenser, of appropriate size and construction to maintain a steady vacuum, and designed to directly utilize the water discharged by the main pumps for condensation of the exhaust steam.

26. The contractor shall furnish and put up Attachments and Appurtenances. all pipes, valves, oil cups, drip pans, fittings and fixtures required to make the construction complete inside the engine room and pump pit, and shall furnish flanges drilled for connection on end of pipes near wall.

27. The various parts of the machinery shall Appearance. be of plain shapes and forms, adapted to their

specific purposes, insuring great strength and reliability with good mechanical effects.

Frame and Fixed Parts.

Expansion.   28.   The frame and foundation of the engines
shall be so designed that changes of temperature
can not alter the distribution of the loads on, or affect
the alignment of the members of, the frame, and,
where necessary, expansion joints shall be used.

Frame.   29.   The frame of the engine shall be designed
to have great stiffness and weight, so that it shall
withstand all working stresses with the minimum
vibration.   All bed plates or sole plates resting on
masonry shall have ample bearing surfaces to safely
distribute the working pressures.

Anchor Bolts.   30.   the machinery shall be substantially and
securely anchored and held in place with a sufficient
number of foundation bolts.

Castings.   31.   All castings shall be designed to avoid
sudden changes of section and of such forms as will
cool uniformly without shrinkage strains.

32.   At all flanges of castings there shall be a
reinforcement, or addition of metal, of at least 30 per
cent. of the regular thickness, which shall extend in
length or height at least twice the total thickness of
the metal at the reinforcement.   All flanges to be of
not less thickness than the total metal at the reinforcement.

33.   All castings must have good sized fillets at
all corners ; no small brackets will be allowed.

Reheaters.   34.   If reheaters are used they shall be designed
and constructed to be absolutely steam tight under
all working conditions to which they will be subjected, and must have proper heating area and space
and facilites for examination, repairs and renewals.

Jackets.   35.   If steam jackets are used they must be
secured to the steam cylinder in such a manner as to
allow free and easy expansion and contraction, without causing internal leakage of joints or derangement
of any description to jackets or cylinders, or undue
strains in any part; and must be arranged to insure
proper circulation of steam and ready removal of the
jacket water.

36.   All flat plates and surfaces acted upon by
water pressure must be substantially proportioned
and strengthened with a sufficient number of heavy
ribs, to make them of ample stiffness and strength to

safely carry the loads to which they will be subjected.

37. All handholes and manholes shall be of ample size, well fitted, and so constructed as to be readily opened and closed.

38. Priming and draining pipes and valves shall be provided for filling and emptying the pump chambers.

39. The condensers must safely stand all work- Condenser. ing stresses to which they may be subjected, without leakage or weakness of any description.

40. The condensers shall be constructed to Examinations and Repairs. give ample facilities and space for the examination, insertion and withdrawal of tubes and packing of joints. The tubes must be provided with perfectly tight and easily removable packings, allowing for expansion and contraction, without injury or leakage.

41. The condensers shall be so arranged that the amount of water passing through, or condensing surface, can be adjusted to suit varying temperatures.

42. Arrangement must be made for proper distribution and circulation of the exhaust steam and condensing water on the cooling surfaces of the condenser, without injurious impingement of the steam or condensing water.

43. All glands and washers used in the condensers shall be made of composition; all bolts and nuts (except stay bolts) used inside the condensers shall be made of Tobin bronze.

44. The condensers must be provided with all necessary auxiliary pipes, valves and tanks.

45. The hot well shall be set at the highest Hot Well. elevation in the pit which the design of the engines will permit.

46. There shall be effectual means and appatus provided for the separation of grease and oil from the condensed water before it is fed to the boilers.

47. The suction and discharge pipes shall be Suction and Discharge Pipes. thirty inches in diameter.

48. For each engine there shall be a single suction or inlet pipe, which shall be attached to the gate valve, furnished by the City of St. Louis, shown in the plans of the pump pits.

49. The discharge pipe for each engine shall be carried up to an elevation of 113.6, and then horizontally through and to a distance of two feet from the outside of the pump pit wall, and shall be provided with a drilled flange for connection to pump main.

Air Chambers.   50. Each engine shall be provided with air vessels of sufficient capacity to insure smooth, easy and equal action of the pumps.

By-pass.   51. Each engine shall be provided with a by-pass pipe, arranged to facilitate draining the pump mains and starting the engines.

Relief Valves.   52. Each engine shall be provided with a pressure relief valve designed and arranged to by-pass the discharge of its pumps when the pressure on the pump mains exceeds 125 pounds per square inch.

53. The pressure relief to be of sufficient capacity to by-pass total discharge of the engine.

54. There shall be platforms or galleries of cast iron plates or wrought iron open work at convenient locations upon the pump and steam ends, which will allow all of the operations necessary in running and maintaining the engines to be performed with the greatest safety and ease.

55. The Contractor shall design, furnish and erect iron stairways, landings and galleries leading from the top gallery down to the bottom of the pump pit, with all intermediate galleries and supporting girders, beams, and composition railings required to make them complete and satisfactory in all respects. All of the above to be made of neat and harmonious proportions, and arranged to leave sufficient space for hoisting and removing the pump chambers and other parts of the machinery without disturbing any beams, bed-plates or other stationary parts, or necessitating the removal of stairways, landings or galleries to any great extent.

Light.   56. The galleries, stairs and platforms shall be arranged to secure as good diffusion of light down the pump pit as possible.

57. The stairs to be made without risers. Tread plates and all gallery plates to be made of a suitable open-work pattern.

All parts of stairs, galleries and platforms shall be accessible for inspection and painting.

Mechanism and Wearing Parts.

Strength and Stiffness.   58. All moving parts shall be of ample strength and of sufficient stiffness to prevent undue vibrations in operation.

Wearing Surfaces.   59. All journals and wearing surfaces shall be of sufficient size and of proper proportion to avoid excessive pressure and heating.

60. When practicable, provision shall be made Counter-bor-
ing. to prevent the wearing of shoulders on either stationary or moving parts at their extreme travel.

61. All stationary journals shall have suitable Journals. boxes, babbitt lined when necessary, and all journals above four inches in diameter shall have provisions for horizontal and vertical adjustment.

62. All glands and guide rings of stuffing Bushings. boxes shall be provided with composition linings forced in and securely held in place, and the glands shall be cupped out to make proper receptacles for lubricants, leakage water, etc.

63. The bodies of all valves, three inches in Valves, etc. diameter and smaller, shall be entirely of composition, but the bodies of valves larger than three inches, may be of cast iron, with composition valve and valve seats.

64. All valves, fittings, fixtures and appurtenances used, shall be of an approved design.

65. The valve motions and starting arrange- Steam End ments of the engines shall be such that each engine can be promptly and safely started and operated by one engineer.

66. The steam distribution valves shall be of Valves. a known reliable type. They shall be well balanced and so designed as to work with the minimum friction, to wear even and steam-tight, and to have proper facilities for refitting and adjustment.

67. The steam valve mechanism shall be of Valve Motion. ample strength and durability, and must be reliable in all its motions and entirely free from any danger of failure, derangement or rebounding. The engine Regulation. and valve mechanism to be provided with an automatic device to prevent racing in case of a broken pump main.

68. The engines shall be fitted with a variable Cut-off. cut-off mechanism so arranged as to be easily and quickly adjusted while the engines are in operation.

69. The running throttle valves of the engines Throttle. shall be of a well-balanced type and operate quickly and easily under full steam pressure.

70. The steam pistons of the cylinders shall Pistons. be provided with Babbitt and Harris piston packing, packing which, in the opinion of the Water Commissioner, is equally efficient.

71. Steam valves above six inches in diameter shall have steel stems provided with Phospho bronze nuts.

72. The area of the suction and discharge valves shall be sufficient to insure proper filling and discharging of the pumps under all conditions, but in no case shall the total suction valve area, or the total discharge valve area of each engine be less than 6 square feet.

Valves.
73. The valves shall be designed and constructed to open and close promptly and quietly, shall be tight and of ample strength, and shall be especially designed for facility of repairs and renewals.

74. All valve stems of stop and gate water valves shall be made of Tobin bronze.

Connecting Pieces.
75. All connecting, piston, plunger and distance rods, and all movable parts must be of ample strength and stiffness to withstand all working stresses.

Guides.
76. The piston rods, plunger and plunger rods, and all reciprocation parts have properly designed guides and crossheads. The crossheads shall have shoes adjustable for wear.

Boxes.
77. All journals and pins of connecting and valve rods, and of all reciprocating and oscillating rods, shall have well proportioned strap or box ends having easily removable composition boxes, Babbitt lined where required, and provided with wedges, keys or bolts for adjustment of wear. Each link or connecting rod shall at the different ends, have provisions for compensation of wear in the same direction

78. All strap or box ends shall be of a shape having great strength and stiffness, holding the composition boxes securely, and giving a neat and workmanlike appearance.

Locked Nuts.
79. All nuts of pillow block caps bolts and follower bolts of pistons, all screw joints of moving parts and all keys shall be provided with a secure locking device.

Fly-Wheel.
80. If a fly-wheel is used, the shafts shall rest in pillow blocks very securely and rigidly supported at ample distances apart.

Air Pumps.
81. The construction of the air pumps must be such that they will at all times perform their work promptly without noise or injurious shocks.

82. The air pump and all accessory pumps required to run the engine, except the boiler feed pump, shall be driven from the main engine.

MATERIALS.

83. All materials used throughout this construction must be of the special class and grade called for in the specifications and designated in drawings, and shall in each case fully stand the specified tests.

84. All castings shall be free from blow holes, Castings. flaws, scabs and defects of any description, and shall be smooth, close grained, sound, tough, and of true forms and dimensions.

85. All casting must be done in accordance with the best modern foundry practice to obtain castings of the very best quality. Castings above 500 pounds in·weight shall be moulded in dry sand or loam. Great care must be taken to make all castings as nearly as practicable of uniform thickness throughout.

86. No plugging or other stopping of holes or defects of castings will be allowed.

87. The cast iron used in the steam cylinders, Cast Iron. the steam distribution valves, the barrels of air pumps and the water plungers shall be close, fine grained, hard and uniform in character and of good wearing qualities. The cast iron used in all other parts of this construction shall be of superior quality, tough and of even grain, and shall possess a tensile strength of not less than 22,000 pounds per square inch. Test bars of the metal 2 inches by 1 inch, when broken transversely, 24 inches between supports and loaded in the center, shall have a breaking load of not less than 2,200 pounds, and shall have a total deflection of not less than 0.35 of an inch before breaking.

88. The test bars shall be cast as nearly as pos- Test Bars. sible to the above dimensions without finishing, but corrections will be made by the Water Commissioner for variations in thickness and width, and the corrected results must conform to the above requirements.

89. If any two test bars, cast the same day, show a tensile strength less than 22,000 pounds per square inch, or do not show the required cross breaking load or deflection, all the castings made from the melting from which the samples were taken may be rejected.

90. All steel castings used in the construction Steel. shall be thoroughly annealed and possess a tensile

strength of 65,000 to 75,000 pounds, and 15 per cent. elongation in two inches.

91. All steel forgings used in this construction shall be equal to forgings manufactured by the Otis Steel Company, Cleveland, Ohio, and have a tensile strength of not less than 75,000 pounds per square inch of section, and show an elongation of 20 per cent. in eight diameters.

Wrought Iron.   92. All of the wrought iron used shall be tough, fibrous and uniform in character, and specimens broken in the testing machine shall show a tensile strength of not less than 50,000 pounds per square inch, with an elongation of 18 per cent. in eight diameters.

93. If any specimen of steel or wrought iron shall not conform to the above requirements, all material of the lot from which the specimen was taken, will be rejected.

Bolts.   94. The Water Commissioner may take at random any wrought iron bolt and nut, and have it broken in a testing machine. If any two bolts shall not fill the above stipulated requirements for wrought iron, the whole lot of that size and make may be rejected; the effective area used in computing the breaking strength, will be the area corresponding to the smallest diameter at the bottom of the threads, when cut in accordance with the U. S. standard.

Rivets.   95. Rivets shall be made from the best refined iron, and must be capable of being bent cold until until the sides are in close contact without sign of fracture on the convex side.

Shapes.   96. All rolled wrought iron shapes shall be free from twists, bends, seams, blisters, buckles, cinder spots or imperfect edges. All sheet and plate iron must be capable of being worked at a proper heat without injury.

Rods.   97. All rods shall be formed in one continuous rolled or forged piece without weld.

Composition.   98. All the composition metal used [excepting for Tobin bronze and hand railing] shall consist of the best quality, new material only, of mixtures specially adapted for the work in each case, and approved by the Water Commissioner.

Phosphor Bronze.   99. All Phosphor bronze used must be homogeneous and uniform in character, and shall have a tensile strength of not less than 30,000 pounds per square inch, with an elongation of 15 per cent. in eight diameters.

100. All Tobin bronze used must be homo- Tobin Bronze. geneous and uniform in character, and specimens broken in a testing machine shall show a tensile strength of not less than 60,000 pounds per square inch, and an elongation of 20 per cent. in eight diameters.

10:. Finished bolts and nuts of Tobin or Phosphor bronze may be tested in the same manner as specified for wrought iron, and if any two bolts shall not fulfill the requirements, the whole lot of that size and make will be rejected.

102. Test specimens and samples of castings, Test Bars. forgings, composition or any other material used in this construction, shall be prepared ready for testing and supplied in the number, shape, finish and sizes required by the water commissioner, and shall be prepared as may be directed at any time during the pouring or working of the materials.

For all material taken by the water commissioner for testing, the following prices will be paid, which shall include the cost of preparing and finishing the test specimens, viz. :

For all wrought iron or steel, the sum of ten cents per pound.

For all composition, the sum of thirty cents per pound.

For all cast iron, the sum of three cents per pound.

All broken material to belong to the city of St. Louis.

103. The Babbitt metal used throughout the Babbitt Metal. construction must be of the following approximate proportions by analysis: 88 per cent. pure tin, eight per cent. antimony, and four per cent. Lake Superior copper.

104. All rubber for valves and gaskets must Rubber. be of a suitable quality, approved by the Water Commissioner before it is used.

105. All other material used in the engines Other materials. and not mentioned in these specifications will be subject to inspection, test and approval by the Water Commissioner before it is used.

CONSTRUCTION.

106. The workmanship and finish of the Workmanship. pumping engines throughout shall be equal to the best American practice, and in every respect satisfactory to the Water Commissioner.

Machine Worked.'   107. All surfaces worked in machine tools must be true and smooth, and accurately conform to the drawings in shape, size and alignment.

108. The bearing surfaces of all sole and bed plates and parts resting on masonry shall be planed.

109. If fly-wheels are used, the parts shall be fitted and fastened together in the most careful and workmanlike manner and the outer circumferences and the sides of the rim shall be turned smooth and true.

Joints.   110. All joints of bed plate and frame to be planed or faced and carefully fitted.

Boring.   111. The steam cylinders shall be bored in a vertical position, perfectly smooth and truly cylindrical, with a boring bar of proper diameter.

Turning.   112. All circular flanges shall be faced on the outer circumference.

113. All centers of lathe work must be made of ample size and carefully preserved.

114. All corners in journals and elsewhere in turned work shall be rounded to proper radii.

Joints.   115. All steam joints shall be made in an approved manner, with a very thin gasket of Jenkins' Usidurian packing.

116. All water joints to be made with rubber or paper gaskets, arranged with special care to prevent blowing out.

117. All seats of steam and water gates must be scraped and ground tight.

Journals.   118. All journals to be turned straight, cylindrical and smooth. Particular attention and care shall be paid to the proper fitting and scraping of all journal boxes, to make the same of an extraordinarily good bearing surface, and accurate fit to their housings or carrying members.

Straps, etc.   119. Straps, gibs, keys, reamed bolts and boxes of all connecting rods must be fitted with the utmost care and accuracy, and finished in a thorough and workmanlike manner.

Scraping.   120. The final fitting marks shall, for all parts, be preserved for examination and must in all cases be satisfactory to the Water Commissioner.

121. All journal boxes, pins, keys and other details of the machinery shall be taken apart at any time during the process of fitting or erecting, when the Water Commissioner so directs, to allow a thorough examination of fit and workmanship.

122. If gear wheels are used in the valve motion of the engines, they shall be properly designed and accurately cut in gear cutting machines. *Gear.*

123. The treads of cams and other parts of the valve motion subject to intermittent or sudden motion and heavy wear shall be of tempered steel or case hardened iron. *Cam Treads, etc.*

124. The tempering or hardening processes must be so conducted that parts will retain their proper size and shapes and have the requisite hardness. *Tempering or Hardening.*

125. All parts of the engines must be well secured and correctly centered with accurately fitted dowel pins, reamed bolts or male and female joints. *Centering.*

126. All flanges must be cast solid, and all bolt holes shall be drilled with perfectly sharpened and centered twist-drills to insure accurate round holes. *Bolting.*

127. All dowel pins must be of proper taper, and well fitted; and where necessary, shall have proper facilities for removal. *Dowel Pins.*

128. All holes intended to receive tapering parts shall be carefully reamed and ground and the tapering parts driven or forced into place. *Taper.*

129. Nuts and bolts and all threads shall be of the U. S. standard, except where special threads are necessary. *Threads.*

130. The threads and shanks of all bolts above $\frac{5}{8}$ inch in diameter shall be cut and turned in the lathe, and the ends of all bolts shall be finished to a neat conical or hemispherical point.

131. The resting surface for nuts and heads of all bolts shall be faced to present a smooth, plane surface, square to the axis of the bolt.

132. Case hardened, finished and polished nuts shall be used in all exposed work above the upper floor level, and also for all parts requiring frequent removal and adjusting. All other nuts and bolt-heads above the upper floor level, and nuts for all stuffing boxes, and at such other places as may be necessary, shall be finished. *FinishedNuts.*

133. Finished Phosphor bronze nuts and rolled Tobin bronze studs and bolts to be used for all fastenings inside the pump chambers, and for all glands of stuffing boxes of the pump end.

134. Cold pressed nuts shall be used for all stationary parts of the pump chambers, and in all cases where not otherwise specified. *Cold Pressed Nuts.*

Hexagonal.   135. All nuts and bolt heads shall be hexagonal in shape and must be faced on top and bottom. The sides shall fit their wrenches accurately.

Keys.   136. All key-ways and keys must be accurately fitted and properly driven or forced into place, and must be of appropriate size and taper.

137. All riveted work shall be specially designed for its particular uses, and executed in a thorough and workmanlike manner.

Calking.   138. All riveted joints subject to pressure shall be thoroughly and neatly calked with a round-nosed tool.

Finishing.   139. All connecting rods, links and valve rods shall be draw-file finished.

140. All bright and specially finished work must be of the highest grade and entirely free from scratches, specks and flaws.

141. All visible composition work shall have a bright finish.

142. All exposed machine worked surfaces of all parts above the upper floor level and of all moving parts, except fly-wheels, shall have a bright finish.

Lagging.   143. The steam cylinders, steam chests, reheaters, steam and distribution pipe and other heated surfaces of the machinery, when necessary, shall be protected by neat mahogany or walnut lagging, securely fastened and held in place by brass bands and button-headed brass screws, or by bright finished false covers.

Covering.   144. All steam pipes and heated surfaces shall be protected with approved non-conductors to the depth of flanges.

145. The material to be used in covering steam pipes, cylinders, reheaters and all protected parts, and the method of its application, shall be subject to approval by the Water Commissioner.

146. No non-conductors, lagging or false covers shall be applied until the construction has been thoroughly tested by working steam pressure and all leakages and defects developed have been thoroughly remedied.

ERECTION.

In Shop.   147. The Contractor shall erect in the shop such parts of the steam and water ends of the engines as may be necessary, in order that the final erection can be carried on with despatch in a thorough and workmanlike manner.

148. The Contractor, shall, at his own expense and risk, transport all parts of the machinery to the pumping station, but will be allowed the use of the power traveling crane in the engine house for erecting. *Transporting.*

149. All foundations and piers required for the support and anchorage of the engines, in addition to that shown in the city's drawings, will be built by the city of St. Louis, to drawings furnished by the contractor. All foundation piers will be built of first-class coursed cut stone masonry and provided with granite capstones of appropriate sizes, and charged to the contractor at $20 per cubic yard. *Masonry.*

150. The contractor shall deliver at the pumping station all bolts, washers, wall boxes, girders, etc., intended to be inserted in the masonry, in ample time to prevent delay during the building of the foundation walls and piers. *Wall Boxes, etc.*

151. The contractor shall be responsible for the proper and exact location of all parts, when placed in accordance with his drawings and templets.

152. The contractor shall do all work necessary to erect, fit and secure the engines in the pump pit upon the foundation piers as completed and built by the city of St. Louis. *In Pit.*

153. Every sole plate, girder, bed plate and casting resting on or secured to masonry, shall be provided with a rust joint of sufficient thickness, carefully driven and packed and consisting of ingredients satisfactory to the Water Commissioner. *Rust Joint.*

154. Great care shall be taken in the erection of the engines to place and secure the various sole and bed plates upon solid, plane and smooth bearings. All joints between stationary details must be made with the utmost accuracy and precision, insuring perfect and permanent alignment. None of the parts shall be unduly strained in lining up. *Bearings.*

155. The contractor shall so conduct his operations as not to interfere with the work of other contractors, and the disposal of his tools and materials during storage and erection will be subject to the approval of the Water Commissioner. *Other Work.*

156. The party of the second part will furnish and set the gate valves of the suction pipes, but the contractor shall pump out all accumulated water in *Water.*

19

the pump pit before commencing erection, and do all necessary pumping during erection of engines.

Protection of Parts. 157. All finished parts must be well protected in shops and during transportation to prevent injury and abrasion.

Damage. 158. All injured parts must be replaced, when in the judgment of the Water Commissioner, refitting will not suffice.

Cleaning up. 159. The contractor shall remove all staging used in erecting the engines, and leave the pump pit, engine room and premises neat and clean.

Damage to Masonry, etc. 160. The contractor shall, at his own cost, make good all damages to masonry, buildings, or other property of the city of St. Louis, occasioned by the contractor or his employes in the transportation and erection of the machinery.

Storage of Machinery Parts. 161. The city of St. Louis will furnish space within its premises for the reception of the various parts of the machinery, but shall not be responsible for the safe keeping of these parts, nor for damage caused to them from exposure or other cause.

PAINTING.

162. . All castings and details must be inspected and approved before painting, and in no case shall the paint or pitch be applied until all surfaces are trimmed and thoroughly cleaned.

Paraffine Varnish. 163. All unfinished iron work not visible from the engine room floor (except where otherwise required) and that above the floor intended to be encased, shall be thoroughly painted inside and out with three coats of No. 1 paraffine varnish, applied hot. The first coat shall be put on at the shop, and the others after erection, excepting for inside surfaces of pumps, pipes, etc., which shall receive two coats at the shop and one after erection.

Oil paint. 164. All unfinished iron work visible from the engine room floor, shall be thoroughly cleaned, rubbed down and painted with four coats of a good quality of paint and strictly pure linseed oil. The first coat shall be put on at the shop and the others after erection.

165. The paint shall be of a grade and color approved by the Water Commissioner, and shall be applied, striped and varnished to his satisfaction.

166. All parts to be covered by non-conductors must be thoroughly cleaned and freed from rust, and painted with three coats of paint of a kind,

color and quality to be determined by the Water Commissioner before application of the non-conductors.

167. All finished and polished surfaces must be kept entirely free from rust until erected and finally accepted. *Finished Iron Work.*

TESTING.

168. After erection has been completed, and before the final painting, a blank flange shall be bolted on the out-door end of the discharge pipe, and the whole construction tested with hydraulic pressure. A force pump shall be connected to the discharge pipe, and a pressure of 200 pounds per square inch applied in such manner as to test the pumps, pump valves, air vessels, discharge pipes, pump rods and the frames of the engines. After this test the engine is to be run to full capacity, discharging through the pressure relief valves for the purpose of testing same ; a further test to be made by suddenly opening gate on pump main to test speed controlling device mentioned in section 67. *Pressure.*

These tests must be conducted by the contractor with great care and in a manner satisfactory to the Water Commissioner.

The contractor shall furnish all labor necessary, and all piping, cocks, valves, gauges, force pumps, flanges and appliances required in the tests.

169. For the purpose of determining the duty of the engines furnished under this contract, there shall be an expert duty test of twenty-four hours continuous run for each engine. These tests shall be conducted by three experts, one to be selected by the Water Commissioner, one by the contractor, and the two thus named to select the third. *Duty Test.*

The duty tests shall be conducted for one engine at a time, unless otherwise ordered by the Water Commissioner.

170. The water of condensation from all steam jackets and reheaters shall be gathered and its weight carefully determined, and it shall be charged against the engines during all of the duty tests.

171. The total weight of water fed to the boilers during the tests, shall be considered the amount of steam used when corrected for entrainment exceeding two per cent.

172. Steam used for running the boiler feed pumps during the duty tests will not be charged against the engines.

Expert Test.   173. The twenty-four hours duty test shall be made with the water in the wet well at an approximate elevation of 110, and shall be conducted by the experts selected in accordance with section 169 of this contract.

Speed.   174. If, in the opinion of the Water Commissioner, the speed of the engines at any time during the twenty-four hours test is such as to jeopardize their safety, he shall have the right to order them run at such reduced speed as will give a smooth and quiet action.

Head (h).   175. The head (h) to be inserted into the formula for computing the duty of the engines during the running test, shall be ascertained by attaching a gauge to the discharge pipe close to where it turns into and runs through the foundation walls of the pit, and by the elevation of the water in the wet well.

176. Any part or detail of the engines showing undue strain or weakness of any description, must be replaced, and all defects developed in these tests shall be corrected by the contractor to the entire satisfaction of the water commissioner.

ADDITIONAL APPLIANCES.

Wrenches.   177. The contractor shall furnish for all sizes of bolts a complete set of wrenches for each engine, accurately fitted to the respective sizes of nuts. The wrenches for all finished nuts about the engines shall have a bright finish and shall be marked with their respective sizes.

178. Each engine shall be provided with one steam gauge, graduated from o to 250 pounds, one vacuum gauge, one suitable steam gauge on each receiver (if such be employed in the construction), and one engine revolution counter; all of them to have brass cases, triple silver plated, and placed convenient for observation. The dials of gauges to be ten (10) inches in diameter.

179. Each of the air vessels of the pumps shall be provided with one glass water gauge of satisfactory design. The hot well for each engine shall be provided with a suitable, permanently attached thermometer of appropriate design.

180. The contractor shall furnish one steam Indicators.
indicator for each steam cylinder and three indicators
for the main pumps, and one indicator for the air
pumps. The indicators shall be the Thompson,
Crosby or Tabor.

181. Each steam cylinder, main and air pumps
of the two engines shall be provided with permanent
piping, fixtures and motion appliances for attaching
and working the indicators. All valves, cocks,
pipes and appliances for the attachment of the indicators to the steam cylinders and pumps shall be
made of composition, of ample size and complete in
every respect.

182. All journals must be provided with sight- Oil Cups,
feed oil cups. There shall also be brass drip pans etc.
or pockets at all journals and oiling places to catch
lubricants.

183. The steam cylinders shall be fitted with
sight-feed lubricators.

184. There shall be valves, pipes and drip
pans at all places where necessary, for receiving and
conveying water from stuffing boxes, etc.

185. The contractor shall furnish an extra set
of suction valves and an extra set of discharge valves
with all parts complete.

REPAIRS.

186. Near the end of the year of probation,
the Water Commissioner will make an examination of
the engines, and any part or detail found to be defective or injured through excessive wear, overstrain,
bad material or faulty design, shall be replaced by
the contractor, at his own cost and expense, to the
satisfaction of the water commissioner.

The said part of the first part further agree   G
that all the work contemplated and described in this
contract and the foregoing specifications, shall be
done in accordance with the general drawings
approved by, and on file in the office of, the board
of public improvements, and with the detail working drawings submitted to and approved by the
Water Commissioner. It is further agreed that the
drawings and specifications form a part of this contract, and that, if any discrepancies appear between
any of the drawings and the specifications, or between
any of the several drawings in themselves, such discrepancies shall be adjusted by the contractor to the
satisfaction of the Water Commissioner. And it is

further expressly agreed that the approval of the general and working drawings shall not in any case relieve the contractor from any of his responsibilities under this contract.

H   The said part of the first part hereby expressly agree that the inspection of materials and workmanship shall not relieve——————————of any of———————obligations to perform sound and reliable work, as herein described. And the said part of the first part further agree to repair or replace any defective part or piece of the pumping engines during one year from the end of the 24 hours running test, at his own cost and expense.

And it is further agreed that during the aforesaid year, the Water Commissioner may make all necessary repairs requiring prompt attention, and that the cost of such repairs shall be borne by the contractor.

I   And it is further agreed that any work not herein specified which may be fairly implied as included in this contract, of which the Water Commissioner shall judge, shall be done by the contractor without extra charge. The contractor shall also do such extra work in connection with this contract as the Water Commissioner may in writing specially direct, and the price for such extra work shall be fixed by the water commissioner, but no claim for extra work shall be allowed, unless the same was done in pursuance of a written order, as aforesaid.

J   The said part of the first part further agree that the work embraced in this contract shall be begun within one week after written notice so to do shall have been given to the contractor by the Water Commissioner, and continued (unless the said commissioner shall otherwise in writing specially direct), with such force and in such manner as to secure its completion within twenty-six months thereafter, the time of beginning, rate of progress, and time of completion being essential conditions of this contract. And the part of the first part further agree that if the pumping engines to be furnished under this contract are not completed at the time above specified, then there shall be retained by said second party, as ascertained and liquidated damages, the sum of fifty ($50.00) dollars per day for every day thereafter until said engines are ready for service.

The party of the second part agrees to have the **K**
pump pits ready for the commencement of the
erection of the engines within twenty months, and
to have the steam ready for testing and running the
engines twenty-three months after the date of the
above notice to begin work.

And the part of the first part further agree **L**
that   shall not be entitled to any claim for any hindrance or delay from any cause whatever in the
progress of the work, or any portion thereof; but
any hindrance or delay occasioned by the party of
the second part shall entitle said part of the first
part to an extension of the time for completing this
contract, sufficient to compensate for the detention,
the same to be determined by the Water Commissioner.

The said part of the first part further agree **M**
that will not sublet the work to be done under this
contract, but will keep the same under   control, and
that will not assign the same by power of attorney
or otherwise, and that will at all times have a representative present where any work is in progress
under this contract.   Whenever it may be desired to
give directions, orders will be given by the Water
Commissioner and obeyed by the contractor's representative who may have charge of the particular
work in reference to which orders are given.   If any
person employed by the contractor on the work
should appear to the Water Commissioner to be
incompetent or disorderly, he shall, upon the requisition of the Water Commissioner, be at once discharged and not again employed.

It is further agreed that if the part of the first **N**
part shall assign this contract, or abandon the work
to be done under this agreement, or shall neglect or
refuse to comply with the specifications or stipulations herein contained, the board of public improvements shall have the right, with the consent of the
mayor, to annul and cancel this contract, and to
relet the work or any part thereof; and such annulment shall not entitle the contractor to recover
damages on account thereof; nor shall it affect the
right of the City of St. Louis to recover damages
which may arise from such failure.

And the said first part hereby agree to pro- **O**
tect and defend and save harmless the said city of
St. Louis against any demand for patent fees on any
patented invention, article or arrangement that may

be used by said first part in the pumping engines furnished under this contract.

P   The said part   of the first part further agree to idemnify and save harmless the City of St. Louis from all suits or actions brought against the said city on account of injuries or damages received or sustained by any party or parties during the construction of the pumping engines, or by or in consequence of any negligence in guarding the same, or any improper materials used in the construction, or by or on account of any act or omission of the said part of the first part or   agents.

Q   The part   of the first part further agree that each engine furnished under this contract shall have a pumping capacity of ten million U. S. gallons in twenty-four hours. The capacity to be at a speed that will insure smooth and quiet action, and to be determined by the experts during the duty test.

R   The part of the first part hereby agree that the pumping engines furnished under this contract shall perform, during a running test of twenty-four hours, a duty of one hundred and twenty-five million foot-pounds per thousand pounds of commercially dry steam.

The part   of the first part further agree that in case either engine fails to perforn a duty of one hundred and 'twenty-five million foot-pounds per thousand pounds of steam, during the working test of twenty-four hours, ———— will pay to the party of the second part, as an agreed measure of damages for lack of efficiency of the engine, in the ratio of $2,500.00 for each one million foot-pounds which the duty falls below one hundred and twenty five million.

In case either engine exceeds, during the twenty-four hours working test, an average duty of one hundred and twenty-five million foot-pounds per thousand pounds of steam, the party of the second part agrees to pay to the part   of the first part, as a reward for the superior efficiency of the engine, an amount to be in the ratio of $1,000.00 for each one million foot-pounds which the duty comes above one hundred and twenty-five million.

S   On condition of the true and faithful performance of all the conditions of this agreement and specifications, the said party of the second part agrees to pay to said part   of the first part the sum of ———— dollars, subject to such additions or deductions as are authorized by the provisions and conditions of this contract, in full payment for all the work and materials, designs and drawings required by this contract, embracing the satisfactory construction and erection of such pumping engines and appurtenances as are herein defined and described in all their parts and requirements.

Payments on account will be made as follows, viz.:

a. On or about the first of each month, the **T** Water Commissioner shall cause an approximate estimate to be made of the value of the materials and word done, based on the total amount to be paid for the engines; from the amount so found he shall deduct 20 per cent. and all sums previously paid or retained under this contract, and certify the remainder as then due. Provided, however, that nothing herein contained shall be construed to affect the right of the City of St. Louis, hereby reserved, to reject the whole or any portion of the work aforesaid, should the said certificates be found or known to be inconsistent with the terms of this agreement, or otherwise improperly given.

b. When the twenty-four hours running test shall have been satisfactorily completed, the Water Commissioner shall make an estimate for the amount of the contract price, less 10 per cent., and all sums retained under this contract.

It is further agreed that the water commissioner shall have charge of and operate the engines furnished under this contract, during the twenty-four hours duty test, and the year following, and that the part   of the first part shall not be relieved or released thereby from any of —— obligations under this contract. .

At the end of said year, the pumping engines and appurtenances, if found to be in good working condition, shall be finally accepted, and the Water Commissioner shall make and certify a final estimate in favor of the first part   and the responsibility of said first part   shall then cease.

The said part   of the first part further agree   **U** that —— shall not be entitled to demand or receive payment for any portion of the aforesaid work or materials, except in the manner set forth in this agreement; nor until each and all of the stipulations hereinbefore mentioned are complied with, and the Water Commissioner shall have given his certificate to that effect. The party of the second part hereby agrees and binds itself to pay the said part   of the first part in cash, the whole amount of money accruing to the said part   of the first part under this contract, excepting such sum or sums as may be lawfully retained under any of the provisions of this contract hereinbefore set forth, upon the giving by the said part   of the first part to the party of the second part a release from all claims and demands whatsoever growing out of this agreement.

V

This agreement is entered into subject to the city charter and ordinances in general, and in particular to the following provisions of Article VI., section 28, of said charter, to wit:

"*a.*" The aggregate payments under this contract shall be limited by the appropriations made therefor.

"*b.*" On ten day's notice the work, under this agreement, may, without cost or claim against the city, be suspended by the board of public improvements, with the approval of the mayor, for want of means or other substantial cause. Provided, that on the complaint of any citizen and tax payer, that any public work is being done contrary to contract, or the work or material used is imperfect or different from what was stipulated to be furnished or done, the said board shall examine into the complaint and may appoint two or more competent commissioners to examine and report on said work, and after such examination, or after considering the report of said commissioners, they shall make such order in the premises as shall be just and reasonable, and what the public interests seem to demand, and such decision shall be binding on all parties. The cost of such examination shall be borne by the contractor, if such complaint is decided to be well founded, and by the complainant if found to be groundless.

W

Ordinance 16,514, approved December 22d, 1891, is hereby made part of this contract, and must be observed in all its provisions, namely:

SECTION 1. All contracts hereafter entered into wherein the City of St. Louis is a party, for the doing of any kind of work or labor for the City of St. Louis, including work on all public buildings, works and enterprises, shall contain the following terms and conditions: (*a*) That the men, persons or laborers who may be employed in the doing, prosecuting, or accomplishment of such work done by the contractor with the City of St. Louis, or any one under him, or any person controlling the said men, persons or laborers, shall not be required to work more than eight hours a day; (*b*) That in case of the violation of such provisions of such contracts, the mayor shall immediately declare such contracts canceled and forfeited, and the work being done under such contracts shall be relet in the manner provided for the letting of such work, and such contractor shall thereafter be ineligible to bid upon such work under such reletting, and the difference in the cost of doing such work under such contract so canceled and forfeited, and under such reletting, shall be sued for on the bond of such contractor so violating such contract.

For the faithful performance of all and singular the terms and stipulations of this contract, in every particular, the said————— part of the first part, as principal, and—————as securities, hereby bind themselves and their respective heirs, executors and administrators, unto the said City of St. Louis, in the penal sum of——dollars, lawful money of the United States, conditioned that in the event the said —————shall faithfully and properly perform the foregoing contract according to all the terms thereof, and shall as soon as the work contemplated by said contract is completed, pay to the proper parties all amounts due for material and labor used and employed in the performance thereof, then this obligation to be void, otherwise of full force and effect, and the same may be sued on at the instance of any material man, laboring man or mechanic, in the name of the City of St. Louis, to the use of such material man, laboring man or mechanic, for any breach of the condition hereof; provided, that no such suit shall be instituted after the expiration of ninety days from the completion of said contract.

In witness whereof, the said————part of the first part, as principal, and—————securities, parties of the first part, have hereunto set their hands and seals respectively, and the City of St. Louis, party of the second part, acting by and through the board of public improvements aforesaid, have subscribed these presents the day and year first above written.

WITNESS:

————————————   —————————————[seal]
————————————   —————————————[seal]
                 —————————————[seal]
                 —————————————[seal]

The City of St. Louis by——————
                President Board of Public Improvements.
Countersigned: ——————————
                                    Comptroller.

CITY COUNSELOR'S OFFICE.
            St. Louis,————————18—
    The foregoing Agreement and Bond are in due
form according to law.

——————————

                                    ·   City Counselor.

MAYOR'S OFFICE.
            St. Louis,————————18—
    I hereby approve of the Securities to the fore-
going Contract and Bond.

——————————

                                                Mayor.
                                    M. L. H.

**169. Complete General Specifications for Water
Tubular Boilers and Settings.** The following complete
general specifications for horizontal water tubular boilers were
used in connection with the engine specifications given in the
previous article, and the contract was let under similar contracting, general, and surety clauses. These portions are omitted
from these specifications for the sake of brevity. They were
prepared by the same gentleman who prepared the specifications in the last article, and are thought to represent an equally
good practice.   ·

1. The work to be done consists in furnishing
designs and plans, material, tools and labor, and
building, transporting and erecting complete in
place, ready for firing, in the boiler-house at Bissell's
Point, eight horizontal water tube boilers, the
boilers to be provided with all necessary valves,
gauges, breechings and connection to underground
smoke flue.

DESIGN.

2. The boilers to be of the type designated as
horizontal water tube boilers, designed and built
with special reference to easy access for cleaning
and repairing of both internal and external surfaces.
The boilers to be designed for natural draft of pres-

ent smoke stack. No stays or obstructions of any kind shall be placed inside of the water tubes.

3. The boilers to be designed for a working steam pressure of 140 pounds per square inch, with a factor of safety of six on minimum sections.

4. Each boiler shall have a total tube heating surface of not less than 3,000 square feet, and a grate area equivalent to 75 square feet of straight grate.

5. The boilers to be provided with smoke preventing furnaces, which shall effectually stop smoke while burning southern Illinois coal at a rate of from twenty (20) to twenty-five (25) pounds per square foot of grate per hour. The furnace shall be some well tested and approved device for prevention of smoke, which does not use a steam jet or a system of brick arches in the fire box.

6. The boilers to be set in four independent batteries, as shown on drawing, each boiler to be provided with walls, settings, valves, gauges, smoke breeching and dampers necessary for operating or repairing independently of other boilers.

7. The fire fronts shall be designed to facilitate firing and removing ashes. The fire doors to be of suitable design to secure the regulation of air admitted to the fire, and prevent radiation through the fire door openings during regular service. The boiler dampers to be arranged to regulate from front of boiler.

8. Each boiler to have an eight-inch stop valve, Fittings. admitting of independent connection to main steam pipe.

To each boiler there shall also be attached, besides the eight-inch stop valve, two three and a half inch improved pop safety valves, placed in such positions that their escape pipes will not interfere with the roof trusses or sky-lights of the boiler house.

9. All steam drums to be made of steel plates of the quality hereinafter specified.

10. The boilers to be set and supported in a manner admitting of expansion and contraction of the same, without injury to the brick work or boilers in any way.

All beams required to support or carry the boilers to be of ample strength, and must be either wrought iron or steel.

There shall be central air spaces in all walls enclosing the boilers.

FITTINGS AND APPURTENANCES.

11. The contractor shall furnish and put in place all necessary valves, steam gauges, water glass gauges, safety valve escape pipes, and all appurtenances, and make connection to steam main, feed and blow-off pipes and underground smoke flue.

Wrenches. 12. For all nuts on the boilers and fittings, there must be furnished well-fitted wrenches.

Steam Gauges and Plugs. 13. The steam gauges shall be attached to the boiler fronts with nickel-plated brass siphon pipe and cocks, in a neat manner, admitting of easy removal.

The feed-water valve of each boiler to be provided with a suitable arrangement for its regulation from the front of the boilers.

Steam gauges to be brass case, nickel-plated, fourteen inches in diameter, maximum pressure 250 pounds, five-pound divisions.

14. Each boiler to be provided with three Bailey's safety copper cap fusible plugs, or other safety plugs of equally good manufacture and satisfactory fusibility.

Drains. 15. There shall be suitable copper spouts and polished brass piping wherever visible, to catch the steam and water from the gauge cocks and glass water gauges, and they shall be piped and connected to the ash box in an acceptable manner.

Valves. 16. Steam valves above six inches in diameter shall have steel stems, provided with phosphor bronze nuts, and the glands of all stuffing boxes shall be of composition.

17. All valves, fittings, fixtures and appurtenances used shall be of the best design.

18. The steam drums and all parts of the boilers and pipes not covered by brick work, and the breechings to be covered with magnesia covering, not less than one and a half inches in thickness, thoroughly secured in place.

19. Hand hole plates must be secured in an approved manner, to insure the greatest possible safety against accidents from breaking of fastenings.

MATERIALS.

20. All material used throughout this construction must be of the special class and grade called

for in the specifications, and shall in each case fully stand the specified tests.

21.   All plates in the boilers to be made of Steel Plates. steel.

The steel plates used in these boilers must be stamped with the maker's name and the tensile strength; to be homogeneous and of uniform quality, to have a tensile strength of not less than 55,000 pounds, nor more than 62,000 pounds per square inch, an elastic limit of at least 30,000 pounds per square inch, and an elongation of at least twenty-four (24) per cent. in eight inches.

Specimens must stand the following bending test, viz. :

To bend double, closing up completely without showing sign of fracture when bent cold, or after having been heated to a cherry red and plunged into water at 70 degrees Fahrenheit.

The water commissioner shall have the right to order test specimens 2x14 inches, to be cut out of any of the plates to be used in the boilers.

22.   All wrought iron for bolts, nuts or other Wrought Iron. purposes shall be double refined, and have an ultimate tensile strength of at least 52,000 pounds per square inch, an elastic limit of 26,000 pounds per square inch, and an elongation of eighteen (18) per cent. in eight inches.

23.   Rivets to be Burden's best, and must be capable of bending cold until the sides are in close contact, without sign of fracture ; and iron used for screw stays, stay bolts and braces to be of best quality of American manufacture.

24.   Tubes to be lap-welded of the best quality Tubes. of American manufacture, of a diameter of 3½ inches or 4 inches, and must stand a satisfactory hammer test.

25.   All castings shall be free from blow holes, Castings. flaws, scabs, and defects of any description, and shall be smooth, close-grained, sound, tough, and of true forms and dimensions.

Great care must be taken to make all castings, as nearly as practicable, of uniform thickness throughout, when not otherwise required.

26.   All cast iron used under steam pressure Iron Castings. shall be of good quality, tough and of even grain, and shall possess a tensile strength of not less than 22,000 pounds per square inch.

Test bars of the metal, two inches by one inch, when broken transversely, twenty-four inches between supports and loaded in the center, shall have a breaking load of not less than 2,200 pounds, and shall have a total deflection of not less than $\frac{35}{100}$ of an inch before breaking.

The test bars shall be cast as nearly as possible to the above dimensions, without finishing, but corrections will be made by the water commissioner for variations in thickness and width, and the corrected results must conform to the above requirements.

27. If any two test bars, cast the same day, show a tensile strength less than is required in these specifications, or do not show the required cross breaking load or deflection, all castings made from the melting from which the samples were taken may be rejected.

Specimens.

28. Test specimens and samples of castings and forgings, or any other kind of material used in this construction, shall be prepared ready for testing and supplied in the number, shape, finish and sizes required by the water commissioner, and shall be prepared as may be directed at any time during the pouring or working of materials.

29. The stamps put upon the steel sheets by the manufacturer must at all times be preserved for identification, and so placed as to be visible on the outside of boilers; if any stamp is cut out in process of manufacture, the water commissioner shall first replace it by a duplicate stamp.

WORKMANSHIP.

30. The best workmanship on these boilers will be exacted, and it must be equal in all respects to that executed in the best boiler works in this country.

31. All holes for bolts, studs and rivets in castings must be drilled. No cored bolt holes will be allowed.

No plugging or other stopping of holes or defects of castings will be allowed.

32. Any rivet which is deformed, cracked, burnt, improperly driven, leaky, or in any way injured, must be cut out and properly replaced.

33. All surfaces of sheets, and other parts to be riveted, must be brought together to close con- · tact and accurately fitted, with bearing surfaces

smooth and clean, and while being riveted to be held firmly in position and alignment without exerting injurious strains upon any portion or detail of the boiler.

34. The use of drift pens, to bring rivet holes to match, or come true and central, will not be allowed in the process of riveting, and must be dispensed with entirely. The utmost accuracy in punching the rivet holes will be exacted. Rivet holes failing to fit, or come fair and true, must be reamed out accurately, and rivets of suitable size used.

35. All sheets of the boilers must be satisfac- Sheets. torily straightened before being planed, bent, flanged, drilled, fitted, etc.

36. All scarfing to be done in a neat and workmanlike manner. Sufficient allowance of material must be made at all places where scarfs are required.

37. The edges of all sheets to be planed to a suitable bevel.

38. All seams to be caulked on both sides Caulking. where accessible.

All caulking to be done in the best manner, with round-nosed caulking tools; great care to be taken not to mar the sheet or rivets.

39. The threads of all studs, bolts, screw Threads. stays, stay bolts and nuts, to be chased with great care and skill, to insure uniformity in pitch and accuracy in fit.

All holes which are to receive bolts, screw stays, studs or stay bolts, to be accurately centered, drilled and tapped, to give a desirable fit and tightness of the threads.

The stay bolts, screw stays and studs to be entered, screwed in and riveted in a careful and workmanlike manner, to insure true and parallel surfaces and an equitable distribution of the stress upon all of the sustaining members.

40. All expanding of tubes and nipples shall be done in a careful and workmanlike manner, and shall be absolutely water-tight under the test pressure.

41. The fire, ash and cleaning doors to be Doors. fitted air-tight to their seating or bearing surfaces.

All holes in the lugs for hinges of the doors used in the construction to be drilled and reamed, to accurately fit the turned pins for same.

20

42. The brick work must be executed in a thorough and workmanlike manner, the brick used to be strictly first-class in every respect. Outside of setting to be laid with stock brick in white mortar; inside, where exposed to heat, to be lined with best quality fire brick.

43. All red bricks to be laid in mortar of approved quality, and all fire brick to be laid in ground fire clay.

FOUNDATIONS.

Foundations. 44. The city will furnish complete foundations for the boilers, the position in the house to be as shown on plans on file in the office of the water commissioner, and the space occupied by each battery of boilers to be not greater than that shown.

GENERAL CLAUSES.

Pressure Test. 45. The boilers shall be tested by the contractor with a water pressure of 210 pounds per square inch, under which they must be water-tight.

Paint. 46. When the boilers shall have been tested to the satisfaction of the water commissioner, they shall be thoroughly scraped, cleaned, dried and painted outside with one coat of linseed oil.

47. The fire front, fire and ash doors, and other cast and sheet iron parts, except grate bars, after approval shall be painted in the shop with one coat of paraffine varnish, and after erection they shall receive another coat of the same.

Erection. 48. The contractor shall, at his own expense and risk, transport the boilers and appurtenances to Bissell's Point, furnish all necessary labor, tools and appliances, and erect the same complete, as above specified.

Every possible and necessary care must be taken in handling and transporting the boilers, to prevent injury of any description to the same.

49. The contractor shall so conduct his work as not to interfere with the operation of any boilers under fire, and the disposal of his tools and materials, during storage and erection, will be subject to the approval of the water commissioner.

50. The contractor shall, at his own cost, make good all damages to masonry, buildings or other property of the city of St. Louis, occasioned by the contractor or his employees in the transportation and erection of the machinery.

51. The city of St. Louis will furnish space Storage. within its premises for the reception of the boilers and details, but shall not be responsible for the safe keeping of the same, nor for damage caused to them from exposure or other causes. ·

52. The city will remove the old boilers and prepare foundations below the floor line for new boilers, contractors to furnish castings to be set in underground flue for smoke connections.

53. The contractor shall get all finished materials on the ground at the earliest possible moment, and proceed with the erection of the same as soon as notified by the water commissioner.

The work of erection in place, ready for firing, shall be carried on continuously, night and day, and the contractor shall provide for that purpose three. complete erecting gangs.

If at any time during the erection the water commissioner shall be of the opinion that the work can be expedited by the employment of additional labor or tools, he shall order the contractor to make such increase in his working force or appliances as he may deem necessary to secure the most rapid progress possible; and it is especially understood and agreed that if the contractor fails to put the required force at work promptly, that the water commissioner shall employ such labor as he may deem necessary, and charge the cost of the same to the contractor.

54. The contractor shall bear the cost of making all repairs necessitated by defective materials, workmanship or design of the boilers and furnaces for the space of one year after the boilers are put into regular operation.   M. L. H.

170. Specifications for an Engine House. The following specifications for an engine house differ from those in the two previous articles inasmuch as they were accompanied by complete detail drawings. The contracting and surety clauses are here omitted, since they would be the same as those given in article 168. This engine house covers three large pump pits, designed for three sets of low service pumping engines, and it is entirely without a floor, nearly the entire space being occupied by the pits. The walls rest directly upon the natural rock, and an electric traveling crane is carried by a track ne..r the top of the two side walls, this crane spanning the entire opening and running the entire length of the building. The side walls, therefore, were made very strong and substantial.

Work to be done. 1. The work to be done consists in building and finishing complete Low Service Engine House at Chain of Rocks, St. Louis City Water Works Extension. The foundation on which the structure will rest is now completed. The work is shown in detail on the following drawings:

No. 1. Elevation of side walls.
" 2. " end walls.
" 3. Longitudinal section.
" 4. Transverse sections.
" 5. Plan below traveler.
" 6. " above "
" 7. Gallery plan.
" 8. Roof plan.
" 9. Roof plan for iron trusses.
" 10. Cut stone courses.
" 11. Details of stone faced door and window openings, terra cotta details.
" 12. Details of cut stone in cornice, fire walls and brick arches.
" 13. Details of windows, doors, ceiling and cast iron door sill.
" 14. Details of large sliding doors and hangers.
" 15. Details of door and window frames.
" 16. Details of sky lights.
" 17. Details of galleries, stairs, ladders, balcony and door sills.
" 18. Details of iron trusses.
" 19. Strain sheet.
" 20. Details of brick cornice, fire walls, etc.

MORTAR.

Sand. 2. All sand for mortar shall be clean, sharp, coarse, Mississippi river channel sand.

Cement. 3. All cement used in the masonry shall be H. H. Meier's Puzzolan cement, put up in well-made barrels.

4. It shall be subject to such tests as may be necessary to fully determine its character, and any cement which, in the opinion of the water commissioner, is unfit for the work herein specified will be rejected.

5. All short weight or damaged barrels of cement, or cement without the maker's brand, will be rejected without test. Samples for testing shall be furnished at such times and in such manner as may be required. On all barrels accepted inspection marks will be placed, and the contractor shall care-

fully preserve these marks and not allow them to be imitated.

6. All cement for use on the works shall be kept under cover, thoroughly protected from moisture, raised from the ground—by blocking or otherwise—and dry until used. The contractor shall keep in storage a quantity of accepted cement sufficient to insure the uninterrupted progress of the work.

7. Cement may be reinspected at any time, and, if found to be damaged or of improper quality, will be rejected. All rejected cement shall at once be removed from the line of work.

8. All mortar used in the masonry shall be **Mortar.** cement mortar, and shall be made of three parts of sand and one part of cement, each of the quality above specified. All mortar shall be made fresh for the work in hand, and any mortar which has begun to set shall not be used.

9. All brick in outer face of walls shall be laid **Colored** in mortar, colored with a red mortar stain that is **Mortar.** even in color and durable, and approved by the water commissioner.

STONE MASONRY.

10. The base, ashler and water table courses **Granite.** shall be of Missouri red granite, sound, free from discolorations, and of even color. All visible rock face shall be free from drill-holes or tool-marks. Base course and water table shall be six-cut work, ashler course, rock face.

11. Base course shall be 12 inches high, 8½-inch bond, with 4-inch by 4-inch chamfer on top. Ashler course shall be 1 foot 4½ inches high, 13-inch bond on the setting bed and 8½ inch bond on the top bed, and cut for iron anchors. Water table shall be 7½ inches high, 6-inch bond, cut for iron anchors and chamfered on top as shown. The ashler and water table course shall be anchored to the brick backing with tarred wrought iron anchors.

12. All of the granite work shall be laid in the most workmanlike and substantial manner, with even and equal joints, ¼ inch thick. Each stone must have perfect and level beds. All joints shall be pointed well and neatly with pointing mortar, colored red. Pointing joints must show equal size throughout, and be struck with pointing tool and straight edge.

13. Eight stones, 2 feet 0 inches by 15 inches by 18 inches, and ten stones, 2 feet 0 inches by 14 inches by 15 inches, shall be furnished and set as directed, to be used as bed stones for roof trusses; said stones shall be of granite, sound in all respects, top and bottom beds dressed true and level.

Sandstone.   14. Window sills, sill courses, belt course, coping, pediments, range work around door and window openings, and all cut stone work above the water table, shall be of Lake Superior red sandstone; fine Crandall finish, laid with equal and even ¼-inch joints in full beds of mortar. All joints shall be without chipping and beds of stone level and perfect. Spalls shall not be used in leveling any portion of the work. Window sills shall be cut with drips and seats, the seats not being cut to exact size until after the frames are set.

15. All the sand stone work shall be cut and set in the very best manner, and the whole cleaned down perfectly, and pointed with red pointing mortar, with concave joints, and backed up as soon as set.

16. The stone must be perfect in all respects, even color, free from all defects or pockets.

17. In cleaning down the work, care must be taken that the joints are rubbed to a level surface.

Limestone.   18. The stone bed course for the track of the crane shall be made of lime stone from approved quarries, dressed smooth on top bed, bush-hammered on face, and with true and parallel beds. This course shall extend the entire length of the building on each side, and it shall be 14 inches wide and 8 inches high, set in a swimming bed of cement mortar. When set same must be leveled *perfectly*, the entire length of the building, taking each side out of wind with the other. Special care must be exercised in cutting and setting this course. See detail sheet No. 20.

BRICK WORK.

Brick.   19. All the exterior faces of the walls, jambs, etc., shall be executed with even-colored dark red and hard brick. All other portions of the brick work executed with strictly red and hard quality. Light red brick shall not be used in any portion of the building, nor will salmon or defective brick be allowed in *any* part of the walls or on the premises.

20. Brick in exterior of walls shall be laid in *Face Walls.*
red mortar, with even and full bed and end joints,
struck with a concave tool, as the work progresses.

21. The standard height for laying all brick *Height of*
shall be 2 courses to 5 inches, unless otherwise *Courses.*
ordered by the water commissioner.

22. Figured thickness of walls will govern. *Thickness of Walls.*

23. The brick in every fifth course shall be *Bond.*
headers, and face work shall be laid to bond with
Flemish bond headers, as directed by the water
commissioner, during the progress of the work.

The different courses shall be slushed, and all
joints thoroughly filled with cement mortar.

All courses shall be laid to a line, front and
rear; plumb, true, straight and level.

24. All arches shall be turned with arch-brick, *Brick Arches.*
ground to proper radiating lines, and the face of
same shall be laid, alternately, $8\frac{1}{2}$ inches and $4\frac{1}{2}$
inches, and backed up with row locks laid with shove
joint. All jambs shall be returned and neatly
pointed. All arches shall be full depth of wall. Turn
brick arches over seats of each truss, as shown on
sheet No. 18.

25. Brick must be thoroughly wet before lay- *How Laid.*
ing, if required. Stone walls shall be well swept
off and sprinkled with water before any brick is laid
on them.

26. Cut a sufficient number of recesses through *Notches in*
the stone foundation walls for passage through same *Foundation.*
of the copper down-spouts, and build them in with
stone work, as shown on sheet No. 1.

27. All frames, anchors, wood, bricks, etc., *Setting Cut Stone above*
that are necessary shall be built in. *Water Table.*

28. All cut stone above the water table shall
be set, and the walls carefully leveled for the reception of the iron trusses. After the walls are built all
sills shall be under-pinned with red mortar.

29. All necessary wood plates for the fastening of tin flashing shall be built in.

30. All terra cotta shall be bonded firmly to the *Setting Terra Cotta.*
brick work and neatly pointed with red mortar at
completion.

31. Two iron I beams shall be built in and *Iron I Beams.*
covered with a $\frac{3}{8}$-inch plate, where shown on sheet
No. 3, in the side wall over the traveler off-set and
above the circle head windows, leaving the wall open
on the under side. so that the traveler can be carried
through this opening. After traveler has been set

in position the opening shall be closed up with brick work, leaving the I beams in the walls, but not exposed.

Cleaning Down.   32.   All exterior brick walls shall be cleaned of all dirt and mortar stains at completion. ·

TERRA COTTA.

Quality.   33.   All the terra cotta letters and border around same, on east and west walls, to be hard burned, best quality red, even in color, and of designs and dimensions shown, free from "flashing" or warping.

Moulds.   34.   The letters shall be first modeled and a plaster mould made, and from the mould the letters shall be pressed.

Fitting.   35.   After terra cotta has been burned it shall be laid out and carefully fitted and shaded and trimmed if necessary, after which each piece shall be lettered to correspond with a setting plan which shall accompany the delivery of all terra cotta. The details for the terra cotta will be found on sheet No. 11.

How Set.   36.   All terra cotta to be set in putty, colored to match, and properly bonded to the brick backing. The bricklayers shall set all terra cotta.

COPPER WORK.

Down-Spouts.   37.   Four 10-inch square down-pipes, 16-ounce copper, to lead water from roof and connect same with sewer, shall be furnished in place. Each down-pipe shall have square copper head of 20-ounce copper, and moulded copper bands of double thickness of 16-ounce copper placed not more than 4 feet apart, and secured with 3-inch copper holdfasts, with rosette heads.

Gutters, etc.   38.   Gutters shall be formed with roofing tin of form and size shown on drawings for same, and constitute a part of the roof-covering, and graded so as to carry the water from the center to the four corners of the building and open into the copper down-spouts.   Tin gutters shall be carefully flashed and counterflashed into the brick fire-walls, and nailed to wood strips provided for said flashing, as provided for in clause No. 44.

Finials.   39.   The copper finials for the skylights shall be furnished and secured in place.

40. Copper drip strips, 1¾ inches wide, 1 inch to project into the mortar joint, and ¾ inch to be exposed and bent to an angle of 30°, as shown on detail sheet No. 12, shall be furnished the bricklayers on the scaffold.   *Drip.*

TIN.

41. The roof shall be covered with roofing plates, standing seam, with joints well tacked, anchored and soldered, using rosin as a flux, and 8-pound soldering coppers, and tin well and closely cleated to roof.

42. The roofing plate used shall be Scott's IX extra coated American roofing tin plate, and must bear a coating of not less than 36 pounds to the box, and must be fully guaranteed, with the maker's name stamped in each sheet, and each sheet must be coated in perfect uniformity and free from "wasters."   *Tin Plate.*

43. The gutters shall be lined with Scott's IX roofing tin, flat seam, and shall conform to the gutter plan as shown on sheet No. 8, and shall be carefully flashed against the brick fire-walls, and be firmly nailed to the wood flashing strips, after which all of this gutter flashing must be well and carefully counterflashed.   *Gutters and Flashing.*

44. The wood flashing strips shall be built into the brick fire-walls 8 inches above the wall edge of gutter at center of roof and 24 inches above at each of the four corners.

45. The flashing around the skylights shall extend against and 8 inches up the wood skylight frame, and shall be finished before the carpenter lays the base.   *Skylight Flashing.*

46. Tin shall be painted, before being laid, with two coats of the best quality of iron oxide, ground in pure linseed oil, on the under side, and must be perfectly dry before laying; and that part of the roof covered with tin shall have two layers of heavy straw building paper laid over sheathing boards before putting down the said tin. Each layer of paper to overlap and be fastened down smooth and flat, and to be kept free from moisture. (See clause No. 51.)   *Paint and Paper.*

LUMBER.

47. All the lumber used in the construction of the building shall be graded as follows:

Carpenter Work.
Purlins—4-inch by 10-inch, yellow pine, long leaf, surfaced three sides and stub moulded.

Lower Roof Sheathing—1½ inches by 4 inches, tongued, grooved and beaded on under side, "B" select, surfaced one side.

Upper Roof Sheathing—⅞ inch by 8 to 10 or 12 inches No. 1 ship lap, surfaced one side.

Skylight Frame—Posts and plates, 5½ in. x 5½ in. yellow pine No. 1.

"   "   Rafters, 4 in. x 4 in. yellow pine No. 1.

"   "   Nailing girths, 2 in. x 6 in. white pine No. 1.

"   "   Outside casing, ⅞ in. "B" select.

"   "   "   base, 1½ in. "B" select.

"   "   Inside casing, ⅞ in. "B" select.

"   "   "   lining, ⅞ in. x 4 in. beaded white pine, "C" select.

"   "   Plinth and base blocks, 1½ in. "B" select.

Flashing strips—2 in. x 4 in., No. 1 white pine.

1st gallery floor girders.—3 in. x 8 in. and 2 in. x 6 in. No. 1 long California yellow pine, surfaced three sides.

1st gallery floor.—1½ in. x 3 in., first and second yellow pine, tongued and grooved.*

Mill Work. Tread board—2 in. x 12 in. white oak.

Window frames—"B" select.

Sash—"A" select.

Doors—"B" select.

Quality.
48.   All lumber must stand *strictly* on grade, kiln dried, free from large loose knots, sap, shakes, rot, stain or any other defects foreign to their respective grades.

CARPENTER WORK.

Roof.
49.   The roof shall be covered with two courses of sheathing.   The lower sheathing shall be white pine, tongued, grooved and beaded, and shall be 1½ inches thick by 4 inches wide; under side dressed and smoothed at the bench to a perfect smooth surface, and fastened to place and left free from hammer-marks or other defects.   Upper sheathing shall be $\frac{7}{8}$ inch by 8, 10 or 12 inches wide, No. 1 ship lap, laid diagonally, and nailed to the lower sheathing.   Care must be taken that nails shall not go through the lower sheathing; the nailing to be in the purlins.

50.   All purlins shall be 4 inch by 10 inch long leaf yellow pine, surfaced three sides, stub moulded, dry, sound and straight grained.   They shall be spaced on centers, as shown on detail sheets Nos. 18 and 9, and secured to upper cord of truss by angles and bolts, as detailed.

* There is no floor proper in the building—only a narrow gallery around the sides.   AUTHOR.

51. Cover the entire roof with two (2) layers of heavy straw building paper, laid over the ship lap sheathing before putting down the tin. Each layer of paper to overlap and be fastened down smooth and flat, and to be kept free from moisture. This work shall be performed by the carpenter, under the direction of the tinner, and laid in such sections only as required to keep in advance of the tinners. The upper sheathing, paper and tin shall be laid as fast as the lower sheathing is nailed in place, so as to protect the ceiling at all times from the weather.

52. Skylights shall be three (3) in number, **Skylights.** and framed in accordance with details for same, as shown on sheet No. 16 (this sheet shows details for the two end lights only; the center light shall be of same construction, but of sufficient length to reach the distance of two truss centers, as per longitudinal elevation and roof plan). The principal posts shall be of 5½ inch by 5½ inch yellow pine, and shall be fastened to purlins with wrought iron anchor straps firmly bolted to both purlins and posts. The upper end of all posts shall be tenoned.

53. Wall plates shall be 5½ inch by 5½ inch yellow pine, and mortised to fit the post tenons, and all fastened together with strap iron anchors and bolts.

54. The hip rafters shall be made of 4 inch by 4 inch yellow pine and dressed four (4) sides, and shall be backed same as for wood sheathing, and upon the top of plates between heels of rafters spike a triangular strip of wood secured rigidly to the plate to receive the thrust of the skylight bars. The center cage shall have the necessary rafters shown on sheet No. 8, and be firmly bolted at apex through a ridge piece of 2 inch yellow pine, top edge of ridge beveled.

55. Nailing girths shall be 2 inch by 6 inch white pine, No. 1, and be firmly spiked to the framing.

56. The outside shall be cased up with ⅞ inch "B" select, to form the finish above the base. A baseboard of 1½ inches thick, "B" select, beveled on top edge. shall run around the entire frame and be firmly nailed to the casing. See clause No. 45.

57. The inside shall be trimmed by casing up the posts with ⅞ inch thick "B" select, fluted and nailing on plinth and base blocks as shown. The inside below the window stool shall be ceiled with

⅞ inch x 4 inch beaded "B" ceiling, nailed on diagonally, with the nails countersunk and finished at the bottom with a 2-inch band mould.

58. (For specifications of skylight roof, see clause No. 112.)

Ceiling.

59. The roof ceiling shall be finished by running a mould along the truss and purlins, forming the different panels, as per detail sheet No. 13.

Gallery Floor.

60. The floor of lower gallery shall be made of 1½ inch by 3 inch tongued and grooved first and second yellow pine in continuous lengths. It shall be blind nailed to girders and the nailing joist.

61. The nailing joist shall be 2 inch by 6 inch No. 1 long leaf yellow pine, and in continuous lengths from bracket to bracket and bolted to channel bar.

62. The floor girders shall be 3 inch by 8 inch No. 1 long leaf yellow pine, surfaced three sides, notched on brackets. All joints shall rest on brackets.

Centers, Templets, etc.

63. The carpenters shall furnish all centers and templets, and shall put up and take down same. The centers and templets shall be made in a proper manner, strong and well braced.

Frames.

64. The carpenter shall set all frames, and verify their plumb after the brick arches are turned.

MILL WORK.

Window Frames.

65. Window frames below the traveler shall be solid frames for top and bottom pivoted sash, and shall be made of form and dimensions called for by the drawings. The lumber used in their construction shall be clear, dry, and sound Wisconsin white pine, "B" select, free from knots or sap. Faces of all frames shall be moulded as per detail.

66. All frames shall be given a heavy coat of paint all around, including back of jambs, and shall be set perfectly plumb; and the sill shall rest on a bed of cement mortar, ¼ inch higher on the inside, so as to make a water tight joint. Casings, mullions, transoms, etc., shall be moulded as shown on details. The frames shall be put together in a strong manner, well and closely nailed, and the stop-beads fastened with 1⅛-inch blued round-headed screws. All the lumber in frames shall be Wisconsin white pine, "B" select, as above specified.

67. The frames shall be provided with moulded stool, terminating with mould against plaster. Mullions and jambs shall be cut down square on stone seats.

68. Stiles, heads, mullions and transoms shall be solid. All circular portions of frames shall be worked in the solid and put together with white lead, so as to break joints throughout, and firmly spiked.. A 2 inch by 4 inch bond strip shall be spiked to all frames (except the two door and the two large window frames), extending from sill to spring of arch for anchoring same to brick backing. The heads of frames shall have wood blocks of 2 inch by 4 inch by 8 inche nailed to same and spaced a distance of 18 inches on centers.

69. The two door and window frames above mentioned shall be anchored to the brick backing with wood blocks of bone dry white pine 4 inch by 12 inch by 12 inch, built in the brick work, and spaced as shown on detail sheet No. 11. The frames shall be bolted to same with ¾ inch by 8 inch lag screws. The frames shall fit in a recess of one inch in the brick work.

70. Inside mould and stools will not be nailed in place until plastering is perfectly dry.

71. Two iron dowels shall be placed in the bottom of each jamb and mullion. The dowels shall be of 1-inch round wrought iron, and sunk 1½ inches in stone sill.

72. All frames above the traveler shall be solid frames for side pivoted sash and for 13-inch walls, and shall be made of "B" select.

73. All window sashes shall be of the form **Sash.** and dimensions called for by the drawings. The lumber used in their construction shall be clear, dry and sound Wisconsin white pine ("A" select), free from knots or sap.

74. All sash shall be moulded and rebated 2¼ inches thick, and divided into lights as shown. Each sash shall be neatly fitted and properly hung with Wollensak's plain bronze sash centers, No. 144, and shall be secured with bronze cupboard turns, and provided with casement rods or shutter holders. No. 8020, p. 1876, "Simmons."

75. All transom sash that are fixed shall be closely fitted and secured in place with heavy coat of white lead in the stop joints, so that all joints shall be water tight.

76. Skylight sash shall be 2¼ inches thick, and divided as shown on sheet No. 16, center pivoted and made water tight.

DoorFrames.

77. The door frames at each end of the building shall have rebated solid plank frames, beaded and moulded on outer face to match window frames. They shall be built in same manner as specified for window frames, using "B" select, and shall be secured to brick work in the same manner as specified for other frames. The frame for the double door, south end, shall be rebated for 2¾ inch doors, and shall have a transom bar 3⅞ inches thick. The single door at north end shall be made without transoms and shall have 1¾ inch rebate.

Doors.

78. The doors at the south end shall be double, and each door shall be hung with four 6 inch by 6 inch real bronze butts, rebated at center joint and beaded.

79. The doors shall be 2¾ inches thick and made of dry "B" select, with stationary sash divided in lights as shown ; lower portions of the doors shall be paneled and moulded as shown by detail sheet No. 13 ; the panels shall be made of tongued and grooved ⅞ inch "B" select, 2 inches wide, with all joints beaded and driven up in white lead.

80. The single door in the north end shall be made of "B" select, 1¾ inches thick. It shall have stationary sash panels, and be built as specified for south door.

81. Doors shall be secured with mortise locks, rebated for double doors, bronze knobs, plates and trimmings, and flush spring bolts, top and bottom of real bronze.

82. The sash in all doors and transoms for same shall be as specified for the window sash, and the transom sash pivoted and hung with the same kind and quality of hardware.

83. The large sliding doors shall be made of the same quality of lumber as specified for the small doors, and in two thicknesses of 1⅞ inches each, making a total of 3¾ inches, framed separately, and put together with white lead, and firmly screwed up with 3 inch screws, countersunk ; they shall have tenoned stiles, rails and muntins ; all tenons shall be double pinned with ½ inch white oak pins. and all shall be bolted together with iron rods, as shown by dot lines on detail sheet No. 14. The panels shall

be made of tongued and grooved "B" select, 2 inches wide and beaded both sides.

84. Small swing doors shall be framed into each large door, making four in all. They shall be hung with three (3) 4 inch by 4 inch real bronze butts, and fitted with Yale mortise locks, with keys to pass.

85. Each pair of the large sliding doors shall have wrought iron drop bars, made of 2 inch x ¾ inch iron, bolted at one end and made to drop into a hook at the other ; with a turned iron hand lift as shown on detail sheet No. 14.

86. An oak brace frame for each door opening, *Oak Brace Frame for Large Doors.* having form, size and radius as shown on detail sheet No. 14, rounded on each edge, shall be secured to the brick work with expansion bolts, in the most substantial manner, and framed into a girder made of 1¾ inch by 7½ inch oak, of length equal to the run of both doors, and firmly bolted to the wall with bracketed bolts, as detailed. All to be put in position before plastering, in the most careful and substantial manner.

HARDWARE.

87. The numbers and pages given for hard- *Windows.* ware refer to Simmons' Catalogue.

88. All pivoted windows shall be hung with Wollensak's plain bronze sash centers, No. 144, and secured with plain bronze cupboard turns, No. 8535, page 1650, and real bronze casement stays, No. 8020, page 1876. Pivoted transoms shall be fitted with similar sash centers, and with Payson's solid grip transom lifts, real bronze, ½ inch by 6 feet, No. 0336.

89. All doors, except large sliding doors, shall *Doors.* be hung with butts, and fitted with Yale mortise locks and keys to pass, and top and bottom bolts for the double doors.

90. The two double doors in south end shall each be hung with four 6-inch by 6-inch real bronze butts, and fitted with rebated mortise Yale lock, plain bronze, for 2¾ inch door, with keys to pass, and real bronze extension flush bolts, plain, same as Yale pattern No. 788E., B. 34, page 1675.

91. The single door in north end shall be hung with four 6-inch by 6-inch real plain bronze butts, and fitted with lock for 1¾ inch door of same kind as specified for double door.

92. The four small swing doors that are built in the large sliding doors shall each be hung with three (3) 4 inch by 4 inch real plain bronze butts, and fitted with lock as specified for the north door.

Finish. 93. The finish of all hardware shall be real plain bronze throughout, and all locks shall be Yale, with keys to pass.

Iron Work for Sliding Doors. 94. Construct for the two sets of large sliding doors a hanger and track as per detail sheet No. 14. The hanger shall be made of wrought iron 11 inches wide and $\frac{1}{4}$ inch thick, bent to correspond with profile shown, and fitted with a steel track-wheel, turned with a groove, and necessary bearings. The hanger shall be firmly bolted to the doors with heavy screw-bolts, as shown.

95. The track for above hanger shall be made of steel Z bars, 4.72 pounds per foot (Carnegie catalogue No. 295), with the short flange ground to fit groove in wheel, and the large flange firmly bolted to oak girder. The track shall be equal to the run of both doors. These doors must be made to run easy and work perfectly.

PLASTER.

96. The side and end walls from the stone foundation to the top of the brick walls in the inside of the building, including window-jambs and stools, shall be plastered with Acme plaster, and given a granulated finish with white sand, applied according to the directions of agent. All walls shall have straight, true surfaces, angles plumb, jambs and stools plastered. Walls shall be laid off to represent stone courses, as directed. Joints shall be marked off when plaster is green, and shall be $\frac{1}{4}$ inch throughout, and cut with clean edges, the joints to continue around window-jambs and to be struck to represent arch stones over openings.

97. A plaster base and wainscot moulding shall be made around the building of height shown, base 12 inches by 1 inch, chamfered on top; wainscot cap, 5 inches by 1 inch, moulded and chamfered top and bottom, as shown on sheets Nos. 3 and 4.

PAINTING AND GLAZING.

98. The contractor shall furnish all material and perform all labor necessary for the proper painting of the building. All sap, knots, etc., of the wood-work shall be covered with a good coat of strong shellac before priming.

99. All wood-work to be painted shall be Priming. primed with French ochre and boiled linseed oil, and all iron-work shall be primed with oxide of iron and boiled linseed oil. All holes and cracks in the wood-work shall be puttied and stopped on the priming coat, and again before applying the last finishing coat.

100. All outside wood-work, usually painted, Outside Wood-work. shall have four (4) coats of pure white lead, ground in linseed oil, and mixed with pure boiled linseed oil. The exterior of all frames, doors, sashes, skylights, etc., shall have the last two (2) coats in colors, as directed by the water commissioner.

101. All tin and galvanized iron shall have, Tin and Galvanized Iron. after completion, three (3) coats of Dixon's Silica Graphite paint, thinned with pure boiled linseed oil; each coat shall be allowed to dry thoroughly before the next is applied. Each coat of paint on the tin roof must be of a different shade, and each shade shall be approved by the water commissioner.

102. All of the inside wood-work, including Inside Wood-work. sash, doors and frames, etc., shall be painted four (4) coats of pure white lead, ground in oil, and mixed with pure boiled linseed oil, brushed on smooth and even, and grained a perfect oak on the last coat, after which it shall receive a heavy coat of coach varnish, evenly flowed on and left in the gloss.

103. The ceiling, including purlins, skylights, Ceiling. etc., shall be painted with four (4) coats of paint of quality specified above, and of such colors as the water commissioner may direct, and each succeeding coat must be of a different shade, as per direction of the water commissioner.

104. The first gallery floor shall receive four Wood Floor. (4) coats of pure boiled linseed oil, and the floor girders shall be painted to match the iron channel bar and brackets.

105. All iron and steel work before leaving Iron Work. the shop shall be thoroughly cleaned from all loose scale and rust, and after inspection be given one good priming coating of pure, raw linseed oil and iron oxide, well worked into all joints and open spaces.

106. In riveted work the surfaces coming in contact shall be painted before being riveted together. Bottoms of bed-plates, bearing-plates, and any parts which are not accessible for painting after erection, shall have two coats of paint.

21

107. After the structure is erected the iron work, both wrought and cast, shall be thoroughly and evenly painted with three additional coats of paint, of quality specified for the wood-work, mixed with strictly pure linseed oil, and each succeeding coat shall be of a different shade, and each shade must be determined and approved by the water commissioner.

Workman-ship.

108. The painter must see that all wood-work is perfectly clean before priming or painting, and putty up all nail heads and other defects, and sand-paper smooth and perfectly prepare all wood before applying a second coat. The whole of the painting work throughout to be done in the best and most workmanlike manner, and all paint and varnish spots must be cleaned off the glass, walls and galleries at the completion of the work, and all left clean and perfect, without exception.

109. All paint must be mixed at the building, and under the direction of the water commissioner, except the priming for the iron work.

Glazing.

110. All of the glass throughout shall be American, double thick, perfectly free from any blemish, flaw or defect. All shall be set in oil putty, carefully tacked with tin glazing tacks, and back puttied.

111. All glass to be cleaned after glazing, and again after painting sash.

SKYLIGHT ROOFS.

Manufacture.

112. The skylight roofs used on this building shall be of the Vaile & Young patent, and shall be adapted to the wood cage construction, as detailed on sheet No. 16.

Bars.

113. The bars shall be of galvanized iron, except the parts exposed to the weather, which shall be of 20-ounce copper, and the said bars must be rigid enough to support the glass without deflection. The apex shall not be finished to a point, but shall be fitted to the square of the size of the copper finial, and said finial shall fit over the apex and cover all joints. This finial shall be made of 18-ounce copper and furnished with the skylights.

114. All bars not resting on rafters shall be wrought iron, encased with galvanized iron.

Gutters.

115. All skylights shall have hanging gutters of 20-ounce copper, with a fall to one corner, and

from this corner the water shall be conveyed to the main roof by means of a copper down-spout, which shall be furnished with the skylight.

116. All glass used in the skylights shall be ⅜ Glass. inch thick and ribbed. It shall be furnished by the manufacturers of the skylights, and it shall be set with special care, and under rigid inspection, and shall be of a continuous length.

ROOF TRUSSES.

117. The castings shall be made from a Cast Iron. superior quality of iron, tough and of even grain, and must conform in shape and dimensions to the drawings. Castings must be clean and perfect, without flaw or sand holes or defects of any kind.

118. With the exception of the bearing plates, Soft Steel. the roof trussing shall be of soft steel throughout.

119. The steel must be uniform in character. The finished parts must be free from cracks on the faces or corners, and have a clean, smooth finish. No work shall be put upon any steel at or near the blue temperature, or between that of boiling water and of ignition of hardwood saw-dust.

120. All tests shall be made by samples cut from the finished material after rolling. All broken samples must show uniform fine grain fractures of a blue, steel grey color, entirely free from a fiery luster or blackish cast. Soft steel shall have an ultimate strength of 54,000 to 62,000 pounds per square inch; an elastic limit not less than 30,000 pounds per square inch, and a minimum elongation of 25 per cent. in 8 inches.

Before or after heating to a light yellow heat and quenching in cold water, this steel must stand bending 180 degrees to a curve, whose inner radius is equal to the thickness of the sample, without sign of fracture.

121. Specimen pieces of a size and form suitable for the testing machine shall be cut from any plate, angle or bar, when directed by the water commissioner.

122. If any specimen shall not conform to above requirements, all the material of the same form and manufacture as the piece from which this specimen was taken will be rejected.

123. All rivets shall be made of soft steel, and the steel for rivets must, under the above bending

test, stand closing solidly together without sign of fracture.

Specimen Bars.   124.   For all material taken by the water commissioner for testing there will be added to the final estimate the following prices, viz.:

For all steel, the sum of five cents per pound.

For all cast iron, the sum of three cents per pound.

All broken material to belong to the party of the second part.

Finish.   125.   The workmanship and finish throughout shall be thorough and of the very best, and any piece or part, however perfect it may be in other respects, if defective in workmanship, will be rejected.

Planed.   •   126.   That part of the bed plate on which rests the three eighth inch bottom plate of the truss shall be planed or faced to a true plane surface. All abutting joints in top and lower chord shall be planed or faced.

Punching.   127.   In punching rivet holes, the diameter of the die shall in no case exceed the diameter of the punch more than one sixteenth inch, and all holes must be clean cut, without torn or ragged edges.

Rivet Holes.   128.   All rivet holes shall be so accurately spaced and drilled or punched that when the several parts are assembled a rivet one-sixteenth inch less in diameter than the hole can be entered hot into any hole without straining the iron by drifting. Occasional variations shall be corrected by reaming.

Rivet Work.   129.   Whenever possible, all rivets must be machine driven. The rivets, when driven, shall completely fill the holes. The rivet heads shall be round and of a uniform size throughout the work. They shall be full and neatly made, and be concentric with the rivet holes, and thoroughly pinch the connected pieces together. The several pieces forming one built member must fit closely together, and when riveted shall be free from twists, bends or open joints. The angle irons forming the top chord must be bent at the different panel points to the proper angle. The lower chord shall have sufficient camber to allow for the deflection of the loaded truss.

Bolts and Nuts.   130.   All bolts and nuts to be made from the best quality of soft steel. The nuts to be hexagonal and the heads square. Heads, nuts and threads to be standard size. All bolts shall have a washer under the heads or nuts, where in contact with wood.

131. All rods with screw ends shall be upset Upset Ends. at the ends so that the diameter at the bottom of the threads shall be one sixteenth inch larger than any part of the body of the bar.

132. All the angles, filling and splice plates Angles, etc. must fit at their ends to the flange angles sufficiently close to be sealed, when painted, against the admission of water, but need not be boat finished.

133. To support and hold purlins in place, short Fastening and pieces of angle iron 3½ inches by 6 inches by ¾ inch Supporting Purlins. shall be riveted to principals with two ¾-inch rivets, and purlins shall be fastened to them by ¾-inch bolts. The contractor shall furnish all bolts, each with one cast iron washer.

134. All the bed plates under fixed and sliding Bed Plates and An- end must be fox-bolted to the masonry with 1¼ chors. inch bolts. The contractor must furnish all bolts, drill all holes and set bolts to place with cement.

IRON GALLERY, LADDERS, ETC.

135. The galleries shall consist of three differ- ent sections, as follows:
1st. A lower or first gallery. (Sheet No. 7.)
2d. An upper or second gallery. (Sheet No. 7.)
3d. A balcony gallery. (Sheet No. 4.)
Details for above galleries will be found on Sheet No. 17.

136. The first gallery shall extend around the First Gallery. entire building on a level with grade (El. 115), and shall consist of brackets, railing, chains, posts, and wood floor.

137. The wood floor shall be 4 feet and 2 inches wide and made of 1½ inch thick by 3 inch wide yellow pine flooring, and shall rest upon two girders and one channel bar, and both girders and channel bar shall be supported by cast iron brackets. The channel bar shall be 6 inches high, weighing 9½ pounds per lineal foot, and to this channel shall be bolted yellow pine nailing joists 2 inch by 6 inch; the bolts shall be ⅜ inch, with round head, nuts and washers, and shall be spaced three to each panel. The two yellow pine girders shall be 3 inch by 8 inch and notched so as to seat on the top of brackets.

138. Brackets shall be cast, according to detail, showed on sheet No. 17, and shall be firmly bolted to the stone-mason work with 1 inch by 10 inch expansion bolts, at top and bottom of each bracket.

The stone walls must be recessed sufficiently to give an even bearing for the backs of all brackets.

139. Railing shall be made of gas pipe and suitable fittings connecting same, made in accordance with details. Top and bottom rails shall be 1½ inch and 2 inch gas pipe; intermediate rails, 1¼ inch gas pipe; principal posts 2½ inch, and intermediate posts 2 inch diameter cast iron.

140. At the angle where the stairs commence this first gallery shall be constructed, on a radius, as shown on gallery plan, sheet No. 7, to make room for said stairway. A round hole must be made in this floor to suit stair column.

141. Suitable chain fastening gates shall be provided at all openings in gallery with suitable hooks, etc., chain to be of wrought iron ½ inch in diameter. There shall be two chains at each opening. See sheet No. 7.

Upper or Second Gallery. 142. The second gallery shall extend across south end of building, and terminate at one end with a spiral staircase, and shall consist of brackets, channels, railings, post and floor. The brackets shall be cast as per detail, shown on sheet No. 17, (scale, ¾ inch), and fastened to brick work by an expansion bolt at the foot and a bearing plate at the head. Upon these brackets shall rest a six inch channel bar weighing 9½ lbs. per lineal foot, and another bar of same size and weight shall be fastened to the brick work by expansion bolts. Upon these two channels the cast floor plates shall take their bearing. The railing, posts, etc., shall be made the same as specified for the first gallery. For a plan of this second or upper gallery, refer to sheet No. 7. The floor plates shall each be cast with three ribs; said ribs shall be spaced on centers, according to the length of the floor plates, and shall be located, one on each extreme edge and one in the center; all 3 inches deep and 1 inch thick.

Stairs. 143. Winding stairs shall consist of cast iron center column, treads, rail and newels.

144. The center column shall be cast ⅞ inch metal and be 7 inches in diameter, terminating at upper end with a newel, as shown on sheet No. 17.

145. The center column shall be supported by two 12 inch steel I beams, 42 pounds per foot, located diagonally across one corner of the stone foundation, with bolts and separators, and set in place before commencing the brick work. The col-

umn shall have a square iron flange on the lower end of 1 inch metal, and said flange must be firmly bolted to the steel I beams.

146. Steps or treads shall be cast without risers, but shall have thimble height of step, cast on each step, with tread nosing continued around.

147. These thimbles shall have freedom figured on drawing, and the vacant space shall be well and thoroughly calked with sulphur.

148. Steps shall be cast of ¾ inch metal, diamond pattern tread. Each step-thimble, bracket and flange shall be cast in one piece, each step being bolted to the next at connections. The first risers shall be housed into the wood floor, if necessary.

149. Stair rail shall be made of 2 inch gas pipe, bent to proper sweep and curve, terminating top and bottom at newels. Newels shall be cast iron ⅝ inch metal. All shall be executed according to drawings, each and every portion put up, bolted and secured in the strongest and most workmanlike manner, and to the satisfaction of the water commissioner.

150. The third or balcony gallery shall be constructed of wrought iron brackets, made of ½ inch by 2 inch metal, and fastened to the brick work with expansion bolts. It shall be provided with an oak tread board. This tread board shall be furnished by the carpenter and put in place by the gallery contractor. *Balcony Gallery.*

151. This balcony shall extend across north end of building, as shown on section plan No. 4.

152. Two wrought iron ladders with ½ inch by 2 inch sides and ¾ inch round rungs, passing through side pieces and riveted, shall be furnished, put in place and properly secured. One ladder to start on the first or lower gallery and extend up and through the balcony gallery as per drawings. One ladder shall be located on the exterior of the building and commence about 10 feet from the ground and extend upwards to and be anchored into the fire-wall coping, as shown on elevation sheet No. 1. The details for these iron ladders will be found on sheet No. 17. *Ladders.*

153. There shall be cast and set in place cast iron door sills for the doors in the north and south ends and the two large doors in each side. *Door Sills.*

154. Sill for the south door shall be 5 feet 8 inches long and 3 feet 4¾ inches wide, ¾ inches thick, and cast in diamond pattern, with door saddle and seats for wood frame drilled for ½ inch expansion bolts.

155. Sill for the north door shall be 3 feet 6 inches long and 3 feet 4¾ inches wide, cast same as specified for south door.

156. Sills for the large doors shall be cast diamond pattern, 1 inch thick, and shall have a square flange on outside and inside edge as shown. These sills shall be cast in three separate sections, as shown and figured in sheet No. 17.

Hand Rail. 157. A hand rail made of 1¼ inch gas pipe shall be provided and put in place and continued along both sides of the building its entire length, 3 feet 6 inches above the traveler I beam. This railing shall project from the wall 6 inches, and be firmly bracketed to the wall at sufficient intervals to insure ample stiffness. The ends shall be secured to the wood window frames. See sheet No. 3.

Traveler Track. 158. The traveler track shall consist of an iron I beam, 8 inches in height, and weighing 34 pounds per lineal foot, Carnegie catalogue, No. 8 C, page 22, extending the entire length of building on each side. It shall be firmly bolted to the stone sill course with ¾ inch expansion bolts, and the space between the web of beam and sandstone sill shall be filled with hard burned brick, laid in the best of cement mortar.

159. Upon the top flange of this 8 inch I beam a flat top steel rail, weighing 52 pounds per lineal yard, shall be bolted, extending the entire length on both sides of the building. This rail must be drilled in each flange, and these flanges bolted with ¾ inch bolts into the flanges of the I beam. The rail shall be connected at joints with fishplates and bolts.

I Beams in Side Walls. 160. Two 8-inch I beams, weighing 34 pounds per foot, with bolts and separators, shall be built in brick work, as shown on plan and specified in clause No. 31, and covered with a ⅜-inch iron plate.

GENERAL CLAUSES.

Finish Complete. 161. All of the materials and work required for the full completion of the building herein specified, to the entire satisfaction of the water commissioner, shall be furnished and done by the contractor, and should anything not mentioned within this specification be necessary to fully complete the work, the same shall be furnished and done without extra charge.

162.   No masonry work of any description *Frost.* shall be laid in freezing weather, except with special permission of the water commissioner.

163.   All unfinished work shall be properly protected from injury by frost.

164..   Any masonry work found damaged by frost shall be taken down and rebuilt at the cost of the contractor.

165.   When the work is completed, the build- *Cleaning up.* ing, substructure and surrounding grounds shall be cleared of all rubbish caused by construction, and left in a neat and presentable condition for immediate use.

166.   Measures shall be taken by the contract- *Public Safety.* or, whether required by city ordinance or not, to insure the safety of the public, by such precautions of fencing, watching, lights, etc., as the exigencies of the case may call for.

167.   The contractor shall furnish, at his own *Erection.* cost and expense, all necessary centering and scaffolding, and remove same at the completion of the work.

168.   Due facilities must be afforded the water commissioner for giving the lines, grades and points, and all stakes or marks given by him must be preserved undisturbed.

169.   The contractor shall keep on the work, accessible at all times, the plans furnished him by the water commissioner, and a copy of these specifications.

170.   At all times, when work is in progress, there shall be a foreman or head workman on the grounds.

171.   Necessary conveniences shall be constructed for the use of the contractor's employees, and during the progress of the work herein specified the contractor shall not use or interfere in any manner with the present buildings, pipes or appurtenances of the waterworks.

172.   The use of the railroad tracks and switches belonging to the waterworks will be permitted to the contractor for the work herein specified at such times only as will not interfere with the delivery, switching and handling of coal cars.

173. Particular care must be exercised in the protection of all finished work as the building progresses, such as exterior projections, cut stone, iron stairs and galleries, etc., which must be fully protected from injury or defacement during the erection and completion of the building.

174. The erection shall be carried on in such manner as will in no way interfere with the erection, completion and operation of the pumping engines or machinery. The extra cost of handling the erection in this manner must be included in the sum bid for the work.

175. The directions of the water commissioner as to the disposition of building materials and location of sheds, temporary buildings, etc., must be strictly observed.

Examination of Work 176. Whenever required by the water commissioner, the contractor shall furnish all facilities and labor to make an examination of any work, complete or in progress, under this contract. If the work so examined is found defective in any respect, or not in accordance with this contract and specifications, the contractor shall bear all expenses of such examination and of satisfactory reconstruction. If the work so examined is found to be in accordance with the contract and specifications, the expense of the examination and reconstruction will be estimated to contractor at a fair price, to be determined by the water commissioner.   M. L. H.

171. General Specifications for Highway Bridges and Viaducts of Iron and Steel.

The following general specifications for highway bridge work have been prepared by Mr. G. Bouscaren, M. Am. Soc. C. E., who has had a very large experience in structural designing of a superior grade of railway and highway bridge work. These specifications were revised and reissued in 1890.

General Clauses.

Plans & Stress Sheets. 1. Structures shall be built in accordance with the general plans exhibited or furnished by the engineer of . . . .

2. Unless stress sheets and plans are also furnished by the engineer, bidders must submit with their proposals, complete stress sheets for the structure and detail plans showing the form and connections of each typical member.

3. The stress sheets must show for each member the total maximum stress or stresses caused by the dead load, the live load, the wind, and the effect of temperature, separately, and the dimensions and area of cross section; also the dead weight assumed in the calculation which must not be less than the actual weight of the structure as built.

4. Complete detail drawings must be submitted for approval of the engineer, and work shall not be commenced until the stresses and details relating thereto have been approved.

5. A copy of every approved stress-sheet and drawing shall be furnished without charge to the engineer within ten days after its approval.

6. All parts of the structure, excepting the floor timbers hereinafter specified under the head of "floor," shall be of iron or steel or both combined, as may be approved by the engineer. The kind of metal to be used for each member, or class of members, must be noted on the stress-sheet. Cast iron may be used in minor details at the discretion of the engineer.   Material.

7. Through bridges shall be built of two trusses, unless otherwise specified, and shall have a clearance above floor of not less than fourteen feet, measuring from top of floor to the lowest point of portals. The depth from center to center of chords of trusses shall not generally be less than one-eighth of the span. The depth of plate girders shall not generally be less than one twelfth of span. The length of span used in calculation of stresses shall be the distance from center to center of end pins for trusses, and the distance from center to center of bearings for plate or lattice girders.   General Dimensions and Provisions.

8. Through spans shall be designed, when practicable, with inclined end posts.

9. Iron trestles and piers shall have, when practicable, a width of base sufficient to give a moment of stability on the masonry, exclusive of the anchorage, greater than the overturning moment of the wind.

10. Provision must be made in all structures for the free expansion and contraction of all parts, corresponding to a variation of 150 degrees Fahrenheit in temperature.

Loads.

All parts of structures shall be proportioned to sustain the stresses produced:

11.   1st. By the weight of the structure itself, considered integrally and separately for each particular member.

12.   2d. By the live load I, II or III, as specified for each particular case on the general plan, considered in positions and conditions, namely:— continuous or discontinuous, standing or moving, giving the greatest results.

13.   3d. By the specified wind pressure, giving the greatest results.

14.   4th. By the effects of a variation of temperature of 150 degrees Fahrenheit.

15.   The bending effect produced on every individual member by the side pressure of the wind and the weight of the member itself shall be considered.

Dead Load.

16.   In determining the total weight of the structure for the purpose of calculating stresses, the weight of the iron shall be assumed at the rate of $1\frac{0}{3}$ pounds per lineal foot of bar of one square inch area. The weight of the steel at the same rate as for iron, with two per cent. added.   The weight of the timber shall be assumed at the rates of five pounds per foot, board measure, for creosoted timber, four pounds for oak and yellow pine, and three pounds for white pine.

Live Load.

17.   The live load shall consist of either class I, II, III or any other load designated by the engineer for each structure.

Load I shall consist of one hundred pounds per square foot of floor, and two concentrated weights of 10,000 pounds each, six feet apart at right angles with the direction of the bridge.

Load II shall consist of eighty pounds per square foot of floor, and two concentrated weights of 7,500 pounds each, six feet apart at right angles with the direction of the bridge.

Load III shall consist of sixty pounds per square foot of floor, and two concentrated weights of 5,000 pounds each, six feet apart at right angles with the direction of the bridge.

18.   The live load per square foot shall be applied to the width in clear, specified for the wagon way, and to the width in clear, specified for the side walks.   The distribution of the uniform load shall be considered continuous or discontinuous, such as may give the largest result.   The concentrated loads shall also be taken in position giving the largest result in conjunction with the uniform load.

To provide for the effect of impact and vibration, additions to the stresses produced by the above specified live load shall be made as follows:

19. Riveted connections of stringers and floor beams, and hangers two feet long or less—50 per cent.

20. Hangers and suspenders over two feet long $25(1+\frac{2}{l})$ per cent., where l=length of hanger or suspender.

21. Floor beams, stringers and other plate girders, $25(1-\frac{d}{125})$ per cent., where d=one half length of girder.

22. Web members of trusses, and trestle posts $25(1-\frac{d}{125})$ per cent., where d=distance of members from center of trusses.

23. Chords of trusses $25(1-\frac{d}{125})$ per cent., where d=one half length of span.

No addition shall be made where d in above formula exceeds 125.

Wind stresses shall be calculated: Wind Pressure.

24. 1st. For a wind pressure of thirty pounds per square foot on the exposed surfaces of floor, of both trusses and railings, and on a moving load surface of six square feet per lineal foot of bridge.

25. 2d. For a wind pressure of fifty pounds per square foot on the exposed surfaces of floor and of both trusses and railings, the direction of wind giving the largest surface being assumed in the calculation and the greatest results shall be taken in the proportioning of parts.

26. Coefficients of friction shall be assumed as follows: Friction.

	When acting to	
	Increase the Strain.	Decrease the Strain.
For wheels sliding on iron or steel rails............................	$\frac{25}{100}$	$\frac{15}{100}$
For plane surfaces of iron or steel..	$\frac{30}{100}$	$\frac{15}{100}$
For plane surfaces of wood on iron or steel..................	$\frac{50}{100}$	$\frac{25}{100}$
For steel rollers between plane surfaces of iron or steel..........	$\frac{5}{100}$	$\frac{1}{1000}$

Description and Dimensions of Parts.

All parts of structures shall be so proportioned, Limits of that the combined effect of temperature and of all Stresses per Square Inch.

the loads specified, except the wind, shall not cause the stress per square inch to exceed the following maximum limits:

27. In tension, $\begin{cases} \text{Rolled bars .. 12,000 pounds.} \\ \text{Plates and} \\ \text{shapes .... 10,000 pounds.} \end{cases}$

28. In compression for lengths less than 50 times the least radius of gyration . 9,000 pounds.

29. In shearing across fibers . 9,000 "

30. On pins closely packed, tension and compression on extreme fibers . . . 18,000 pounds.

31. On bearing surfaces . . 15,000 "

32. The bearing surfaces of pins and rivets shall be reckoned from the diameter, not from the semi-circle.

The stress per square inch in compression shall be reduced with the ratio of diameter to length of member whose length exceeds fifty times the least radius of gyration, according to the following formulæ:

33. For members with square bearings . . . . $R = \dfrac{9,000}{1 + \dfrac{1^2}{36,000 r^2}}$

34. For members with square bearing at one end, and pin bearing at the other . . . . . . $R = \dfrac{9,000}{1 + \dfrac{1^2}{24,000 r^2}}$

35. For members with pin bearings . . . . $R = \dfrac{9,000}{1 + \dfrac{1^2}{18,000 r^2}}$

36. For top flange of rolled I beams between supports . . . $R = \dfrac{10,000}{1 + \dfrac{1^2}{5,000 b^2}}$

37. For top flange of built I beams between supports . . . $R = \dfrac{9,000}{1 + \dfrac{1^2}{5,000 b^2}}$

Where R=Modulus of allowable stress per square inch of cross section.
l=Length in inches of member between support.

r=Least radius at gyration of cross section.
b=Breadth of top flange of girder in inches.

38. In tension . . . . . 14,000 pounds. For Steel.
39. In compression for lengths
less than fifty times the
least radius of gyration 12,000 "
40. In shearing . . . . . . 10,000 "
41. In bending on pins closely
packed . . . . 22,000 "
42. On bearing surfaces . . 18,000 "
43. For compression steel members whose
lengths exceed fifty times the least radius of gyration,
the stress per square inch shall be determined by the
same formulæ as prescribed for iron members, but a
sufficient number of tests shall be made on full size
members to determine the value of the constants in
the formulæ, with a factor of safety of 4.

44. On extreme fibers in bend- For Wood.
ing, tension and compres- (Oak & Yellow Pine).
sion . . . . 1,200 pounds.
45. On bearing surfaces transversely to fibers . . . 400 "
46. Members subjected to alternate tensile and Alternate Stresses.
compressive stresses shall be designed and proportioned to resist both.

The moduli of allowable stress per square
inch of such members, shall be:

47. In tension
$$
\begin{cases}
\text{For iron}
\begin{cases}
\text{For rolled bars} & 12,000\left(1-\tfrac{1}{2}\tfrac{s}{S}\right) \\
\text{For plates and shapes} & 10,000\left(1-\tfrac{1}{2}\tfrac{s}{S}\right)
\end{cases} \\
\text{For steel} \quad 14,000\left(1-\tfrac{1}{2}\tfrac{s}{S}\right)
\end{cases}
$$

48. In compression for
lengths less than fifty
times the least radius
of gyration .
$$
\begin{cases}
\text{For iron} & 9,000\left(1-\tfrac{1}{2}\tfrac{s}{S}\right) \\
\text{For steel} & 12,000\left(1-\tfrac{1}{2}\tfrac{s}{S}\right)
\end{cases}
$$

When s and S are respectively the smallest and largest
of the two maximum stresses regardless of sign.

49. For compression members, whose lengths
exceed fifty times the least radius of gyration, the
constant in the numerators of the formulæ, used to
determine the moduli, shall be:

For iron . . . . . . 9,000 $\left(1-\tfrac{1}{2}\tfrac{s}{S}\right)$

For steel . . . . . . P $\left(1-\tfrac{1}{2}\tfrac{s}{S}\right)$

Where P represents the numerical value derived from the results of tests on full size members.

50. In shearing of riveted connections:

For iron . . . . . . 9,000 $\left(-\frac{s}{2S}\right)$

For steel . . . . . 10,000 $\left(1-\frac{s}{2S}\right)$

Wind Stresses. 51. An addition of twenty-five per cent. to all specified limits of stress per square inch, shall be allowed for wind stresses. These limits shall apply wherever the wind stresses are added to the stresses due to other loads.

52. Independently of the wind stresses, lateral struts shall be proportioned to resist the resultant of an initial stress of 10,000 pounds per square inch on all rods attached to them. No lateral or sway rods shall be less than one inch diameter.

Eye Bars and Upset Rods. 53. The eye and threaded parts of all bars and rods must not be less in strength than the body of the bar. The shape, size and mode of manufacture of the heads of eye-bars shall be subject to the approval of the engineer. Welding the head to the body of the bar will not be allowed.

54. In upset-rods, the area of section at base of thread shall exceed sectional area of rod by not less than seventeen per cent.

55. Long tension members shall be supported at suitable intervals, to avoid rattling and undue stress by bending.

Compression Members. 56. When practicable, compression members in trusses shall be designed with pin bearings. Long compression members, supported at intervals by lateral bracings, such as top chords of trusses and trestle posts, shall be considered and calculated as columns with pin bearings.

57. The thickness of metal in compression where the stress per square inch is 9,000 pounds, shall not be less than one sixteenth$\left(\frac{1}{16}\right)$ of the distance between supports in line of stress, and one thirtieth $\left(\frac{1}{30}\right)$ of distance between supports at right angles to line of stress, nor less than one eighth$\left(\frac{1}{8}\right)$ of distance from edge of plate or flange to line of support. When the stress per square inch is less than 9,000 pounds, the limits aforesaid can be increased proportionately, but the thickness of metal shall in no case be less than one quarter$\left(\frac{1}{4}\right)$ inch when both faces are accessible for painting and five sixteenths $\left(\frac{5}{16}\right)$ inch when one face only is accessible.

58. The ratio of length to diameter shall not exceed fifty.

59. In built posts and struts, the segments shall all be of one length without break, and shall be rigidly riveted together to act collectively as one solid body.

60. The sectional area of rivets in one segment in a distance from the end as short as the specified minimum spacing of rivets will allow, shall be sufficient to resist the entire stress on that segment.

61. In lattice work, the distance between rivets shall not be more than length of segment of equal strength per square inch as the column itself. The sizes and stiffness of the lattice bars shall be proportioned to the weight and spacing of the segments joined by them, so as to resist bending from rough handling of the finished member, and all stresses that they may be subject to, from eccentricity of the line of pressure, and from the inclination or curvature of the segments.

62. Pins shall be proportioned to resist the bending as well as the shearing forces acting upon them. The limits of stress specified for shearing and for the pressure on bearing surface of holes, shall determine the number and size of rivets. The sectional area of rivets before driving shall be taken for the effective shearing area. The allowable stress per square inch on hand driven field rivets shall be only three fourths of the specified limits for shearing. *Pins and Rivets.*

63. Shall be so proportioned that the top and bottom flanges will resist the bending moments without considering the web, and the web will resist the shearing forces without considering the flanges. *Plate Girders.*

64. The thickness of web of girders shall not be less than one fourth inch. When shearing stress on same exceeds $\dfrac{9{,}000}{1+\dfrac{d^2}{6{,}000\,t^2}}$ (Where t = thickness of web, and d = vertical depth between flanges), stiffeners must be provided at intervals not greater than depth of girder. Stiffeners of sufficient strength to resist the shear shall be provided at the ends and at all points of concentrated load. Webs of plate-girders shall be spliced with a plate on each side. The top and bottom flanges of plate girders shall have the same gross area. When flange plates are

22

used, the flange angles must be as large and heavy as practicable.

Connections & Attachments. 65. Of all members shall be so designed, that the stress on each member can be correctly calculated. The lines of stress shall coincide with the lines of center of gravity of members, and intersect at the joint point. The strength of all connections shall be at least equal to that of the member or members which they are designed to connect. This shall be demonstrated by testing, if required by the engineer.

Bed Plates and Friction Rollers. 66. Bed-plates and bearing plates shall be truly planed on all sliding and rolling surfaces, and shall be so proportioned that the maximum pressure per square foot on masonry will not exceed 30,000 pounds. They shall be securely anchored against upward and sideway motion.

67. The rollers shall be of steel, and not less than two inches in diameter; they shall be truly dressed to a smooth finish. The pressure in pounds per lineal inch on friction rollers shall not exceed $700 \sqrt{d}$, (d) being the diameter of rollers in inches. The rollers and rolling surfaces of bed-plates shall be protected by approved wrought iron casings to keep out foreign matter.

When practicable, adjacent ends of consecutive spans shall have a common bed-plate.

Workmanship and Details of Construction.

General. 68. All workmanship shall be first-class in every particular.

69. As far as practicable, all parts shall be accessible for inspection and painting.

70. All members shall be free from undue twists and bends.

71. All parts working together as one member of the truss, shall be uniformly stressed.

72. Tensile stress shall be avoided in a transverse direction, and shearing stress in a direction parallel to the fibers of the iron.

Pin Holes. 73. Shall be bored, not punched, exactly perpendicular to the center lines of stress, and not more than one fiftieth ($\frac{1}{50}$) inch larger than the diameter of the pin.

Eye Bars. 74. Shall be straightened before boring. Bars working together shall be bored in one operation, piled and clamped together, and at the same temperature. The eye shall be in the center of the

head and on the center line of the bar. A discrepancy in length from center to center of eye, exceeding $\frac{1}{20000}$ of the length of the bar will not be allowed.

75. All forging shall be done at the tempera- Forging.
ture best suited to the kind and quality of the metal.
No work shall be done on iron or steel at a black
heat.

76. Steel eye bars, upset rods, and all pieces Annealing.
of steel which have been partly heated, or bent cold,
must be properly annealed.

77. Shall be turned true to size, and straight. Pins.
They shall be turned down to a smaller diameter at
the ends for the thread and driven in place with a
pilot-nut, when necessary to save the thread. There
shall be a washer under each nut.

78. No discrepancy in length of pins through
the bearing parts will be allowed.

79. The several members attached to the same
pin shall be so arranged as to produce the least
bending moment on the pin; they shall be held and
closely packed in position by filling rings between
them. Fillers shall be of wrought iron.

80. Abutting ends shall be planed or turned, Abutting
in a plane perpendicular to the line of stress. They Joints.
shall be in contact throughout, and held in position
by suitable splices.

81. All segmental joints in riveted work shall Riveted
be square and truly dressed, and in contact through- Work.
out. They shall be fully spliced, no reliance being
placed upon the contact of abutting parts. Sheared
edges of steel plates shall have not less than one
quarter inch of metal removed by planing.

82. In the effective area of riveted members,
pin, bolt and rivet holes shall be counted out for
tension; bolt and pin holes shall be counted out for
compression. Rivet holes shall be assumed to be
one eighth inch larger than diameter of rivets.

83. No tensile stress shall be allowed on rivets.
Rivets shall be used in preference to bolts for all
rigid connections to resist shearing. Where bolts
must be used in place of rivets, they shall be in
double shear; the holes shall be drilled or reamed
and the bolts shall be turned to fit tightly in their
holes.

84. Rivet holes shall be accurately spaced and
shall be drilled or reamed to fit exactly opposite to
each other without drifting. The space between the
edge of a piece and the edge of rivet holes shall be

such that the iron will not crack nor split by punch-. ing. It shall not be less than one and a half diameters of rivet. The pitch of rivets shall not be less than three diameters of rivet nor more than sixteen times the thickness of plate.

85. Rivet holes in steel, if punched, shall be reamed not less than one eighth inch larger than the die sides of the punched holes, and the sharp edge of holes under the rivet heads shall be trimmed.

86. When practicable all riveting shall be done by a machine capable of holding on to the rivet, after the upsetting is complete.

87. Rivets when driven, shall completely fill the holes.

88. Rivet heads shall be full size, well formed and concentric to the holes.

89. No loose rivets will be allowed.

90. All rivet holes for field riveting shall be reamed in place.

Adjustment.   91. All members requiring adjustment, shall be provided with adjusting screw-threads, nuts and check-nuts convenient of access. Where sleeve-nuts are used, they must be open so that the threaded lengths of rods engaged at each end can be verified.

Counter and Lateral Bracing.   92. When practicable, counter rods will be dispensed with, by designing all web members to resist counter, as well as direct stress. Preference will be given to a system of lateral bracing designed to resist compression as well as tension.

Portals.   93. The end posts shall be rigidly connected by riveted portals of approved design, as deep as the specified clearance above the floor will allow.

Vertical Lateral Bracing.   94. Shall be provided between posts at each panel in deck bridges, and in through bridges where practicable. This bracing shall be proportioned to resist the unequal loading of trusses, the effect of the wind on the moving load being taken into consideration.

95. The top chord of through girders or pony-trusses shall be rigidly braced sideways at the ends and at suitable intermediate points.

Bolts, Washers and Nuts.   96. Washers and nuts shall have a uniform bearing. All nuts shall be easily accessible with a wrench for the purpose of adjustment, and shall be effectively checked after the final adjustment. No round headed bolts will be allowed. All bolts through wood must be provided with wrought iron washers under head and nut: the use of more than one washer under the head or nut to make up for deficiency in length of thread, will not be allowed.

97. The plan of floor shall be of the kind A or B, or any other description specified for each particular case by the engineer. *Floor.*

98. The stringers shall be of iron or steel, *Floor A.* spaced not more than 3 feet 6 inches from center to center under the wagon-ways and capped with nailing pieces 5 inches thick of creosoted wood. The flooring of the wagon-ways shall be made of two thicknesses of plank, 3 inches thick each for "load I," $2\frac{1}{2}$ inches thick each for "load II," and 2 inches thick each for "load III," the under flooring shall be creosoted, and the top flooring of white oak shall be laid transversely, breaking joints with the under flooring. The top flooring pieces shall not exceed 8 inches in width. The sidewalk floors shall consist of 2x6 inch planks nailed on creosoted joists laid transversely on the iron or steel stringers and spaced not more than 2 feet from centers. The joists shall be fastened to the stringers so as to allow for the longitudinal motion of expansion and contraction. The under flooring of wagon-way shall be fastened to the nailing pieces with wrought spikes $7x\frac{7}{16}$ inches for load I, $6x\frac{3}{8}$ inch spikes for load II, and $5x\frac{5}{16}$ inch spikes for load III; the top flooring shall be fastened with 60 penny nails of the best quality for load I, 40 penny for load II, and 30 penny for load III. The flooring of the sidewalks shall be fastened with 30 penny nails of the best quality.

99. The joints shall be of creosoted wood, not *Floor B.* less than 3 inches thick, laid longitudinally with the line of the bridge for the wagon-way and sidewalks, and supported by the floorbeams. The flooring shall be made as for Plan "A" and fastened in the same manner, but the flooring of the sidewalk shall be laid transversely with the line of the bridge.

100. The joists of consecutive panels shall lap by each other on the floorbeams; they shall be bolted to the floorbeams so as to allow for the motion of expansion and contraction, and shall be fastened to each other by two five-eights ($\frac{5}{8}$) inch bolts or lag-screws; they shall be braced between floorbeams by rows of bridging spaced not more than 6 feet apart.

101. Guard-timbers not less than 8x8 inches shall *Floor General.* be securely bolted to the floor on each side of the wagon-way, to give a clearance of 24 inches between

the inside faces of the guard and the inside faces of the widest part of the truss for through bridges, and the inside lines of sidewalks for deck bridges and viaducts.

102. All floor timbers, guards and railings shall extend over all piers and abutments, and make suitable connection with the embankments at the ends of the structure. Cast iron aprons or cover-joints shall be provided at the free ends of spans if required.

103. Stresses due to the friction of the floor on the stringers and floorbeams, shall be considered and treated the same as stresses due to regular load.

104. Should one or more street railroad tracks be required on the bridge, the contractor shall furnish all necessary material for the same, and lay the track conformably to the instructions of the engineer. The tread of the rails must not deviate more than six inches from the center line of the stringers.

105. The floor of the sidewalks shall extend to and connect with the floor of the wagon-way so as to leave no open space between them. Openings shall be left at suitable intervals in the floor, if required by the engineer, to dispose of the sweepings on the bridge and of the drainage.

106. All framing shall be done to a close fit, and in a thorough and workmanlike manner. No open joints or filling shims will be allowed. All under and top flooring must be passed through the planer so as to secure a uniform thickness.

Railings.

107. Substantial iron railings of approved design, not less than four feet high, shall be provided for the outward lines of sidewalks. The railings shall be supported directly by the floorbeams and braced laterally with outside stays riveted thereto. Intermediate stays fastened to the stringers shall be provided at intervals not more than ten feet. The top flange of railings shall be proportioned to resist a transverse horizontal thrust of not less than one hundred pounds per lineal foot.

Camber.

108. The camber measured on the center line of the pins of chords shall not be less than one twelve hundredth ($\frac{1}{1200}$) of the span. The camber line shall not deviate from an arc of a circle, more than one quarter ($\frac{1}{4}$) of an inch at any place.

Anchorage.

109. All bridges and viaducts shall be sufficiently anchored to the masonry, to resist displacement by the strongest wind specified. The base of

all piers and trestles shall, when practicable, be sufficient to avoid tension under the most unfavorable circumstances of load and wind, but sufficient anchorage shall, nevertheless, be provided to resist · not less than one half ($\frac{1}{2}$) of the overturning moment of the wind.

110. All the necessary drilling and dressing of masonry shall be done, and all the necessary fastenings and anchorage provided and put in by the contractor without extra allowance.

Quality of Material.

111. Shall be double rolled, tough, ductile, Wrought Iron. uniform in quality, and shall have a limit of elasticity of not less than 26,000 pounds per square inch.

When tested in specimens of uniform sectional area of at least one half ($\frac{1}{2}$) square inch, cut out of the full size finished piece, it shall stand without breaking, the following tensile stresses and elongations in the distance of 12 diameters.

112. For bar iron 4 sq. in. and less in sect'al area, } 52,000 pounds.per square inch; elongation, 20 per cent.

113. For bars over 4 square inches in area, a reduction will be allowed in the strength per square inch and elongation of specimen of 500 pounds and one half per cent. respectively for every additional square inch of area of the bar, down to the limits of 48,000 pounds per square inch, and 16 per cent. elongation.

114. For all shapes and plates 24 in. wide and less, } 50,000 pounds per square inch; elongation, 15 per cent.

115. For plates over 24 inches wide, } 48,000 pounds per square inch; elongation, 10 per cent.

116. The ultimate strength of full size tension bars shall not be less than 45,000 pounds per square inch, with elongation not less than 10 per cent. in ten feet, measured on any part of the bar. Their elastic limit shall not be less than 25,000 pounds per square inch.

Specimens as above shall bend cold without sign of fracture on the convex side.

117. For bar iron, 180 degrees around a circle whose diameter is equal to the thickness of the specimen.

118. For shapes and plates, 90 degrees around a circle whose diameter is equal to twice the thickness of the specimen.

119. Full size pieces shall bend cold 90 degrees without sign of fracture around a circle whose radius is equal to the thickness of the piece. For bars, angles, shapes and plates 24 inches wide and less, the radius shall be equal to twice the thickness, and for plates more than 24 inches wide to four times the thickness. Full size rivet iron shall bend cold and set flat on itself 180 degrees without sign of fracture on the convex side.

120. Iron plates rolled in a universal mill shall be used in preference to others, when practicable.

Steel.

121. Shall be tough, ductile, uniform in quality and incapable of tempering; it shall not contain more than one tenth ($\frac{1}{10}$) of one per cent. of phosphorus.

122. Test pieces three fourths ($\frac{3}{4}$) inch in diameter, cut out of the finished pieces, shall stand without breaking, a tensile stress not less than 64,000 pounds and not more than 70,000 pounds per square inch, with an elongation not less than twenty per cent. in a length of 12 diameters. The same allowance as for iron specimens shall be made in the strength and elongation of the specimens according to the sizes of the bars from which they are cut, down to the limit of 60,000 pounds and 16 per cent. They shall bend cold and set flat on themselves without sign of fracture. Heated uniformly at a low cherry red and cooled in water at 82 degrees Fahrenheit, they shall bend around a circle one and one half ($1\frac{1}{2}$) inches in diameter, 180 degrees, without fracture.

123. Full size pieces shall bend cold 90 degrees without sign of fracture, around a circle whose radius is equal to the thickness of the pieces, and shall have an ultimate strength of not less than 56,000 pounds per square inch, with elongation not less than ten per cent. in ten feet measured on any part of the bar; their elastic limit shall not be less than 33,000 pounds per square inch.

124. Steel for rivets shall have in test pieces an ultimate strength of from 56,000 to 62,000 pounds per square inch, with 25 per cent. elongation; it shall stand the quenching test at a light yellow heat. Full size rivet bars shall bend cold and set flat on

themselves without sign· of fracture on the convex side.

125. All.steel plates must be rolled in a universal mill.

126. Shall be of the best quality of tough gray metal. A cast bar five feet long, one inch square, four feet six inches between supports, shall bear, without breaking, a weight of 550 pounds suspended at the center. **Cast Iron.**

127. Castings shall be smooth, well-shaped, free from air-holes, cold shorts, cracks, cinders and other imperfections. All finished pieces of ·iron and steel shall be smooth, free from injurious seams or flaws, blisters, buckles,, cinder spots and imperfect edges. Hammering and heating for the purpose of straightening will not be allowed. **General.**

128. The timber shall be of the sound heart wood of long-leaf yellow pine or white oak. It shall be sawed true and out of wind, full size, free·from wind shakes, large or loose knots, decay, brash or sap-wood, worm holes, ·or any defect impairing its strength or durability. **Timber.**

129. All timber shall be inspected and accepted by an authorized inspector before being used.

130. Shall be prepared in the following manner unless otherwise specified. **Creosoted Timber.**

1st. By a thorough seasoning of the wood at a temperature not to exceed 230 degrees Fahrenheit in a vacuum of twenty-four inches of mercury, a sufficient length of time being used in this operation to avoid the cracking or splitting of the timber.

2d. By the injection into the wood under a pressure of not less than 150 pounds per square inch, of not less than ten pounds of heavy creosote oil to each cubic foot of timber.

All framing and trimming shall be done before injection.

131. The creosote oil used shall generally be solid at a temperature of 50 degrees Fahrenheit, and entirely liquid at a temperature of 100 degrees Fahrenheit. It shall contain not less than five per cent. of tar acids. It shall contain not less than twenty-five per cent. of constituents that do not distill over at a temperature of 600 degrees Fahrenheit. It shall be free from water, ammonia, naphtha, and other **Quality of the Oil.**

impurities, and shall be subject to the acceptance of the engineer.

Inspection and Tests.

132. An expert inspector appointed by the engineer will inspect the material, supervise the work at the shops, the work of erection and all tests to be made. All finished parts of the structure shall be inspected and weighed by him before shipment.

133. All facilities for inspection, testing, and weighing, shall be furnished by the contractor free of charge.

The following tests shall be made by the inspector at the expense of the contractor:

134. 1st. For every lot of 50,000 pounds or less of the same kind (iron or steel) and the same class (bars, angles, plates or other shapes) of material:

Three specimen tensile tests.

Three specimen bending tests.

One full size bending test.

One specimen quenching test, if the material be of steel.

135. 2d. All additional tests required to duplicate any of the above, by reason of faulty material or manipulation in the first tests.

136. All other tests required by the engineer, shall be made by the inspector, and paid for at cost, less the scrap value of the material, if the test proves satisfactory. If the test is not satisfactory, the contractor shall receive no compensation.

137. Failure to stand the foregoing tests, or a discrepancy in weight of materials of more than 2½ per cent. shall be a sufficient cause for rejection.

138. The acceptance of any material or finished member by the inspector, shall not prevent the subsequent rejection of the same if found defective after delivery, and the contractor shall replace the rejected material or member without extra compensation.

139. Before the final estimate is paid, a thorough test of the structure shall be made by the engineer, by loading each span with the nearest equivalent load obtainable to that described under the head of "loads," distributed and moving in such a way as the engineer may see fit. The load will be allowed to remain on the structure any length of

time deemed necessary by the engineer. Each span shall not deflect under such a load more than one eighteen hundredth ($\frac{1}{1800}$) of its length, and shall return to its original camber, when the load is removed. There shall be no permanent changes in the position or condition of any part of the structure as a result of the test.

Paint.

140. All iron and steel before leaving the shop shall have all loose scales scraped off, and shall be thoroughly coated with boiled linseed oil. All planed or turned surfaces shall be coated with white lead mixed with tallow.

141. All inaccessible surfaces shall be painted before being put together with two coats of red lead or other metallic paint, approved by the engineer. After erection, the entire structure, excluding timbers, shall be painted with two coats of the same paint.

142. No painting shall be done in wet or freezing weather.   ·

143. All depressions in the erected structure where water is liable to collect, shall be drained by suitable drain holes or filled with approved waterproof mastic.

Travel and River Navigation.

144. The contractor shall conduct all his operations so as not to impede travel on the road or street for which the bridge is designed, nor travel and operation of trains on any road, street or railroad, crossing under or above the bridge.

145. When rivers are navigable, they shall at all times during the construction and erection of the structure be kept open for navigation.

146. All staging, coffer-dams, and other temporary structures used in the construction of the bridge, as also the old bridge, if any exists, shall be removed by the contractor.

Risks.

The contractors shall assume all risks from floods and storms, damage to persons and properties, and casualties of every description, and shall furnish all materials, tools, machinery and labor incidental to, or in any way connected with the manufacture, transportation, erection, and maintenance of the structure until its final acceptance, without additional compensation.

Additional Clauses.

———————————————————

———————————————————

———————————————————

G. B.

172. Advertisement, Instructions, Specifications and Proposals for Improving St. Mary's Falls Canal. The following complete set of papers illustrate the practice of the U. S. Engr. Corps, and also describe one of the finest pieces of masonry construction in this country. The work was done under Col. O. M. Poe, of the U. S. Engr. Corps, with Mr. E. S. Wheeler, M. Am. Soc. C. E., as Chief U. S. Asst. Engr. The lock gates, buildings, and operating machinery were no part of this contract.

No————————

PROPOSAL OF

———————————————————

———————————————————

Opened January 27, 1891,

FOR FURNISHING ALL MATERIALS, ETC., AND BUILDING THE MASONRY OF A LOCK

AT ST. MARY'S FALLS CANAL, MICHIGAN.

Under Acts of Congress of August 11, 1888, and September 19, 1890.

EXTRACT FROM ACT OF CONGRESS,

Approved Sept. 19, 1890.

Improving St. Mary's River at the Falls, Michigan: Continuing improvement on new locks and approaches, nine hundred thousand dollars: Provided, That such contracts as may be desirable may be entered into by the Secretary of War for materials and labor for the entire structure and approaches, or any part of the same, to be paid for as appropriations may from time to time be made by law.

ADVERTISEMENT.

UNITED STATES ENGINEER OFFICE, ⎫
34 West Congress Street, ⎬
DETROIT, MICH., November 28, 1890. ⎭

Sealed proposals, in triplicate, for furnishing all materials and labor and building the Masonry of a Lock at St. Mary's Falls Canal, Michigan, will be received at this office until 2 o'clock, p. m., January 27, 1891, and then publicly opened.

Preference will be given to materials of domestic production or manufacture, conditions of quality and price (import duties included) being equal.

Attention is invited to Acts of Congress, approved February 26, 1885, and February 23, 1887, Vol. 23, page 332, and Vol. 24, page 414, Statutes at Large.

The Government reserves the right to reject any or all proposals; also, to award the contract upon other considerations than the price.

For further information apply at this office or to the U. S. Engineer Office, Sault Ste. Marie, Michigan.

O. M. POE, Col. Corps of Engineers,
Bvt. Brig. General, U. S. A.

GENERAL INSTRUCTIONS FOR BIDDERS.

1. All bids must be made in triplicate, upon printed forms to be obtained at this office.

2. The guaranty attached to each proposal must be signed by two responsible guarantors, to be certified as good and sufficient guarantors by a United States District Attorney, Collector of Customs, or any other officer under the United States Government, or responsible person known to this office.

3. When a firm bids, the individual names of the members should be written out, and should be signed in full, giving the Christian names; but the signers may, if they choose, describe themselves in addition as doing business under a given name and style as a firm.

4. Anyone signing a proposal as the agent of another, or of others, must file with it legal evidence of his authority to do so.

5. All signatures must have affixed to them seals of wax or wafer.

6. The place of residence of every bidder, with postoffice address, county and state, district or territory, must be given after his signature, which must be written in full.

7. All prices must be written, as well as expressed in figures.

8. Alterations by erasures or interlineations should be explained or noted in the proposal over the signature of the bidder.

9. A firm will not be accepted as a surety, nor will a partner be accepted as a surety for a co-partner, or for a firm of which he is a member.

10. An officer of a corporation will not be accepted as surety for such corporation.

11. A contract will not be awarded to a corporation until it shall have furnished satisfactory evidence of its legal capacity to enter into the same.

12. The bidder must satisfy the United States of his ability to do the work for which he bids.

13. Reasonable grounds for supposing that any bidder is interested in more than one bid for the same item will cause the rejection of all bids in which he is interested.

14. The United States reserves the right to reject any or all bids, and to waive any informality in the bids received; also to disregard the bid of any failing contractor known as such to the Engineer Department.

15. Contingent upon such appropriations as may from time to time be made by law, payments will be made upon monthly estimates, but ten (10) per cent. will be reserved from each payment until the completion of the contract.

16. The contract, which the bidders and sureties promise to enter into, shall be in its general provisions in the form adopted and in use by the Engineer Department, blank forms of which can be inspected at this office, and will be furnished, if desired, to parties proposing to put in bids. Parties making bids are to be understood as accepting the terms and conditions contained in such form of contract.

17. The bond required under the contract will not be greater than one fourth nor less than one fifth the estimated amount of the contract.

18. Transfers of contracts, or of interests in contracts, are prohibited by law.

19. In the form for proposal, the materials to be furnished and the work to be done are itemized for the purpose of comparing the bids, and as a basis for the monthly estimates; but if the contract be awarded it will be as a whole.

20. Any bid in which the prices stated for the several items seem to be "unbalanced" may be rejected on that account alone.

21. The character of the materials proposed will be considered, and if it be deemed to the interest of the United States, for this or any other reason, to accept any proposal other than the lowest in price, the right to do so is expressly reserved.

22. No advantage shall be taken of any error or omission in the following specifications, as full information will be given upon application.

23. Envelopes containing proposals must be endorsed, "Proposals for Lock at St. Mary's Canal, Michigan," and addressed to COLONEL O. M. POE, Corps of Engineers, Detroit, Mich.

SPECIFICATIONS FOR CONSTRUCTING THE MASONRY OF A LOCK
IN THE ST. MARY'S FALLS CANAL, MICHIGAN.

I.—Description.

1. The work required by these specifications is the building of the main walls, miter walls, and stairways of the 800 foot Lock at Sault Ste. Marie, Michigan, and the furnishing of all material, labor and appliances needed for this purpose.

2. The general character of the proposed work is similar to that of the lock now in use, which will be called the lock of 1881 in these specifications; and the general outline of the work will be as shown in the drawings to be seen at the U. S. engineer office, 34 West Congress street, Detroit, Mich.   The location of the work and of all parts of it will be made under the direction of the U. S. agent in charge, who will also furnish the contractor with detailed drawings of the various parts of the work as they become necessary for the construction.

3. The United States will be responsible for the maintenance of the coffer-dam that surrounds the site of the lock, and for keeping the lock-pit free from water.

4. For landing and storing material and carrying on the work, the contractor will have the use of such portions of the U. S. canal lands and piers as may from time to time be designated by the U. S. agent in charge.

5. Bidders are requested to visit the locality of the proposed work and obtain from personal investigation the information necessary to enable them to make intelligent proposals, as the United States will not be responsible for any lack of accurate information on the part of the contractor, regarding the work.

II.—Stone.

6. The masonry of the lock will be of two kinds, cut stone and backing.   Each bidder shall state the location of the quarries from which he intends to obtain each kind of stone, and shall submit with his proposal, at his own expense, a sample of each kind of stone; each sample to have the name and

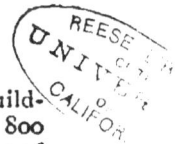

location of its quarry, and the name of the bidder plainly
marked thereon; each sample shall be a "six-inch" cube, one
face showing natural fracture, and the other faces showing different styles of finish. The samples submitted with the proposals will be retained.

7. Each bidder, when requested, shall also furnish, at
his own expense, "two-inch" cubes for testing purposes, from
such portions of his quarries as the engineer officer in charge
may direct; and all stones delivered by the contractor shall be
of as good quality as the samples and test cubes furnished by
him.

8. Stones quarried after October 15 shall not be laid in
the work until the following season, and no stone shall be
quarried between November 1 and April 15. When placed in
the work all stones shall lie on their natural beds. No lewis
holes shall have an opening larger than 1 inch by 4 inches, and
only one lewis hole will be permitted in each stone. Lewis
holes, dog holes, letters, or marks of any kind shall not appear
on show surfaces; but, in the top face of each coping stone, one
lewis hole will be permitted. All lewis holes shall be thoroughly filled with stiff cement mortar of the same quality as
that in which the stone is set. Stones having defects concealed
by cement or otherwise will be rejected.

9. CUT STONE.—The cut stone shall be uniform in
appearance, and of at least as good quality as the cut stone of
the lock of 1881. It shall be free from all defects that, in the
judgment of the engineer officer in charge, will impair its
strength or durability. Sandstone will not be accepted. No
stone shall be used that weighs less than 148 pounds to the
cubic foot. The cut stone shall conform approximately in
dimensions to the bill of materials appended to these specifications, but exact drawings of the stones will be furnished the
contractor when he needs them. The character of workmanship, style of finish, chamfer, etc., shall be the same as that of
the corresponding cut stone of the lock of 1881, except where
otherwise specified below.

10. The stones for the miter walls shall be cut to ⅜ inch
joints throughout. The coping stones shall be cut to ⅜ inch
joints, except for the bed joints that overlie the backing, which
shall not exceed one inch in thickness. The stones for the face
of the main walls, and the facing of the well holes and passages
for the gears for operating the gates and valves shall be cut to
⅜ inch joints on their upper and lower beds, except for such
portions of the bed joints as overlie the backing, which shall
not exceed 1 inch in thickness; the vertical joints of these
stones shall be cut to ⅜ inch for 18 inches from the face, and
the remainder of each of these joints shall not exceed 1 inch in
thickness. No joint shall vary in thickness from the dimensions

specified, and the dimensions of all stones shall be such that the center of each vertical joint shall not vary more than $\frac{1}{10}$ inch on either side of a vertical line through the center of the lowest corresponding joint. Any excess in width of stretchers or length of headers in the face walls will be paid for as "backing stone."

11. All beds and joints shall be fair and true, out of wind and have an equal bearing throughout; the faces of all the walls shall be out of wind, and true to line both horizontally and vertically; and the top surface of each course of stone and of the coping of each wall shall be horizontal throughout, and of the levels given by the U. S. agent in charge.

12. All edges on show lines shall be perfect, and no stones will be accepted that are in any way marred or nicked. Show faces shall be cut true and even, without depressions of any kind; but within the draft lines the show faces of stones in courses below the upper 12 courses of the main walls (coping not included), may be rough finished, with no depression below the plane of the face, and no projection greater than 1 inch above the plane of the face; the face stones of the gate recesses, however, shall be finished throughout, as those in the lock of 1881. All cut stones shall be laid by clamps or by the lewis, and all lewis holes in cut stone kept over winter shall be carefully protected from the weather by the contractor.

13. BACKING.—The backing stones shall be of some good quality approved by the engineer officer in charge. They shall have an average area of bed of at least 8 square feet; the smallest shall have an area of bed of at least 5 square feet, except when the U. S. agent in charge may occasionally specially authorize the use of a smaller stone, and no stone shall have an overhanging top. Backing stones shall be at least 8 inches thick and shall have their faces approximately parallel and their joints at right angles to the faces. No irregular or three-cornered stones will be accepted. The bed joints of the backing shall not exceed one inch in thickness. No vertical joints shall, at any point, exceed four inches in thickness, and the average shall not exceed two inches.

14. About 4,000 cubic yards of stone, now piled on the U. S. canal lands, shall be taken by the contractor from where it now lies and laid in the work as backing. The U. S. agent in charge will designate the stones to be so taken, and the contractor shall furnish only the difference between the total amount of backing required and the amount of this supply.

III.—*Sand, Cement and Concrete.*

15. SAND.—All sand used in mortar or concrete shall be clean, sharp and silicious. It shall be subject to the approval

23

of the U. S. agent in charge as to fineness, and shall be screened and washed if required by him. The cost of all sand used in the work shall be included in the contractor's prices for laying masonry and concrete.

16. CEMENT.—All cement shall be well packed in strong barrels of standard size, lined with paper, so as to be reasonably secure from air. It shall be delivered as directed by the U. S. agent in charge, in a U. S. warehouse at Sault Ste. Marie, Michigan. As soon as possible after its delivery it will be inspected, and, if accepted, the United States will be responsible for its preservation until it is issued at the warehouse to the contractor to use in the work.

17. Tests of cement will be made at a summer temperature of 60 degrees Fahrenheit or above. In making tests samples may be taken from each and every barrel delivered, and a failure of one tenth of the samples tested shall be sufficient cause for the rejection of all barrels in the same lot as those from which the samples have been taken. The amount of cement accepted will be such that the United States will not be responsible, at any time, for more cement than will be needed during the succeeding month, and only so much cement will be issued to the contractor as will be needed from day to day for immediate use in the work. Should the contractor delay the work at any time by an insufficient delivery of cement the United States reserves the right to annul so much of the contract as relates to the delivery of cement, and to purchase elsewhere all the cement needed to complete the work; and all increase in cost to the United States resulting from such purchase shall be deducted from the percentages retained under the contract; and thereafter the contractor shall make no claim against the United States on account of any delays caused by cement not being furnished to him as rapidly as it is needed in the work.

18. The U. S. agent in charge shall direct where Portland Cement mortar is to be used, and where Natural Cement mortar is to be used. Each shall be thoroughly mixed dry of equal parts by measure of cement and sand, and only enough clean water added afterwards to form a paste that can be readily handled with a trowel. The mortar shall be thoroughly mixed and used before it has begun to set, and if required by the U. S. agent in charge, the mortar beds shall be protected from the sun.

19. PORTLAND CEMENT.—This cement shall be of the best quality of hydraulic Portland; no barrel shall weigh less than 375 pounds gross, and the average gross weight per barrel shall not be less than 400 pounds. At least 95 per cent. of the cement, by weight, shall pass through a sieve with 2,500 meshes to the square inch. The cement shall stand a tensile stress of at least

400 pounds to the square inch when mixed neat, and exposed one day in air and six in water. and when mixed in proportions by weight, of one part cement to three sand, and exposed in the same manner, it shall stand a tensile stress of at least 125 pounds per square inch. A cement that cracks or checks when made into thin cakes or that begins to set in less than 30 minutes will not be accepted.

20. AMERICAN NATURAL CEMENT.—This cement shall be of the best quality of natural hydraulic cement, of domestic production. The average gross weight per barrel shall not be less than 300 pounds. At least 90 per cent. of the cement, by weight, shall pass through a sieve with 2,500 meshes to the square inch. The cement shall stand a tensile stress of at least 60 pounds per square inch when mixed neat, and exposed one hour, or until set, in air, and the remainder of the 24 hours in water, and at least 100 pounds per square inch when mixed neat, and exposed one day in air and six in water, and when mixed in proportions, by weight, of one cement to one sand, and exposed for seven days as described above, it shall stand a tensile stress of at least 50 pounds per square inch. A cement will not be accepted that cracks or checks when made into thin cakes, or that, when made into stiff mortar, sets hard enough in less than 30 minutes, to bear a weight of one pound on a wire one twenty-fourth of an inch in diameter.

21. CONCRETE.—The concrete for foundations shall be composed, by measure, of one part cement, one part sand and four parts broken stone. The concrete for filling sha'l be composed, by measure, of one part cement, two parts sand and four parts broken stone. The stone for the concrete may be obtained from such of the stone in the spoil banks of the lockpit as may be designated by the U. S. agent in charge, who will also designate the kind of cement to be used in each case. It shall be broken by the contractor to pass in every way through a ring 2 inches in diameter, and shall be screened and washed clean before being used.

But the proportions of any or all of the component materials of the concrete, either for foundation or for filling, may be varied by the engineer officer in charge, or additional stone in the form of pebbles, boulders of the locality crushed to pass through a half-inch ring, or crushed quartz from stamp mills may be required by the same authority, and when so required the changes shall be made, or the additions shall be supplied by the contractor without any increase in the price per cubic yard to be paid for the concrete in place.

22. The cement and sand shall first be thoroughly mixed dry, the proper quantity of clean water shall then be mixed in and the clean moistened stone shall then be added to the mass

and the whole thoroughly mixed. The amount of water added s all be such that no sign of water shall appear until the ramming of the concrete, as specified below, is nearly finished. The concrete shall be thoroughly rammed in place before its cement has begun to set.

IV.—Construction.

23. The contractor must clean all broken stone and rubbish from the rock upon which the concrete foundation is to rest, and must thoroughly clean and scrub it before any concrete is placed upon it.

24. The concrete foundation shall extend under the bottom course of masonry of all walls, but shall not be so laid as to interfere with the subsequent laying of the lock floor timbers. The top surface of the concrete foundation shall be horizontal throughout, and of the levels given the contractor by the U. S. agent in charge. The concrete shall be deposited in layers not more than 6 inches in thickness, and thoroughly rammed, as soon as laid, with rammers weighing not less than 35 pounds.

25. Unless specially directed otherwise by the engineer officer in charge, the walls shall be carried up in such a manner that not more tnan three courses shall be unfinished at a time. The stones shall be prepared for their respective places they are to occupy in the work before they are brought on the walls; moving stones about on the walls will not be permitted, and no dressing, except by special permission of the U. S. agent in charge, will be allowed on any stone after it is in the wall.

26. Generally the backing shall be laid in courses not exceeding two feet in height. and must be leveled off with the top surface of each course of the face stone. But the engineer officer in charge may occasionally permit the use of backing stone of more than two feet in thickness, in which case the course shall not exceed four feet in thickness and must be leveled off with the top surface of every second course of face stone. The bottom stones of each course of backing shall break all vertical joints at least 9 inches with the top stones of the course below. Whenever possible, in each course, all stones shall break joints at least 9 inches with all stones adjacent to them. The backing shall not be laid in advance of the face stones.

27. No cut stone shall be laid after dark. Every stone, both cut stone and backing, shall be laid in a full bed of mortar, and shall be carefully settled in place in a manner satisfactory to the U. S. agent in charge. All vertical joints shall be completely filled with mortar. The spaces in vertical joints due to irregularities of form of the backing stones. shall all be filled solid with selected hammer shaped stones and spalls, carefully laid and settled in mortar, but no spalls or wedges of any kind

will be permitted in horizontal joints.   The use of grout is prohibited.

28. Where the work is in progress the masonry and concrete shall be kept wet and free from dirt.   All stones shall be washed clean just before they are brought on the wall, and the beds and joints of all stones shall be moistened with water just before they are laid.

29. All irons that must be built in or attached to the masonry, such as gate anchorages and pivot p'ates, snubbing hooks, miter wall bolts, etc., will be furnished the contractor at the site by the United States.   The contractor must place all these irons in the positions indicated by the detailed drawings that will be furnished him at the time, and, as no additional compensation will be allowed for this work, its cost must be included in the prices bid for laying masonry.

30. Before filling of any kind is placed behind the walls, the exposed joints of the backing must be rough pointed by the contractor, with stiff cement mortar.   The cut stone masonry will be pointed by the United States before the completion of the contract, and the contractor is not to interfere in any way with the employees of the United States engaged in this work.

31. Neither masonry nor concrete shall be laid from November 15 to April 15 inclusive, nor at any other time during freezing weather.   The U. S. agent in charge will inform the contractor when the laying of masonry or concrete shall be stopped on this account.   No holes that will hold water shall be left in the walls during winter.   All portions of the walls remaining unfinished over winter must be thoroughly protected by the contractor by suitable covering against the action of frost.   Before laying any masonry in the spring, all mortar that has been injured by frost shall be removed from all the joints by the contractor, and they shall then be properly refilled by him without additional compensation.

32. The space between the back of the walls and the rock face of the lock-pit shall be filled with concrete, laid as described for the concrete foundations and carried up simultaneously with the walls.

33. Except where indicated by the U. S. agent in charge, the space behind the walls, above the concrete filling, shall be filled with material taken from such parts of the spoil banks of the lock-pit as may be indicated by the U. S. agent in charge.   It shall be placed in horizontal layers not exceeding one foot in thickness, and after being well dampened with water shall be thoroughly rammed with rammers weighing not less than 35 pounds each.   Between the south wall and the lock of 1881, all depressions below the level of the coping of the two locks shall be filled to this level; and behind the north wall the top surface of the filling shall be on a level with the

coping, and shall extend back to a line parallel to the face of the north wall and 50 feet from it. All slopes of the filling shall be at least 1 on 2. No stones will be permitted in the filling of a greater volume than one cubic foot, and the filling shall be carried up simultaneously with the walls.

34. The contractor shall be responsible for the preservation of the slopes of the lock-pit until the completion of work under the contract, and immediately thereafter he shall remove all of his machinery, all rubbish and loose material, from between the walls.

V.—General.

35. The approximate estimate of materials to be furnished, and the work to be done under these specifications, is as follows:

Portland cement to be delivered, 22,000 bbls. (more or less).

Natural cement to be delivered, 75,000 bbls. (more or less).

Cut stone to be delivered, 20,000 cubic yards, solid measure (more or less)..

Cut stone to be laid, 20,000 cubic yards, solid measure (more or less).

Backing to be delivered, 55,000 cubic yards, solid measure (more or less).

Backing to be laid, 59,000 cubic yards, solid measure (more or less).

Concrete to be laid, 5,000 cubic yards, measured in place (more or less).

Earth to be filled behind walls, 70,000 cubic yards, measured in place (more or less).

36. Bidders will state the price per unit for the several items separately, but the aggregate determined from the prices and the above approximate quantities will be considered as one bid. No claim shall be made against the United States on account of any excess or deficiency, absolutely or relatively, in the amounts as stated above. Sufficient material shall be furnished, and sufficient work done, to complete the masonry, etc., as specified, and bidders are requested to make the estimate of quantities for themselves. The contractor's prices for the various items shall cover all costs of labor, appliances, and materials, and all expenses of whatever nature (except superintendence by U. S. agents), that may arise during the progress of the work. The best quality of materials and workmanship will be required and the cost to the United States will be but one element in determining the award of the contract.

37.   The work will not be finally accepted until the contractor shall have made good any injury that may have been done to work included in any previous estimates, and the contract shall have been completed.   The United States will not be responsible for the safety of the employees, plant, or materials used by the contractor, nor for any damage done by or to them from any source or cause whatever.   While in the canal, the contractor's plant shall be subject to the rules which govern vessels passing through it, and to this extent shall be under the orders of the superintendent of the canal.   •

38.   All the work under these specifications shall be carried on under, and in conformity with, the direction of the U. S. agent in charge.   Should any changes in any of the plans be made by the engineer officer in charge during the progress of the work, the contractor shall conform to them, and a fair allowance will be paid for any changes, which, in the judgment of the engineer officer in charge, materially increase the cost of the work.   No "extras" of any kind will be paid for, however, unless they have been specially ordered in writing, and their price stated in writing, by the engineer officer in charge. before work on them has begun.   All rejected material shall be removed from the work immediately by the contractor at his own expense.

39.   Contingent upon such appropriations as may from time to time be made by law, payments will be made upon monthly estimates of the amounts of work and materials accepted during the preceding month.   The monthly estimates will be made as follows:

1.   Portland cement delivered and accepted.

2.   Natural cement delivered and accepted.

3.   Seventy-five per cent. of the cut stone delivered and accepted (before it is laid).

4.   Twenty-five per cent. of the cut stone delivered and accepted (after it is accepted in the wall).

5.   Laying of cut stone after it is accepted in the wall.

6.   Backing delivered and accepted (after it is laid).

7.   Laying of backing, after it is accepted in the walls.

8.   Laying of concrete, after it is accepted in place.

9.   Earth filling behind walls, after it is accepted in place.

For all cut stone and backing items, payment will be made only for the volume actually occupied by the stones in the completed wall.   No payment will be made for the volume lost in cutting stones to their proper shapes.

Until the completion of the contract, the contractor shall be responsible for the preservation of all previously accepted work or material, the cement while in custody of the United States alone excepted, and each monthly estimate to be diminished first by the ten per cent. to be retained from each payment, and second, by the estimated value of all previously accepted work or material injured or wasted during the preceding month.

40. In case it be deemed advisable, at the request of the contractor, to extend the contract or modify it in any manner, all cost to the United States resulting therefrom may be deducted from the final payment, and only the remainder will be paid to the contractor.

41. Work under the contract shall be begun on or before May 15, 1891, and entirely completed on or before November 15, 1893. The total amount of the contractor's monthly estimates by the end of the first season's work under the contract shall be at least one fifth of the total amount of the approximate estimate given above; and by the end of the second season's work, the total of the contractor's monthly estimates shall be at least three fifths of the same approximate estimate. In case the total estimates for the work done by the end of any season shall be less than specified above, the United States shall have the right to annul the contract and finish the work, and the contractor and his bondsmen shall be liable for any increase of cost to the United States over that proposed and agreed upon for the entire work as specified. In case the contract is annulled as described above, all amounts that may be due the contractor at the time shall be forfeited to the United States.

42. The River and Harbor Act of September 19, 1890, in making the appropriation for continuing the work of improving St. Mary's river, provides:

" * * * *That such contracts as may be desirable may be entered into by the Secretary of War for materials and labor for the entire structure and approaches, or any part of the same, to be paid for as appropriations may from time to time be made by law.*"

The amount of funds available at any time for work under the contract will be obtained by deducting from the total funds on hand for improving St. Mary's river, the estimated amounts necessary for all the contingencies of engineering, superintendence, etc., etc.

43. If any person employed by the contractor on any part of the work, or upon any work pertaining thereto, should appear to be incompetent or objectionable, he shall be discharged immediately upon the requisition of the engineer officer in charge, and such person shall not again be employed in connection with the work or any part of it.

In case of differences arising between the contractor and the U. S. agent in charge, in regard to the work or to these specifications, appeal may be made to the engineer officer in charge, and his decision shall be final.

No advantage shall be taken of any error or omission in the foregoing specifications, as full information will be given upon application.

APPROXIMATE BILL OF CUT STONE REQUIRED FOR 800 FOOT LOCK AT ST. MARY'S FALLS CANAL, MICHIGAN.

No. of Pieces.	Dimensions of Each Piece.			Cu. Ft. in Each Piece.	No. of Pieces.	Dimensions of Each Piece.			Cu. Ft. in Each Piece.
	Ft.	Ft.	Ft.			Ft.	Ft.	Ft.	
36	2	5	5½	55	6	2	6½	8½	111
7752	2	3	6	36	8	2	5	7½	75
186	2	6	6½	78	10	2	3	4¼	26
348	2	3	7	42	4	2	5	5¼	53
102	2	4¾	6	57	32	2	5½	6	66
194	2	3	6¾	40	2	2	6¾	7½	102
138	2	4	6	48	4	2	3½	7	49
166	2	3	6¼	33	6	2	4½	6¾	60
186	2	3½	6	42	16	2	3½	5	35
276	2	3	6½	39	8	2	4	7½	60
152	2	4¼	6	51	36	2	4½	5	45
230	2	5	6	60	8	2	2¾	7	39
38	2	2½	6	30	61	2	3	4½	27
10	2	5	7	70	16	2	4	5½	44
12	2	5	6½	65	28	2	3	3½	21
227	2	4½	6	54	172	2	2	4	16
12	2	3¼	6	39	190	2	2¾	5	28
64	2	3¾	6	45	77	2	2¾	6	33
66	2	6	7¼	87	12	2	2½	4	20
36	2	3	5½	33	8	2	2	5¼	21
30	2	6	6	72	4	2	2	2½	10
273	2	3	5	30	10	2	3	3	18

APPROXIMATE BILL OF CUT STONE REQUIRED FOR 800 FOOT LOCK AT ST. MARY'S FALLS CANAL, MICHIGAN.

No. of Pieces.	Dimensions of Each Piece.			Cu. Ft. in Each Piece.	No. of Pieces.	Dimensions of Each Piece.			Cu. Ft. in Each Piece.
	Ft.	Ft.	Ft.			Ft.	Ft.	Ft.	
6	2	3	4	24	4	2	4½	5½	50
8	2	2	5¾	23	4	2	3	8¼	50
6	2	3	5¾	34	26	2	2¾	4¾	26
2	2	3¼	5½	36	28	2	3¾	4½	33
34	2	2	5	20	21	2	4	4½	36
8	2	2	2¾	11	14	2	3	8¾	53
4	2	2	3½	14	14	2	3	7¾	47
4	2	2½	6¼	31	12	2	4½	4½	41
12	2	6½	9¼	120	4	2	3¼	8¾	57
74	2	5¾	6	69	8	2	3¼	7	46
74	2	3	5¼	32	14	1½	3¼	6	29
2	2	4½	7¼	65	11	2½	5	6	75
2	2	3½	6¾	47	10	2½	3	5	38
14	2	4	6½	52	4	3	3¼	6½	63
100	2	3	7½	45	7	3	3	7¾	70
2	2	6	7	84	6	3	3½	3½	37
2	2	6	6¾	81	6	3	3	3½	32
14	2	8	8½	136	5	3	3½	7¼	76
14	2	7½	7½	113	5	3	3½	6¼	66
98	2	2¾	4½	25	2	3	3½	7½	79
91	2	5¾	7½	86	12	3¼	3½	6	68
4	2	6	8½	102	28	3½	3½	6	74

APPROXIMATE BILL OF CUT STONE REQUIRED FOR 800 FOOT LOCK AT ST. MARY'S FALLS CANAL, MICHIGAN.

No of Pieces.	Dimensions of Each Piece.			Cu. Ft. in Each Piece.	No of Pieces.	Dimensions of Each Piece.			Cu. Ft. in Each Piece.
	Ft.	Ft.	Ft.			Ft.	Ft.	Ft.	
2	3½	4½	7½	118	52	1½	3	5	23
2	3½	4	4½	63	8	1½	3	5¼	24
2	3½	4½	6	95	64	1½	3	4½	20
2	3½	5½	7½	144	36	1½	3	7½	34
2	2¾	3½	4½	43	12	1½	3	5½	25
2	3½	8	8½	238	8	1½	5¼	6	47
2	3½	7½	7½	197	2	1½	5	7	53
50	⅔	1⅙	8	6	2	1½	4	7	42
2	⅔	1 5/12	8	8	8	1½	3	6¾	30
1052	1½	3	6	27	4	1½	4	7½	45
42	1½	3	7	32	6	1½	4	6	36
22	1½	5	6	45	2	1½	3½	5	26
8	1½	4¼	6	39	4	1½	4	4½	27
28	1½	3·	6½	29	4	1½	3	4	18
30	1½	3½	6	32	6	1½	4½	5	34
8	1½	2½	6	23	2	1½	6	6	54
30	1½	3¾	6	34	134	1¾	3	5	26
2	1½	2½	4½	17	132	1¾	5	6	53
36	1½	4½	6	41	2	1¾	5	5	43
2	1½	2½	7½	28					
14	1½	6	7¼	65	Total No. Stone......13,904				Approximate.
4	1½	4½	7	47	Total Vol. Cu. Yds....19,688				

NOTE.—The linear dimensions are given to the nearest quarter of a foot, and the volume to the nearest cubic foot. They describe a prism from which the required finished stone may be cut.

Payments are not to be made by this bill, but upon the volume of stone laid in the wall after being cut to the proper dimensions.

PROPOSAL.

TO COLONEL O. M. POE,

Corps of Engineers, U. S. A., Detroit, Mich.

_____189—.

SIR—In accordance with your advertisement of November 28, 1890, inviting proposals for furnishing all material and labor and building the Masonry of a Lock, at Saint Mary's Falls Canal, Mich., and subject to all the conditions and requirements thereof, and of your instructions to bidders and specifications dated November 28, 1890, copies of all of which are hereto attached, and, so far as they relate to this proposal, are made a part of it, we (or) I propose to furnish all materials, appliances and labor, and do the work as specified, at the prices named below, namely:

Deliver at Sault Ste. Marie, Mich., twenty-two thousand (22,000) barrels (more or less) of Portland Cement at the rate of ——— (——) dollars and ——— (——) cents per barrel.

Deliver at Sault Ste. Marie, Mich., seventy-five thousand (75,000) barrels (more or less) of Natural Cement, at the rate of ——— (——) dollars and ——— (——) cents per barrel.

Deliver at Sault Ste. Marie, Mich., twenty thousand (20,000) cubic yards (more or less), solid measure, cut stone, at the rate of ——— (——) dollars and ——— (——) cents per cubic yard.

Lay in the lock walls twenty thousand (20,000) cubic yards (more or less), solid measure, cut stone, at the rate of ——— (——) dollars and ——— (——) cents per cubic yard.

Deliver at Sault Ste. Marie, Mich., fifty-five thousand (55,000) cubic yards (more or less), solid measure, backing stone, at the rate of ——— (——) dollars and ——— (——) cents per cubic yard.

Lay in the lock wall fifty-nine thousand (59,000) cubic yards (more or less), solid measure, backing stone, at the rate of ——— (——) dollars and ——— (——) cents per cubic yard.

Lay in the foundations, etc., of the lock, five thousand (5,000) cubic yards (more or less) concrete, at the rate of ——— (——) dollars and ——— (——) cents per cubic yard measured in place.

Fill behind walls seventy thousand (70,000) cubic yards (more or less) earth, at the rate of ——— (——) dollars and ——— (——) cents per cubic yard, measured in place.

We [or] I make this proposal with a full knowledge of the kind, quantity, and quality of the articles required, and of the work to be done, and, if it is accepted, will, within ten (10) days after receiving written notice of such acceptance, enter into contract, with good and sufficient sureties, for the faithful performance thereof.

WITNESS:

. ...(Signature). ....[SEAL]
                         (Address) ....
....(Signature)......... ... .[SEAL]
                         (Address). ....
.. ..(Signature). ....[SEAL]
                         (Address) ....
....(Signature).... ....[SEAL]
                         (Address).. ....
.... (Signature)...... ....[SEAL]
                         (Address) ....
....(Signature)................. ....[SEAL]
                         (Address) ... .
                         (Sign in Triplicate.)

GUARANTY.

We, ——, of ——, in the state of ——, and ——, of ——, in the state of ——, hereby guarantee and bind ourselves and each of us, our and each of our heirs, executors and administrators, to the effect that if the bid of —— herewith accompanying, dated ——, 189—, for furnishing all materials and labor, and building the Masonry of a Lock at Saint Mary's Falls Canal, shall be accepted, in whole or in part, within sixty (60) days from the date of the opening of proposals, the said bidder ——, will, within ten (10) days after being notified of such acceptance, enter into a contract with the United States in accordance with the terms and conditions of the advertisement, and will give bond with good and sufficient sureties for the faithful and proper fulfillment of the same. And in case the said bidder —— shall fail to enter into contract within the said ten (10) days with the proper officer of the United States, and furnish good and sufficient bond for the faithful performance of the same according to the terms of said bid and advertisement, we and each of us hereby stipulate and guarantee, and bind ourselves and each of us, our and each of our heirs, executors and administrators, to pay unto the United States the difference in money between the amount of the bid of the said bidder —— and the amount for which the proper officer of the United States may contract with another party to furnish said materials and labor and build the Masonry of a

Lock, as specified, if the latter amount be in excess of the former, for the whole work covered by the proposal.

WITNESSES:

——————————————————— ——————————————————[SEAL.]

——————————————————— ——————————————————[SEAL.]

Dated ....189..
[Executed in Triplicate.]

JUSTIFICATION OF THE GUARANTORS.

STATE OF————————
COUNTY OF———————— } ss.

I, ————, one of the guarantors named in the within guaranty, do swear that I am pecuniarily worth the sum of two hundred thousand dollars, over and above all my debts and liabilities.

(Signature of Guarantor.)——————————————————————

Before me,

(Signature of Officer administering oath, with seal, if any.)

————————————————————

STATE OF————————
COUNTY OF———————— } ss.

I, ————, one of the guarantors named in the within guaranty, do swear that I am pecuniarily worth the sum of two hundred thousand dollars, over and above all my debts and liabilities.

(Signature of Guarantor.)——————————————————————

Before me,

(Signature of officer administering oath, with seal, if any.)

————————————————————

CERTIFICATE.

I, ————, do hereby certify that ———— and ————, the guarantors above named, are personally known to me, and, that, to the best of my knowledge and belief, each is pecuniarily worth, over and above all his debts and liabilities, the sum stated in the accompanying affidavit subscribed by him.

(Signature of certifying official.)————————————————

NOTE.—The certificate may be given separately as to each guarantor, and modified accordingly.

O. M. P.

173. Complete Specifications and Contract for Dam No. 5, Southborough, of the Boston Water-works, July, 1893.

ADVERTISEMENT.

TO CONTRACTORS.

Sealed proposals addressed to the Boston Water Board, and endorsed "Proposals for building Dam No. 5 in the Town of Southborough," will be received by the Boston Water Board, at their office, City Hall, Boston, Mass., until 12 o'clock M., of Monday the seventeenth day of July, 1893, and at that time · will be publicly opened and read.

Each bidder must make a personal examination of the location of the dam.

All bids must be made upon blank forms, to be obtained of the City Engineer, Boston, must give the prices proposed, both in writing and in figures, and be signed by the bidder, with his address.

Each bid is to be accompanied by a certified check for two thousand dollars ($2.000), payable to the City of Boston, said check to be returned to the bidder unless he fail to execute the contract, should it be awarded to him.

A bond for one hundred thousand dollars ($100,000) will be required for the faithful performance of the contract, the sureties to be residents of Massachusetts, and satisfactory to said Boston Water Board.

The person or persons to whom the contract may be awarded will be required to appear at this office with the sureties offered by him or them, and execute the contract within six days (not including Sunday) from the date of notification of such award, and the preparation and readiness for signature of the contract; and in case of failure or neglect so to do, he or they will be considered as having abandoned it, and the check accompanying the proposal shall be forfeited to the City of Boston.

All bids will be compared on the basis of the Engineer's estimate of quantities of work to be done, which is as follows:—

(*a*)   14,000 cubic yards soil excavated and placed in spoil-banks.

(*aa*)   13,900 cubic yards soil excavated from spoil-banks and placed on dam.

(*b*)   1,610 square yards sodding.

(*bb*)   5 acres seeding.

(*c*)   230,000 cubic yards earth excavation (trenches, embankments, and backfilling).

(*cc*)   10,000 cubic yards rehandling of excavated materials.

(*d*)   13,400 cubic yards rock excavation.
(*e*)   2,000 feet board measure timber work.
(*cc*)   2,000 feet board measure timber work (tongued and grooved).
(*f*)   800 barrels Portland cement.
(*g*)   14,000 cubic yards concrete masonry.
(*gg*)   800 cubic yards concrete masonry.
(*h*)   9,270 square yards plastering.
(*i*)   256 cubic yards brick masonry.
(*j*)   7500 cubic yards paving. *
(*k*)   10,100 cubic yards riprap.
(*l*)   5,400 cubic yards broken stone.†
(*m*)   22,200 cubic yards rubble-stone masonry.
(*n*)   13,300 square feet facing stone masonry (broken ashlar work).
(*o*)   3,650 cubic yards facing stone masonry (range work).
(*p*)   320 linear feet coping.
(*q*)   290 cubic yards dimension stone masonry.
(*r*)   4,110 square feet hammered work.
(*s*)   1,000 cubic yards masonry laid in American cement mortar 1 to 1, an additional price per cubic yard.
(*t*)   1,000 cubic yards masonry laid in Portland cement mortar 1 to 1, an additional price per cubic yard
(*u*)   1,000 cubic yards masonry laid in Portland cement mortar 1 to 2, an additional price per cubic yard.
(*v*)   1,000 cubic yards masonry laid in Portland cement mortar 1 to 3, an additional price per cubic yard.
(*w*)   1,575 linear feet of walk.

These quantities are approximate only, and the Boston Water Board expressly reserves the right of increasing or diminishing the same, as may be deemed necessary by its Engineer.

Plans can be seen, and specifications and forms of proposal and contract obtained, at the office of the City Engineer, City Hall, Boston.

The Boston Water Board reserves the right to reject any or all bids, should it deem it to be for the interest of the City of Boston so to do.   ROBERT GRANT,
                           JOHN W. LEIGHTON,
                           THOMAS F. DOHERTY,
                                 Boston Water Board.

OFFICE OF BOSTON WATER BOARD,
  CITY HALL, BOSTON, JULY 1, 1893.

* 3,200 cubic yards if riprap is used.   † 2,800 cubic yards if riprap is used.

PROPOSAL

The undersigned hereby declares that'he has carefully examined the annexed form of contract and specifications and the drawings therein referred to, and made an inspection of the site of the proposed dam, and will provide all necessary machinery, tools, apparatus, and other means of construction, and do all the work and furnish all the materials called for by said contract and specifications and the requirements under them of the Engineer, for the following sums, to wit:

(*a*) For the removal of soil excavated and placed in spoil banks, including all incidental work, the sum of——— ($——) per cubic yard.

(*aa*) For the removal of soil taken from spoil banks or from other places and placing on the slopes of the embankment, including all incidental work, the sum of——— ($——) per cubic yard.

(*b*) For sodding, including all incidental work, the sum of——— ($——)per superficial square yard.

(*bb*) For seeding, including all incidental work, the sum of——— ($——) per acre.

(*c*) For earth excavation, including its disposal in embankments and refilling, or as otherwise ordered by the engineer, and all incidental work, the sum of——— ($——) per cubic yard.

(*cc*) For rehandling of excavated materials from spoil banks and placing, including all incidental work, the sum of ——— ($——) per cubic yard.

(*d*) For rock excavation, including its disposal and all incidental work, the sum of——— ($——) per cubic yard.

(*e*) For permanent timber work, except tongued and grooved timber, placed, including all incidental work, the sum of——— ($——) per thousand feet B. M.

(*ee*) For permanent timber work, tongued and grooved, placed, including all incidental work, the sum of——— ($——) per thousand feet B. M.

(*f*) For Portland cement ordered by the engin er, delivered where ordered on the work, in barrels containing 400 pounds, including all incidental work, the sum of——— ($——) per barrel.

(*g*) For concrete masonry, in place, formed of five parts of broken stone or screened gravel, to one part of cement, and made with American cement mortar mixed in the proportion of one part of cement to two parts of sand, including all incidental work, the sum of——— ($——) per cubic yard.

24

(*gg*) For concrete masonry, in place, formed of three parts of broken stone or screened gravel to one part of cement, and made with American cement mortar mixed in the proportion of one part of cement to two parts of sand, including all incidental work, the sum of——— ($——) per cubic yard.

(*h*) For plastering all concrete walls with Portland cement, including all incidental work, the sum of——— ($——) per superficial square yard.

(*i*) For brick masonry, laid in Portland cement mortar mixed in the proportion of one part of cement to two parts of sand, and including all pointing, centering, etc., and removing the same, and all incidental work, the sum of——— ($——) per cubic yard.

(*j*) For paving in place, including all incidental work, the sum of——— ($——) per cubic yard.

(*k*) For riprap in place, including all incidental work, the sum of——— ($——) per cubic yard.

(*l*) For broken stone in place (other than that used in making concrete and the walk), including all incidental work, the sum of——— ($——) per cubic yard.

(*m*) For rubble-stone masonry, laid in American cement mortar, mixed in the proportion of one part of cement to two parts of sand, including all incidental work, the sum of——— ($——) per cubic yard.

(*n*) For face work of broken ashlar, in addition to the price paid per cubic yard as rubble, including pointing in neat Portland cement, and all incidental work, the sum of——— ($——) per superficial square foot.

(*o*) For facing stone masonry of range stones laid in American cement mortar mixed in the proportion of one part of cement to two parts of sand and pointing in neat Portland cement, including all incidental work, the sum of——— ($——) per cubic yard.

(*p*) For coping laid in place, and pointed in neat Portland cement, including all incidental work, the sum of ——— ($——) per linear or running foot.

(*q*) For dimension stone masonry laid in American cement mortar mixed in the proportion of one part of cement to two parts of sand, including pointing in neat Portland cement, centering, etc., and all incidental work, the sum of——— ($——) per cubic yard.

(*r*) For fine hammer dressing (six cut work) the sum of ——— ($——) per superficial square foot.

(*s*) For all kinds of masonry laid in American cement mortar mixed in the proportion of one part of cement to one part of sand, in addition to the prices per cubic yard hereinbefore stipulated to be paid for the same class of masonry laid in American cement mortar mixed in the proportion of one part of cement to two parts of sand, the sum of——— ($———) per cubic yard.

(*t*) For all kinds of masonry laid in Portland cement mortar mixed in the proportion of one part of cement to one part of sand, in addition to the prices per cubic yard hereinbefore stipulated to be paid for the same class of masonry laid in American cement mortar mixed in the proportion of one part of cement to two parts of sand, the sum of ——— ($———) per cubic yard.

(*u*) For all kinds of masonry laid in Portland cement mortar mixed in the proportion of one part of cement to two parts of sand, in addition to the prices per cubic yard hereinbefore stipulated to be paid for the same class of masonry laid in American cement mortar mixed in the proportion of one part of cement to two parts of sand, the sum of ——— ($———) per cubic yard.

(*v*) For all kinds of masonry laid in Portland cement mortar mixed in the proportion of one part of cement to three parts of sand, in addition to the price per cubic yard hereinbefore stipulated to be paid for the same class of masonry laid in American cement mortar mixed in the proportion of one part of cement to two parts of sand, the sum of——— ($———) per cubic yard.

(*w*) For building walk, including all incidental work, the sum of——— ($———) per linear or running foot.

(*x*) For all extra work done by written order of the Boston Water Board, its actual reasonable cost to the Contractor, as determined by the City Engineer, plus fifteen per cent. of said cost.

Accompanying this proposal is a certified check for two thousand dollars ($2,000), which it is agreed shall become the property of the city of Boston, if, in case this proposal shall be accepted by the Boston Water Board, the undersigned shall fail to execute a contract with said city under the conditions of this proposal within the time provided for by the advertisement for proposals; otherwise said check shall be returned to the undersigned.

No member of the city council, and no person in any office or employment of the city of Boston is directly or indirectly interested in this proposal or in any contract which may be made under it, or in expected profits to arise therefrom; and this proposal is made in good faith without collusion or connection with any other person bidding for the same work.

Name———————

Address—————————

Date—————1893.

CITY OF BOSTON.

BOSTON WATER WORKS.

CONTRACT AND SPECIFICATIONS FOR BUILDING DAM
NO. 5, IN THE TOWN OF SOUTHBOROUGH.

This Agreement, made and concluded this
————day of————in the year one thousand eight
hundred and ninety-three, between the City of Boston, by its Boston Water Board, of the first part,
and————————————in the State of————————
of the second part:

Commencement of Work. **A.** *Witnesseth*, That for and in consideration of
the payments and agreements hereinafter mentioned,
to be made and performed by the said party of the
first part, and under the penalty expressed in a bond
bearing even date with these presents and hereunto
annexed, the said party of the second part agrees
with the said party of the first part to commence the
work herein required to be done, within fourteen
days after the signing of this contract and to proceed
with the work in such order and at such times,
points, and seasons, and with such force as may,
from time to time, be directed by the engineer, and
at his own proper cost and expense, to do all the
work and furnish all the materials called for by this
agreement, in the manner and under the conditions
hereinafter specified.

Completion of Work. And the said party of the second part hereby
agrees to complete all the work called for under this
agreement, in all parts and requirements and in full
conformity with the plans and specifications on or
before November 1, 1896; provided, however, that
the water board shall have the right at their discretion to extend the time for said completion of the
work. It is further agreed that the permitting of
said party of the second part to go on and finish
said work after the time specified for its completion
shall not operate as a waiver of any of the rights of
said city under this contract.

Referee. **B.** To prevent all disputes and litigation it is further agreed, by and between the parties to this contract, that the city engineer of Boston (meaning
thereby the individual at any time holding the position or acting in the capacity of the engineer of the
Boston Water Board) shall be referee in all cases to
determine the amount or the quantity of the work
which is to be paid for under this contract, and to

decide all questions which may arise relative to the fulfillment of this contract on the part of the contractor, and his estimates and decisions shall be final and conclusive; also that said engineer, by himself, or by assistants and inspectors, acting for him, shall inspect the work to be done under this agreement to see that the same is done strictly in accordance with the requirements of the specifications hereinafter set forth.

C. The parties further agree that wherever in this contract the words defined below are used, they shall be understood to have the meanings herein given:

The term "water board" shall mean the Boston Water Board, or any board or committee duly authorized to represent the city of Boston in the execution of the work covered by this contract. *Water Board.*

The word "engineer" when not further qualified, shall mean the said city engineer or his properly authorized agents, limited by the particular duties entrusted to them. *Engineer.*

The word "contractor" shall mean the person or persons, co-partnership or corporation, who have entered into this contract as party of the second part, or his or their legal representatives. *Contractor.*

D. It is further agreed that the quantities of work to be done and materials to be furnished, as given in the accompanying notice to contractors are only for the purpose of comparing the bids offered for the work under the contract on a uniform basis; and it is hereby agreed that the Boston Water Board expressly reserves the right to increase or diminish the above mentioned quantities, or any of them, as may be deemed necessary by the engineer.

E. The plans and specifications are intended to be explanatory of each other; but should any discrepancy appear, or any misunderstanding arise as to the import of anything contained in either, the parties hereto further agree that the explanation and decision of the city engineer shall be final and binding on the contractor; and all directions and explanations required, alluded to, or necessary to complete any of the provisions of this contract and specifications and give them due effect, shall be given by the said engineer. Corrections of errors or omissions in drawings or specifications may be made by the said engineer, when such corrections are necessary for the proper fulfillment of the intention of such drawings or specifications, the effect of such corrections *Plans.*

to date from the time that the said engineer gives
due notice thereof to said contractor.

Alterations. **F.** It is further agreed that the city engineer may
make alterations in the line, grade, plan, form, position, dimensions, or material of the work herein
contemplated, or of any part thereof, either before
or after the commencement of construction. If such
alterations diminish the quantity of work to be done,
they shall not constitute a claim for damages, or for
anticipated profits on the work that may be dispensed with; if they increase the amount of work,
such increase shall be paid for according to the
quantity actually done, and at the price established
for such work under this contract; or in case there
is no price established, it shall be paid for at its
actual reasonable cost as determined by the city
engineer, plus fifteen per cent. of said cost.

SPECIFICATIONS FOR BUILDING DAM NO. 5 OF THE

SUDBURY RIVER WORKS.

Plans. **G.** 1. The contractor is to furnish all the material
and do all the work necessary to build a dam on the
Stony Brook branch of the Sudbury river near the site
of Nichol's Mill, so called, in the town of Southborough, Mass. The dam to be in accordance with
plans marked Dam No. 5, dated June 16, 1893,
signed by William Jackson, city engineer, and filed
in the office of the city engineer, City Hall, Boston.
The work will also be built in conformity with these
specifications.

These plans show only the general character of
the work, and during its progress such working plans
will be furnished from time to time by the engineer
as he may deem necessary.

Borings. The character of the materials to be met with,
as shown on said plans, is the result of such examinations as the city of Boston has been able to make;
but no guarantee is made as to the accuracy of the
borings or test pits or the representations on the
plans.

General De- 2. The dam is to be built partly of masonry
scription. and partly of earth, approximately on the lines shown;
but if the character of the materials or circumstances
arise which render it advisable to change the location
of the dam or to change the plans of the dam the
city of Boston expressly reserves the right so to do
without payment of damages to the contractor, but

all work actually completed will be paid for as per prices bid for the whole work.

The earth embankments will contain plastered concrete core walls. Water-tight material will be placed next these walls on the water side. The embankments will be protected from wash by linings of riprap or paving. A walk will be built on the top of the dam, and other slopes and surfaces covered with soil as directed. The embankments will be separated from the masonry overfall by heavy wing walls. A gate-house with wells and appurtenances as shown will be built next to the north wing wall. The "masonry portion" of the dam will be about 300 feet in length and will be a solid mass of rubble masonry faced with range stones laid in courses.

Where the rock is of poor quality or for other reasons, it may seem to the engineer to be desirable, the core walls both in the center of the embankments and under the masonry section may be carried down deep into the rock.

3. The work to be done in a general way consists in stripping the site of the dam; building up the embankments in layers, and in paving or otherwise protecting their surfaces; doing all blasting, rock and timber work; constructing all masonry; building in all iron work in connection with brick or other masonry; laying pipes through the dam; doing all pumping or other temporary work in connection with the permanent work, and delivering over to said city of Boston the whole structure in a complete condition with the masonry all pointed and with the dam ready to be put into service in accordance with the plans and these specifications. *Work to be Done.*

All work during its progress and on its completion must conform truly to the lines, grades, and levels to be determined and given hereafter by the engineer, and due facilities and such assistance and materials as he may require must be furnished by the contractor without extra charge, and the engineer's marks must be carefully preserved. The work must also be built in accordance with the plans and directions which shall be given by him from time to time, subject to such modifications and additions as said engineer shall deem necessary during the prosecution of the work, and in no case will any work which may be performed, or any materials furnished in excess of the requirements of this contract or of the plans or *Lines,Grades, Levels,Plans.*

of the specifications, be estimated and paid for, unless such excess shall have been ordered by the water board as hereinafter set forth.

Tools and Implements. The contractor is to furnish all temporary flumes, all materials and all tools, implements, machinery, and labor necessary or convenient for doing all the work herein contracted for, with safety to life and property in accordance with this contract, and within the time specified herein; he will be required to construct and put in complete working order the work herein specified, and is to perform and construct all the work covered by this agreement; the whole to be done in conformity with the plans and these specifications; and all parts to be done to the satisfaction of the city engineer.

Soil. 4. The soil is to be removed from the grounds where the dam, embankments, and other works are to stand. Wherever directed by the engineer said soil to be hauled and put in spoil banks, to remain until required to be placed over the finished surfaces of slopes or embankments. The quantities of soil removed will be measured in the spoil banks and paid for as stipulated in article Q. item (a).

The slopes of the embankment are to be covered with soil taken from the spoil banks; if any additional soil is needed for the work, it shall be obtained and taken from such grounds as may be designated by the engineer, and deposited wherever ordered by him; all soil removed from the spoil banks, or from such grounds as the engineer may designate, shall be measured in excavation. It will be rolled or otherwise compacted, and paid for as stipulated in article Q, item (aa).

All surfaces which are required to be afterwards sodded or seeded are to be covered with soil at least twenty-four inches in thickness.

Sodding and Seeding. 5. The embankments of the dam, and such other surfaces as may be designated by the engineer, are to be sodded or seeded with grass seed.

All the surfaces to be sodded or seeded are to be carefully graded and particular care taken to make a true and even bearing for the sods to rest on.

Sods. The sods to be of good quality of earth covered with heavy grass, sound and healthy, and not less than one foot square, and generally of a uniform thickness of three inches. These sizes may be altered by the engineer during the progress of the work. The sods will be cut with a bevel on all sides, so that when laid they will lap at the edges; to be properly set so as to have a full bearing on their whole lower surface; to be padded down firm with a spade or wooden bat made suitable for the purpose; each sod is to be pinned with one wooden pin, not less than fifteen inches long, so as to be secured to the ground beneath it, and to be so laid that the upper surface shall conform to the true slope of the bank or ground and to the lines given by the engineer. No lean, poor, or broken sods will be allowed in the work, but on the outside edges of the bank sods may be cut to such size and shape as will make a proper finish to the same. The engineer may alter all the above sizes during the progress of the work.

The sodding that shall have been laid shall be well and carefully sprinkled with water as often as the engineer shall deem necessary.

6. The engineer may specify the kind, quality, Seeding. and amount of seed to be used on all surfaces ordered to be seeded, and he may also direct the manner of seeding, including rolling and watering:

EARTH EXCAVATION AND EMBANKMENT.

7. Earth excavation is to be made for the foundations, center walls, etc., and for any grading that may be required either above or below the dam, or for any other work in connection with the dam, structures, or appurtenances which the engineer may order, but no payment will be made for earth or other excavation unless specifically staked out and ordered by the engineer. The price bid for excavation will cover all excavations by the contractor for his own convenience or for temporary or protecting work, none of which will be measured or estimated by the engineer.

8. Earth excavation is to be made in accord- Excavation. ance with the lines established by the engineer, and the price herein stipulated for earth excavation— article Q, item (c)—is to include the work of clearing and grubbing the ground of all trees, stumps, bushes, and roots, and burning or otherwise disposing of the same; of sheeting and bracing and supporting and maintaining all trenches and pits during and after excavation; of all pumping, ditching and draining; of clearing the excavation of all wood or other objectionable materials, of selecting the materials, and of hauling and of disposing of the exca-

vated materials in making embankments, in filling,
refilling, and wasting; of rolling and watering, and
all other labor and expenses incidental to the handling of the excavated materials.

Spoil Banks.   9. Whenever, in the opinion of the engineer,
the material excavated from the pits and trenches
can not properly be disposed of in embankment or
for other work at one hauling, it shall be deposited
in spoil banks, and paid for under article Q, item
(c), and if subsequently ordered to be used in the
work, it shall be paid for a second time under article
Q, item (cc).

Measure-   10. All earth work paid for under article Q,
ment. items (c) and (cc), shall be measured in excavation.

Embank-   11. The embankments for the dam shall start
ment. from a well prepared base, stepped on sloping
ground, and shall be carried up in horizontal layers
not exceeding four inches in thickness; every layer
to be carefully rolled, either with heavy grooved
rollers, or steam rollers, and to be well watered.
The earth to be well rammed with heavy rammers
at such points as can not be reached by the rollers.
Special care shall be required in ramming the earth
close to the center wall, which shall always be kept
at least two feet higher than the adjoining embankment, unless otherwise permitted. The embankments of the dam shall be kept at an uniform height
on both sides of the masonry during construction,
and at no time will the down-stream half of the dam
be allowed to be higher than the up-stream portion.

At all times the earth embankment must be
kept three feet above the "masonry portion" of the
dam.

Watering.   12. Ample means shall be provided for watering the banks, and any portion of the embankment
to which a layer is being applied shall be so wet,
when required, that water will stand on the surface.

The contractor shall furnish at his own cost the
necessary steam pumping plant and force-main for
forcing water into a tank situated on the side hill, at
least fifty feet above the top of the dam when completed. From this tank a three-inch distribution
pipe, fitted with gates and hose connections, will
lead lengthwise over the dam to supply water wherever it may be needed. If the engineer approves,
some other method of equal efficiency for the furnishing of water may be substituted for the above

plant. This work is included in the price to be paid for earth excavation.

13. All the grounds covered by the dam and by the borrow pits shall be cleared of all soil, stones, trees, stumps, or other organic or perishable matter, which shall be deposited at such points as shall be designated. If the borrow pits are, in the opinion of the engineer, sufficiently near the dam, the soil or other useful materials may be removed to the spoil banks and measured, otherwise they will not be measured. Stumps and other vegetable substances shall be burned. *Clearing and Grubbing.*

14. The surfaces of embankments shall be dressed smoothly to line and grade to receive the soil or broken stones supporting the paving or riprap.

15. The earth used for the embankments shall be free from perishable material of all kinds, and from stones larger than three inches in diameter, and it shall be of a quality approved by the engineer. The portion of the embankment next to the core-wall on the up-stream side of the dam and the refilling of all trenches will be composed of hard-pan or other fine, compact, or selected material approved by the engineer, who shall decide upon the quality and character of the earth to be used at various places, and it must be selected and placed in accordance with his orders. *Quality of Earth.*

16. All excavation and disposal in embankments and refilling of earth, hard-pan, and other materials, shall be classified and estimated as earth excavation, and paid for at the price hereinafter stipulated, article Q, item (c). *Classification.*

ROCK EXCAVATION.

17. Rock excavation is to include the excavation of all solid rock which can not, in the opinion of the engineer, be removed by picking, and of bowlders of one cubic yard or more in size; the price hereinafter specified—article Q, item (d)—to be paid for rock excavation shall include the work of hauling and disposing of the same in spoil banks or other places.

18. Rock excavation shall be measured in excavation, and estimated for payment in accordance with the lines given by the engineer. No excavation outside of these lines will be estimated. *How Measured.*

19. Rock is to be excavated for the foundations of the dam, core-walls, and gate-house, and wherever the engineer may order.

Steps.
20. In the wall and pipe trenches and in the foundation for the gate-house or other structures, the rock is to be shaped roughly in steps or other form that may be ordered by the engineer.

The price bid for rock excavation is to include the cost of supporting and maintaining the excavations, of pumping and draining, of disposing of the excavated materials as ordered by the engineer, and all other incidental expenses.

Explosives.
21. All rock excavation in the wall trenches and at any other place designated by the engineer is to be made with explosives of a moderate power, under his directions, and not with high explosives. Black powder may be ordered by him to be used in special cases.

22. All rock surface intended for masonry foundation must be freed from all loose pieces, and be firm and solid, and prepared as directed by the engineer.

FOUNDATION WORK.

23. The foundation work for the centre walls of the dam and for other structures is to be extended to such depth and in such a manner as shall be ordered by the engineer. In bad bottom, sheet piling, tonged and grooved, may be ordered to be driven or placed on one or more sides of the work. If the material of excavation is such, in the opinion of the engineer, as to require especial precaution, the trenches for the centre wall and for other structures may be ordered extended to a great depth, beyond the indications of the plans. The position of the bed rock being uncertain, it is impossible to indicate the bottom of the core-wall with accuracy, and it is distinctly understood that the lines for the foundation shown on the plans are not guaranteed by the city to be correct.

PROTECTIVE WORK.

24. The contractor will be required at his own expense to take care of all water which may come down the stream during the progress of the work, and to make good any damage done to the dam from freshets or other action of the water or the elements.

TIMBER.

25. Timber may be ordered used for platforms, for permanent sheet-piling, and for other permanent

uses. It shall be of the sizes and placed in the manner ordered by the engineer.

26. All timber and lumber so used shall be spruce, sound, straight grained, and free from all shakes, loose knots, and other defects. that may impair its strength and durability. The price b.d for timber shall cover all incidental expenses incurred for labor, or for tools or materials used in placing, securing, and fastening it.

27. No payment shall be made to the contractor for lumber used for bracing, sheeting, scaffolding, and other temporary purposes.

28. All sheeting and other timber work in the trenches and pits shall be removed unless it is ordered left in, in which case such timber shall be paid for as herein stipulated—article Q, item (e)—for permanent timber work.

29. The timber to be used for sheet-piling in the foundations and other places may be ordered tongued and grooved. Such timber shall be furnished and placed as ordered, and the price hereinafter stipulated—article Q, item (ee)—for tongued and grooved timber is to cover the cost of placing, driving, securing, and fastening the same. *Tongued and Grooved Timber.*

MASONRY.

30. All masonry, except where otherwise specified, shall be laid in hydraulic cement mortar, and shall be built of the forms and dimensions shown on the plans, as directed by the engineer from time to time, and the system of bonding ordered by the engineer shall be strictly followed.

31. All beds and joints must be entirely filled with mortar, and the work in all cases shall be well and thoroughly bonded.

32. Care must be taken that no water shall interfere with the proper laying of masonry in any of its parts. *Water.*

33. All means used to prevent water from interfering with the work, even to the extent of furnishing and placing pipes for conducting the water away from points where it might cause injury to the work, must be provided by the contractor at his own expense. *Pipes.*

34. Under no circumstances will masonry be allowed to be laid in water.

35. All iron-work, except the sluice-gates, is to be built in the masonry without other compensa- *Iron-work.*

tion than the price herein stipulated to be paid per
cubic yard of masonry. The pipes, special castings
and other iron work will be furnished and deliverd
by the city on the site of the dam, and must then be
carefully protected, handled and laid by the contractor in a thorough manner as directed by the engineer.

Freezing Weather. 36. No masonry is to be built between the
15th of November and the 15th of April, or in freezing weather, except by permission of the engineer.

All masonry to be amply protected from the
action of frost during the winter. The contractor
will be required to make good any damage resulting
from frost on any portion of the work.

Sprinkling. 37. All fresh masonry, if allowed to be built
in freezing weather, must be covered and protected
in a manner satisfactory to the engineer, and during
hot weather all newly-built masonry shall be kept
wet by sprinkling water on it with a sprinkling pot
until it shall have become hard enough to prevent
its drying and cracking, and if necessary canvass
coverings must be provided.

Cement. 38. American cement and Portland cement
are to be used. The American cement must be in
good condition and must be equal in quality to the best
Rosendale cement. It must be made by manufacturers of established reputation, must be fresh and
very fine ground, and in well-made casks. The
Portland cement must be of a brand equal in quality
to the best English Portland cement. To insure its
good quality, all the cement furnished by the contractor will be subject to inspection and rigorous
tests; and if found to be of improper quality, will
be branded and must be immediately removed from
the work; the character of the tests to be determined
by the engineer. The contractor shall, at all times,
keep in store at some convenient point in the vicinity
of the work, a sufficient quantity of cement to allow
ample time for the tests to be made without delay
to the work of construction. The engineer shall be
notified at once of each delivery of cement. It shall
be stored in a tight building, each cask must be
raised several inches above the ground, by blocking
or otherwise.

39. Cement is generally to be used in the form
of mortar with an admixture of sand, and when so
used, its use is included in the price herein stipulated

for the various kinds of masonry. For the foundation work, however, Portland cement may be ordered by the engineer in exceptionally wet and difficult places, to be used with or without any admixture of sand for grouting seams or for such other purposes as he may direct. The cost of placing said cement will be paid by the city, the price to be paid to be estimated by the engineer unless otherwise stipulated. Such· cement is to be paid for per barrel of .four hundred pounds, furnished and delivered by the contractor at the place where it must be used. .See article Q, item (*f*).

40. All mortar shall be prepared from cement **Mortar.** of the quality before described, and clean, sharp sand. These ingredients shall be thoroughly mixed dry, as follows: The proportion of cement ordered, by measure, with the ordered proportion of sand, also by measure; and a moderate dose of water is to be afterwards added to produce a paste of proper consistency; the whole to be thoroughly worked with hoes or other tools. In measuring cement it shall be packed as received in casks from the manufacturer. The mortar shall be freshly mixed for the work in hand, in proper boxes made for the purpose; no mortar to be used that has become hard or set. If the mortar ingredients are mixed at some distance from the work, water shall not be added until the mortar has been brought to the dam and is ready for use.

41. The price herein stipulated for the various kinds of masonry is contingent on the use of a mortar made of a mixture of one part in a volume of American cement to two parts of sand. Additional prices are herein stipulated for the use of mortars formed with a different mixture of cement and sand. Article Q, items (*s*), (*t*), (*u*), (*v*).

42. The concrete shall be formed of, sound **Concrete.** broken stones or screened gravel stones not exceeding two inches at their greatest diameter. All stones in any way larger are to be thrown out. The materials to be cleaned from dirt and dust before being used; to be mixed in proper boxes, with mortar of the quality before described, in the proportion of five parts of broken stone to one part of cement; to be laid immediately after mixing, and to be thoroughly compacted throughout the mass by ramming till the water flushes to the surface; the amount of water used for making the concrete to be approved or

directed by the engineer. The concrete shall be allowed to set for twelve hours, or more, if so directed, before any work shall be laid upon it; and no walking over or working upon it shall be allowed while it is setting. Article Q, item (*g*).

43. Whenever ordered by the engineer the concrete shall be formed of broken stone not exceeding one inch at their greatest diameter, used in the proportion of three parts of broken stone to one part of cement. Article Q, item (*gg*).

Plastering.   44. The up-stream faces of all core-walls, and such other surfaces as the engineer may direct, will be thoroughly plastered with a half inch coat of Portland cement plastering put on in two portions as follows: Next the concrete a thick coating of Portland cement mortar will be put on, mixed in the proportion of one part of cement to one of sand, rubbed to a uniform surface and left rough; over this will be smoothly spread with trowels a coat of neat Portland cement which shall be thoroughly worked to make a perfectly water-tight surface. All plastering will be measured and paid for by the square yard of superficial surface as per article Q, item (*h*).

45. The bricks shall be of the best quality of hard-burned bricks; burned hard entirely through, regular and uniform in shape and size, and of compact texture. To insure their good quality, the bricks furnished by the contractor will be subject to inspection and rigorous tests, and if found of improper quality will be condemned, the character of the tests to be determined by the engineer. They are to be culled before laying at the expense of the contractor, and all bricks of an improper quality shall be laid aside and removed; the engineer to be furnished with men for this purpose by and at the expense of the contractor.

Brick Masonry.   46. All brick masonry shall be laid with bricks of the quality before described and in Portland cement mortar mixed one part of cement to two of sand. No "bats" shall be used except in the backing, where a moderate proportion (to be determined by the engineer) may be used, but nothing smaller than "half bricks." The bricks to be thoroughly wet just before laying. Every brick to be completely imbedded in mortar under its bottom and on its sides. Care shall be taken to have every joint full of mortar and all joints shall be pointed.

47. All centering shall be made, put up, and Centering. removed in a manner satisfactory to the engineer.

48. All stone masonry is to be built of sound, Stone clean quarry granite stone of quality and size satis- Masonry. factory to the engineer; all joints to be full of mortar, unless otherwise specified.

49. Paving is to be laid without mortar, and is Paving. to be used for portions or the whole of the slopes of the dam embankments, and at any other place that may be designated.

50. This work is to be measured in accordance with the lines shown on the drawings or ordered during the progress of the work. The stones used must be roughly rectangular; all irregular projections and feather edges must be hammered off. No stone will be accepted which has less than the depth represented on the plans or ordered. Each stone used must be set solid on the foundation of broken stone or earth and no interstices must be left.

51. After the slopes which are to receive the Broken paving have been dressed, a layer of broken stone, Stones. nine inches thick or less, is to be spread as a foundation for the paving wherever ordered. The broken stone must be sound and hard, not exceeding two inches at the greatest diameter. Broken stones may be used also wherever the engineer may direct, and paid for under this head. Article Q, item (*l*). The cost of the broken stone used for making concrete is included in the price hereinbefore stipulated for concrete laid.

52. Riprap instead of paving may be used for covering a large portion of the dam slopes, and Riprap. wherever the engineer may order. It shall be made of stone of such size and quality and in such manner as he shall direct, and must be roughly laid by hand. It will generally be put on in thick layers, and if found cheaper will probably be substituted for paving on the lower slopes of the dam below the berm.

53. Rubble-stone masonry is to be used for the Rubble. central part of the dam, for the wing-walls of the earth embankments, for the gate-house, and wherever ordered by the engineer.

It shall be made with sound clean stones of compact texture, free from loose seams and other defects. They must have roughly rectangular forms, and all irregular projections and feather edges must be hammered off before the stones are set. The beds

25

must be good for materials of this class and must present such even surfaces that when lowering a stone on the surface prepared to receive it, there may be no doubt that the mortar will fill all spaces.

After the bed-joints are thus secured, a moderate quantity of spalls can be used in the preparation of suitable surfaces for receiving other stones. No spalling up under a stone after it is laid will be allowed, neither will any grouting or filling of joints be allowed after the stone is set. Especial care is to be taken to have every stone entirely surrounded by mortar.

The quality of the beds is to regulate, to a large extent, the size of the stones used, as the difficulty of forming a good bed-joint increases with the size of the stones. Various sizes must be used.

Generally the largest stones are not to measure more than twenty cubic feet, and they are to be used in the proportion of about twenty-five per cent. of the whole, but they must be omitted partially or entirely if their beds are not satisfactory. It is expected that one quarter of the stones used will be of such size that two men can handle them. The balance to be composed of intermediate sizes. Regular coursing to be avoided.

Broken Ashlar. 54. The exposed faces of the wing walls, retaining walls, and of any other rubble work that the engineer may designate, are to be made of broken ashlar with joints not exceeding one half inch in thickness; the stones not to be less than 12 inches deep from the face, and to present frequent headers. The joints shall be pointed with neat Portland cement. This face work is to be paid for by the square foot of the superficial area for which it is ordered in addition to the price paid per cubic yard of rubble-stone masonry, but the right is reserved to change this masonry to range work, should it be for the interest of the city so to do. Article Q, item (n).

Rangework. 55. The outer faces of the masonry dam, and if found best the gate-chamber and any other masonry that may be d signated, are to be made of range stones, as shown on the plans, the stones to be of unobjectionable quality, sound and durable, free from all seams and other defects, and of such kind as shall be approved by the engineer. They shall be pointed with neat Portland cement.

All beds, builds, and joints are to be cut true to a depth of not more than 4 inches, and not less than

3 inches from the faces and to surfaces allowing of one. half inch joints at most; the joints for the remaining part of the stones not to exceed 2 inches in thickness at any point.

56.   All cut arrises to be true, well defined, and **Arrises.** sharp.

57.   Where this class of masonry joins with dimension stone masonry the courses must correspond, and the joining with arches and other dimension stone masonry must be accurate and workmanlike.

Each course to be composed of two stretchers **Bond.** and one header alternately, the stretchers not less than 3 feet long nor more than 7 feet long.

58.   The rise of the courses may vary from bot- **Courses.** tom to top from 30 inches to 15 inches in approximate vertical progression, and the width of bed of the stretchers is not to be at any point less than the height nor less than 24 inches.   The headers are not to be less than 4 feet in length.

This class of masonry, including the headers, is **Measure-** to be estimated at 30 inches thick throughout.   In no **ment.** case are the tails of the headers to be estimated.

59.   The coping of the wing walls will be **Coping.** classed as coping stone masonry.   The surfaces will be rough pointed to the circular forms given.   The capping stones to the posts will be estimated as dimension stone with hammered surfaces.

60.   The prices herein stipulated for range and **Prices.** broken ashlar stone masonry are to cover the cost of pointing, of cutting chisel drafts at all corners and angles in the work, and of preparing the rock faces; but if any six-cut work is ordered in connection with this class of masonry it shall be paid for at the prices hereinafter stipulated for such work.   Article Q, item (r).

61.   The face bond must not show less than 12 inches lap unless otherwise permitted.

62.   The pointing of the faces of all masonry **Pointing.** in the dam, gate-house, and wings to be thoroughly done with neat Portland cement after the structures are completed, every joint to be raked out therefor to a depth of at least 2 inches, and if the engineer is satisfied that the pointing at any place is not properly done it must be taken out and done over again.   The cement is to be mixed in small quantities and applied before its first setting.

Dimension Stone Masonry. 63. Dimension stone masonry must be made of first-class granite of moderately uniform color, free from all seams, discoloration, and other defects, and satisfactory to the engineer. The stones shall be cut to exact dimensions, and all angles and arrises shall be true, well defined, and sharp. All beds, builds, and joints are to be dressed for the full depth of the stone, to surfaces allowing of one quarter ($\frac{1}{4}$) inch joint at most. No plug-hole of more than 6 inches across or nearer than 3 inches to an arris is to be allowed, and in no case must the aggregate area of the plug-hole in any joint exceed one quarter of its whole area.

The stone shall be laid with one quarter ($\frac{1}{4}$) inch joints, and all face joints shall be pointed with mortar made of neat Portland cement, applied before its first setting. All joints to be raked out to a depth of two inches before pointing; the cost of pointing to be included in the price stipulated for cut stone masonry.

Rock-face. 64. In rock face work the arrises of the stones enclosing the rock face must be pitched to true lines; the face projections to be bold, and from 3 to 5 inches beyond the arrises. The angles of all walls or structures having rock faces are to be defined by a chisel draft not less than $1\frac{1}{2}$ inches wide on each face.

Hammered Work. 65. In fine hammered work the face of the stones must be brought to a true plane and fine dressed, with a hammer having six blades to the inch.

For fine hammer-dressing (six-cut work) the price stipulated in article Q, item (r), per superficial square foot of dressing will be paid in addition to the price per cubic yard of masonry.

Grooves. 66. No payment will be made for cutting grooves and recesses other than the price paid for the dressing of their surfaces, which are to be fine hammered.

Walk. 67. The contractor will build a walk upon the top of the earthen embankments. It will be 8 feet wide and 1 foot in depth, composed of broken stone 9 inches in depth and a thin layer of selected screenings and binding gravel (as ordered). The surface will be moistened and rolled with a hand roller as directed. The broken stone screenings and gravel used in this walk will not be included in any other measurement. Payment will be made for the finished walk according to the number of linear or running feet it may contain.

GENERAL CLAUSES.

68. If any person employed by the contractor on the work should appear to the engineer to be incompetent, or to act in a disorderly or improper manner, he shall be discharged immediately on the requisition of the engineer, and such person shall not be again employed on the work.

Incompetent Workmen.

69. Any materials condemned or rejected by the engineer or his representatives may be branded, or otherwise marked, and shall, on demand, be at once removed to a satisfactory distance from the work.

Materials Branded.

70. Any unfaithful or imperfect work which may be discovered before the final acceptance of the work shall be corrected immediately, and any unsatisfactory materials delivered shall be rejected on the requirement of the engineer, notwithstanding that they may have been overlooked by the proper inspector. The inspection of the work shall not relieve the contractor of any of his obligations to perform sound work, as herein prescribed; and all work, of whatever kind, which, during its progress and before it is finally accepted, may become damaged from any cause shall be removed, and replaced by good and satisfactory work.

Imperfect Work.

71. Whenever the contractor is not present on any part of the work where it may be desired to give directions, orders will be given by the engineer to, and shall be received and obeyed by, the superintendent or foreman who may have charge of the particular work in relation to which the orders are given.

Orders Obeyed.

72. In all the operations connected with the work herein specified, all laws or regulations controlling or limiting in any way the actions of those engaged on the works, or affecting the methods of doing the work or materials applied to it, must be respected and strictly complied with; and during the progress of the work the contractor shall provide such precautions as may be necessary to protect life and property.

Laws.

73. After the completion of the work the contractor is to remove all temporary structures built by him, and all surplus materials of all kinds from the site of the work, and to leave them in neat condition.

Clearing up.

Subletting. **H.** The contractor agrees that he will give his personal attention to the fulfillment of this contract; and that he will not sublet the aforesaid work, but will keep the same under his control, and that he will not assign, by power of attorney or otherwise, any portion of the said work, unless by and with the previous consent of the water board, to be signified by endorsement on this agreement.

Ways and Means. **I.** The contractor shall furnish the necessary scaffolding, ways, and all necessary means and conveniences for the transfer of the material to its proper place and for its erection. And it is also to be understood that the city shall not be held responsible for the care or protection of any materials or parts of the work until its final acceptance.

Access. **J.** It is further agreed that the engineer, or his authorized agent and assistants, shall at all times have access to the work during its progress; and he shall be furnished with every reasonable facility for ascertaining that the work being done is in accordance with the requirements and intention of this contract.

Alteration. **K.** Should it be found desirable by the water board to make alterations in the form or character of any of the work, the said water board may order such alterations to be made, defining them in writing and drawings, and they shall be made accordingly; *provided*, that in case such changes increase the cost of the work, the contractor shall be fairly remunerated; and in case they shall diminish the cost of the work, proper deduction from the contract price shall be made; the amount to be paid or deducted to be decided by the city engineer.

Extra Work. **L.** The contractor hereby agrees that he will do such extra work as may be required by the water board for the proper construction or completion of the whole work herein contemplated; that he will make no claims for extra work unless it shall have been done in obedience to a written order from the said water board or their duly authorized agent; that all claims for extra work done in any month shall be filed in writing with the engineer before the fifteenth of the following month; and that, failing to file such claims within the time required, all rights for pay for such extra work shall be forfeited. The price to be paid for all extra work done shall be its actual reasonable cost to the contractor, as determined by the city engineer, plus fifteen per cent.

M. The contractor is to use such appliances for the performance of all the operations connected with the work embraced under this contract as will secure a satisfactory quality of work and a rate of progress which, in the opinion of the engineer, will secure the completion of the work within the time herein specified. If, at any time before the commencement or during the progress of the work, such appliances appear to the engineer to be inefficient or inappropriate for securing the quality of the work required or the said rate of progress, he may order the contractor to increase their efficiency or to improve their character, and the contractor must conform to such order; but the failure of the engineer to demand such increase of efficiency or improvement shall not relieve the contractor from his obligation to secure the quality of work and the rate of progress established in these specifications.

Appliances.

N. The said contractor further agrees that if the work to be done under this contract shall be abandoned, or if at any time the engineer shall be of the opinion, and shall so certify in writing to the water board, that the said work is unnecessarily or unreasonably delayed, or that the said contractor is willfully violating any of the conditions or agreements of this contract, or is not executing said contract in good faith, or fails to show such progress in the execution of the work as will give reasonable grounds for anticipating its completion within the required time, the said water board shall have power to notify the said contractor to discontinue all work, or any part thereof, under this contract; and thereupon the said contractor shall cease to continue said work, or such part thereof, as the said water board may designate; and the said water board shall thereupon have the right, at their discretion, to contract with other parties for the delivery or completion of all or any part of the work left uncompleted by said contractor, or for the correction of the whole or any part of said work. And in case the expense so incurred by said water board is less than the sum which would have been payable under this contract if the same had been completed by the said contractor, then the said contractor shall be entitled to receive the difference; and in case such expense shall exceed the last said sum, then the contractor shall, on demand, pay the amount of such excess to the said city, on notice from the said water board of the excess so due; but such excess to be paid by the contractor shall not exceed the amount of the security for the performance of this contract.

O. The said contractor further agrees that the said water board may, if they deem it expedient to do so, retain out of and amounts due to the said contractor sums sufficient to cover any unpaid claims of mechanics or laborers for work or labor performed under this contract; *provided*, that notice in writing of such claims, signed by the claimants, shall have been previously filed in the office of the city clerk.

P. The said contractor further agrees that he will indemnify and save harmless said city from all claims against said city, under chapter one hundred and ninety-one of the Public Statutes of Massachusetts, and any laws passed since the Public Statutes, with reference to liens on buildings and lands, for labor done and materials furnished under this contract, and shall furnish the said water board with satisfactory evidence, when called for by them, that all persons who have done work or furnished materials under this contract, for which the said city may become liable, and all claims from the various departments of the city government, or private corporations, or individuals, for damage of any kind caused by the construction of said work, have been fully paid or satisfactorily secured; and in case such evidence is not furnished, an amount necessary and sufficient to meet the claims of the persons aforesaid shall be retained from any moneys due, or that may become due, the said contractor under this contract, until the liabilities aforesaid shall be fully discharged or satisfactorily secured.

The said contractor further agrees that he will indemnify and save harmless the said city from all suits or actions, of every name and description, brought against the said city for or on account of any injuries or damages received or sustained by any person or persons, by or from the said contractor, his servants or agents, in the construction of said work, or by or in consequence of any negligence in guarding the same, or any improper materials used in its construction, or by or on account of any act or omission of the said contractor or his agents; and the said contractor further agrees that so much of the money due him under and by virtue of this agreement as shall be considered necessary by the said engineer may be retained by the said city until all such suits or claims for damages as aforesaid shall have been settled, and evidence to that effect furnished to the satisfaction of the said engineer.

Q. And the said contractor further agrees to receive the following prices as full compensation for furnishing all the materials, and for doing all the work contemplated and embraced in this agreement; also, for all loss or damage arising out of the nature of the work aforesaid, or from the action of the elements, or from any unforeseen obstruction or difficulties which may be encountered in the prosecution of the same: and for all risks of every description connected with the work; also, for all expense incurred by or in consequence of the suspension or discontinuance of said work as herein specified, and for well and faithfully completing the work, and the whole thereof, in the manner and according to the plans and specifications, and the requirements of the engineer under them, to wit:

(*a*) For the removal of soil excavated and placed in spoil banks, including all incidental work, the sum of ―――― ($――――) per cubic yard.

(*aa*) For the removal of soil taken from spoil banks or from other places and placing on the slopes of the embankment, including all incidental work, the sum of ―――― ($――) per cubic yard.

(*b*) For sodding, including all incidental work, the sum of ―――― ($――) per superficial square yard.

(*bb*) For seeding, including all incidental work, the sum of ―――― ($――) per acre.

(*c*) For earth excavation, including its disposal in embankments and refilling, or as otherwise ordered by the engineer, and all incidental work, the sum of―――― ($――) per cubic yard.

(*cc*) For rehandling of excavated materials from spoil banks, and placing, including all incidental work, the sum of―――― ($――) per cubic yard.

(*d*) For rock excavation, including its disposal, and all incidental work, the sum of―――― ($――) per cubic yard.

(*e*) For permanent timber work, except tougued and grooved timber, placed, including all incidental work, the sum of―――― ($――) per thousand feet B. M.

(*ee*) For permanent timber work, tongued and grooved, placed, including all incidental work, the sum of——— ($———) per thousand feet B. M.

(*f*) For Portland cement ordered by the engineer, delivered where ordered on the work, in barrels containing four hundred pounds, including all incidental work, the sum of——— ($———) per barrel.

(*g*) For concrete masonry, in place, formed of five parts of broken stone or screened gravel, to one part of cement, and made with American cement mortar mixed in the proportion of one part of cement to two parts of sand, including all incidental work, the sum of——— ($———) per cubic yard.

(*gg*) For concrete masonry, in place, formed of three parts of broken stone or screened gravel to one part of cement and made with American cement mortar mixed in the proportion of one part of cement to two parts of sand, including all incidental work, the sum of——— ($———) per cubic yard.

(*h*) For plastering all concrete walls with Portland cement, including all incidental work, the sum of——— ($———) per superficial square yard.

(*i*) For brick masonry, laid in Portland cement mortar mixed in the proportion of one part of cement to two parts of sand, and including all pointing, centering, etc., and removing the same, and all incidental work, the sum of ——— ($———) per cubic yard.

(*j*) For paving in place, including all incidental work, the sum of ——— ($———) per cubic yard.

(*k*) For riprap in place, including all incidental work, the sum of ——— ($———) per cubic yard.

(*l*) For broken stone in place (other than that used in making concrete and the walk), including all incidental work, the sum of ——— ($———) per cubic yard.

(*m*) For rubble-stone masonry, laid in American cement mortar mixed in the proportion of one part of cement to two parts of sand, including all incidental work, the sum of ——— ($———) per cubic yard.

(*n*) For face work of broken ashlar, in addition to the price paid per cubic yard as rubble, including pointing in neat Portland cement, and all incidental work, the sum of ——— ($———) per superficial square foot.

(*o*) For facing stone masonry of range stones laid in American cement mortar mixed in the proportion of one part of cement to two parts of sand and pointing in neat Portland cement, including all incidental work, the sum of——— ($———) per cubic yard.

(*p*) For coping laid in place, and pointed in neat Portland cement, including all incidental work, the sum of ——— ($———) per linear or running foot.

(*q*) For dimension stone masonry laid in American cement mortar mixed in the proportion of one part of cement to two parts of sand, including pointing in neat Portland cement, centering, etc., and all incidental work, the sum of ——— ($———) per cubic yard.

(*r*) For fine hammer dressing (six-cut work) the sum of ——.— ($———) per superficial square foot.

(*s*) For all kinds of masonry laid in American cement mortar mixed in the proportion of one part of cement to one part of sand, in addition to the prices per cubic yard hereinbefore stipulated to be paid for the same class of masonry laid in American cement mortar mixed in the proportion of one part of cement to two parts of sand, the sum of——— ($———) per cubic yard.

(*t*) For all kinds of masonry laid in Portland cement mortar mixed in the proportion of one part of cement to one part of sand, in addition to the prices per cubic yard hereinbefore stipulated to be paid for the same class of masonry laid in American cement mortar mixed in the proportion of one part of cement to two parts of sand, the sum of ——— ($———) per cubic yard.

(*u*) For all kinds of masonry laid in Portland cement mortar mixed in the proportion of one part of cement to two parts of sand, in addition to the prices per cubic yard hereinbefore stipulated to be paid for the same class of masonry laid in American cement mortar mixed in the proportion of one part of cement to two parts of sand, the sum of ——— ($———) per cubic yard.

(*v*) For all kinds of masonry laid in Portland cement mortar mixed in the proportion of one part of cement to three parts of sand, in addition to the price per cubic yard hereinbefore stipulated to be paid for the same class of masonry laid in American cement mortar mixed in the proportion of one part of cement to two parts of sand, the sum of ———— ($——) per cubic yard.

(*w*) For building walk, including all incidental work, the sum of ———— ($——) per linear or running foot.

(*x*) For all extra work done by written order of the Boston Water Board, its actual reasonable cost to the contractor, as determined by the engineer, plus fifteen per cent. of said cost.

R.   And it is agreed that payment for the work embraced in this contract shall be made in the following manner:

A payment will be made, on or about the first day of each month, of 85 per centum of the value of the work completed in place by the contractor on the fifteenth of the previous month, as estimated by the engineer.

Provided, however, that the making of such payment may be deferred from month to month, when, in the opinion of the engineer, the value of work done since the last estimate for payment is less than one thousand dollars.

The said contractor further agrees that he shall not be entitled to demand or receive payment for any portion of the aforesaid work or materials, until said work shall have been completed to the satisfaction of the city engineer, and the said city engineer shall have given his certificate to that effect; whereupon the said city will, within forty days after such completion, and the delivery of such certificate, pay the said contractor the whole amount of money accruing to the said contractor under this contract, excepting such sum or sums as may be lawfully retained by said city.

Provided, that nothing herein contained be construed to affect the right hereby reserved of the said water board to reject the whole or any portion of the aforesaid work, should the said certificate be found or known to be inconsistent with the terms of this agreement, or otherwise improperly given.

S.   The parties hereto further agree that this contract shall be in writing, and executed in triplicate, one of which triplicates shall be kept by the said engineer, one to be delivered to the city auditor of said Boston, and one to the said contractor; that this contract shall be utterly void as to the said city if any person appointed to any office, or employed by virtue of any ordinance of said city, is either directly or indirectly interested therein.

And the said contractor, further agrees that he will execute a bond in the sum of one hundred thousand dollars ($100,000) and with such sureties as shall be approved by the said Boston Water Board, to keep and perform well and truly all the terms and conditions of this contract on his part to be kept and performed and to indemnify and save harmless the said water board as herein stipulated.

T.   And it is also to be understood and agreed that, in case of any alterations, so much of this agreement as is not necessarily affected by such alterations shall remain in force upon the parties hereto.

U.   And the said contractor hereby further agrees that the payment of the final amount due under this contract and the adjustment and payment of the bill rendered for work done in accordance with any alterations of the same, shall release the city from any and all claims or liability on account of work performed under said contract or any alteration thereof.

In Witness Whereof, the parties to these presents have hereunto set their hands the year and day first above written.

The City of Bos-
ton, by its Boston
Water Board.

SIGNED in the presence of

Know all Men by these Presents,

That we———————————————— are held and firmly bound unto the CITY OF BOSTON, in the sum of———————————————— dollars, to be paid to the CITY OF BOSTON, or its certain attorney, its successors and assigns, for which payment, well and truly to be made, we bind ourselves, our heirs, executors, and administrators, jointly and severally, firmly by these presents.

The Condition of this obligation is such that if the above-bounden——————————————— shall well and truly keep and perform all the terms and conditions of the foregoing contract for building Dam No. 5, in the town of Southborough on—— part to be kept and performed, and shall indemnify and save harmless the said CITY OF BOSTON, as therein stipulated, then this obligation shall be of no effect; otherwise it shall remain in full force and virtue.

In Witness Whereof, we hereto set our hands and seals on this——day of——in the year eighteen hundred and ninety-three.

....

....

....

....

SIGNED AND SEALED in presence of

....

....

....

....

174. SPECIFICATIONS FOR THE STRUCTURAL IRON WORK

OF A

HOTEL BUILDING,

TO BE ERECTED ON THE SOUTHWEST CORNER OF 34TH STREET

AND 5TH AVENUE FOR JOHN JACOB ASTOR.

H. J. HARDENBERGH,	PURDY & HENDERSON,
Architect,	Consulting Engineers,
New York.	New York and Chicago.

MAY, 1895.

In order to understand the business relations involved in the following specifications, some explanation of them is necessary.

Messrs. Purdy and Henderson, the consulting engineers, are under contract with Mr. H.J.Hardenbergh, architect, to furnish those parts of the plans and specifications for the building which relate to the iron and steel frame work. They are also under contract with Mr. Downey, the agent of the owner, to prepare all the shop drawings, to supervise the inspection, to superintend the erection of the steel frame work, to check all bills rendered by the contractor for this portion of the work. and, in general, to see that all the contracts relating to this part of the building are faithfully fulfilled. The contract for the iron and steel work was let on a pound basis erected. A separate set of specifications were prepared for the inspection of the work, and also one for the use of the computers and draftsmen in preparing detail plans. It will thus be seen that the consulting engineers are under contract to do a great deal more in this matter than is usually expected of the architect, and much more, therefore, than the architect could afford to pay for, if all this service had to be remunerated out of his professional fees. In the most common practice, the owner checks his own bills, pays the contractor for the shop drawings and divides the remaining portion of this additional service with the architect. Only a small portion of the additional fee paid the engineer by this arrangement is consequently an added expense. It is important that consulting engineers should make contracts with the owner for the additional detail work and supervision as well as with the architect for the preparation of the general plans. This kind of a double connection is desirable and likely to secure the most satisfactory service.

The steel construction described in these specifications is that for a new hotel adjoining The Waldorf on the north, corner 34th street and Fifth avenue, New York city. The building is in plan 350 feet by 100 feet, and is sixteen stories high above the sidewalk, with basement and sub-basement, extending 35 feet below ground. It is the largest steel con-

structed building ever designed, containing over 10,000
tons of structural iron. The exterior of the building is
finished with stone to the height of three stories above
the sidewalk, and with brick, with terra cotta trimmings,
above that line. The construction involves many unusual
conditions, such as a ballroom on the second story· 100
feet long, and 85 feet wide, with vaulted ceiling reaching to
the fifth floor. The floors above this great room, and also the
roof, are carried on two trusses extending through four stories,
the total load carried by the two being about nine million
pounds. The columns in the walls around this ballroom are
from 60 to 70 feet in length, and some of them carry over three
million pounds each. There is also a large dining room on the
first floor, which necessitates the use of very heavy trusses, and
difficult and costly work. The spaces between the columns
are unusually long, 35 and 40 feet being common, thus requiring an unusual quantity of plate girder work. In several other
places in the building, rooms extend through two stories, and
the roof on three sides has a Mansard slope fifty feet in height,
with large towers on the three street corners. All the details
have been worked out with great care, and the business relations of the engineer of construction to both the architect and
the owner are considered ideal.

Specifications Explained: These specifications are supplemental to the contract already entered into for the construc-
tural iron and steel work of this building, between ————
——, parties of the first part,* and——————, parties of the
second part. They are the specifications referred to in the first
clause of said contract, and which are to be considered a part
of that contract.

These specifications are intended to cover all the structural
iron work in said building. They are intended to co-operate
with the drawings for the same, both those furnished by the
architect and those furnished by the engineers as hereinafter
specified, and what is called for by either, is as binding as if
called for by both. They are intended to describe and provide

---

* Mr. Astor's agent, who stands as the party of the first part in these specifications,
is Mr. John Downey, and he is so named in various parts of the document.

for a finished piece of work. The contractor will understand
that the steel construction herein described is also to be complete
in every detail, and in every portion of the work, and all material entering into it is to be first-class, and he will be expected
to thoroughly understand the construction and to fully inform
himself in regard to any points that he may not clearly understand, for what is herein intended to be described, viz.: The
complete and perfect construction of the building is the thing
required. When necessary or desirable, he must apply to the
architect or the engineers for further details or specifications
during construction or before proceeding with his work.

Requirements Outlined: This contractor must furnish and
set all the iron and steel shown or referred to in these specifications and called for by the said drawings hereinbefore referred
to, and when the erection is completed, he must remove all the
materials used in performing the work. He must furnish in all
cases the exact sections, weights and kinds of material that are
called for, and he must follow exact details, methods and instructions called for by these specifications and said drawings.
He must set the iron work as fast as may be considered practical in the judgment of the architect, always keeping at least one
story in advance of the masonry. He will be expected to give
this work his personal supervision, or have a capable man at
all times to take care of it. He must also do all the cutting and
fitting that may be required in his work to receive the work of
other contractors.

Reference in Case of Dispute: Should any difference of
opinion or dispute arise in relation·to the meaning of these
specifications, or of the said drawings furnished by either the
architect or the engineers as hereinafter specified, reference
must be made to the engineers, but if their decision is not satisfactory appeal may be made to the architect, whose decision
on all such points shall be final and conclusive.

Drawings: The general dimensions, arrangement and sections required for the structural iron work herein specified, are
shown on the general structural iron drawings prepared and furnished by the architect, and included in pages ——————
to ————,inclusive.

The sections given are those of the Carnegie Steel Company's manufacture. In general, these drawings are made to
scale, but scale dimensions must never be used. These drawings, together with these specifications, are the property of the
architect, to whom all copies must be returned on the completion of the work. Detail or shop drawings, including drawings
of every part and piece of the work, with all the lists, schedules,
indexes, erection plans or other directions necessary for the
proper manufacture, finish and erection of the work covered

26

by these specifications, and the said general drawings prepared by the architect, will be made and furnished by the engineers.

Blue prints of the shop drawings, lists and schedules, as many copies of each as are necessary, but not more than five, will be furnished to the contractor for his use in the manufacture of the material. Another complete set of these prints, together with one complete set of prints of the erection drawings, will be furnished to the contractor for use in erection. One complete set of all the drawings, plans, lists and schedules will be furnished to the inspector. All the above-mentioned prints will be furnished by the engineers, free of expense. Additional prints of any of these drawings may be taken by said contractor or inspector, if desired, at their own expense, but originals taken from the office for that purpose must be promptly returned.

Orders: All materials required to be furnished or work to be done under these specifications or by the said general structural iron drawings, prepared by the architect, will be ordered by the engineers from time to time with the shop drawings, lists, schedules, etc., for the same, as fast as they can be prepared, and the contractor for the structural iron work must order no material and perform no work under these specifications until he has received the said detail drawings, lists and schedules for the same. Bolts or other material used temporarily for erection purposes are not included in this specification.

Extras and Bills: No additional work or material, over and above what is called for by said detail drawings, lists and schedules, prepared and furnished as hereinbefore provided, will be allowed unless ordered by the architect in writing. When said detail drawings, lists and schedules are received by the contractor, they must be immediately examined to determine whether the material and work called for by the same may be properly classified in the price classification contained in the contract hereinbefore referred to, and of which these specifications are considered a part; or, in any supplemental agreement that may be made to said contract. In case either or both may not be properly classified, in said price classification, the engineers must be promptly notified of the fact in writing, and a copy of such notification must also be sent to the architect. If no reply, verbal or written, to such notification is received within three days, a second notification must be sent the same as the first, but, in any case, the work called for by such detailed drawings, lists or schedules must proceed without delay, unless the contractor shall receive written instructions to the contrary from the architect or engineers.

No bill for extra work ordered by the architect as herein provided, or not called for by said drawings, lists or schedules will be approved by the architect unless it is rendered immediately upon the completion and acceptance of said work. All bills for material or work not properly included in the price classification hereinbefore referred to, must be made separate from the bills for work and material properly covered by said price classification. All bills must be made sufficiently in detail to permit of their ready verification. The originals of all bills must be sent to the engineers, Purdy & Henderson, and exact duplicates must, at the same time, be sent to John Downey, parties of the first part in the contract hereinbefore referred to.

Building Laws: This contractor must comply with all municipal or corporation ordinances and the laws and regulations relating to buildings in the city of New York.

Risks: This contractor will be liable and responsible for any damage to life, limb or property that may arise or occur to any party whatever, either from accident or owing to his negligence, or that of his employees during the operations of constructing or completing the works herein specified.

Rubbish: This contractor must remove from the premises all rubbish arising from his operations as the work proceeds and at completion of same.

Signs: No signs of any description will be allowed to be placed on or about the building or premises.

Co-operation and Cleaning Up: This contractor must cooperate with the contractors for the other parts of the building, so that when completed it shall be in accordance with the architect's design and a complete and perfect piece of work. He must arrange and carry on his work in such a way that the other contractors shall not be delayed, subject always to the architect. When his work is finished he must remove from the premises all the tools, apparatus, machinery, scaffolding, and the debris pertaining to his part of the work, and leave the job free from all obstruction.

Kind of Material Required: All material required for the trusses, and all the material required for the flanges of riveted girders must be open hearth steel.

All other material required for riveted members, and the beams and channels used in the floors with their connections, may be made of Bessemer steel, unless in special cases, it shall be otherwise specified.

Pins over five inches in diameter must be of forged steel.

All machine driven rivets must be of steel.

Tie rods, bolts, anchors, lateral ties and all hand driven rivets must be of wrought iron.

Bearing plates in masonry, bases under columns, separators, brackets under plates, and filler blocks more than 1½ inches thick, must be made of cast iron.

Shoes for trusses and column blocks where required must be made of cast steel.

Character and finish of materials: All the steel used in this building must comply with the following specifications:

	Medium Steel.	Soft Steel.
Maximum ultimate strength in lbs. per sq. in..	68,000	60,000
Minimum ultimate strength in lbs. per sq. in..	60,000	
Minimum elastic limit in lbs. per sq. in........	32,000	30,000
Minimum percentage of elongation in 8 inches..	22%	26%

Test pieces of medium steel must bend cold 180° about a diameter equal to the thickness of the piece without any sign of fracture on the convex side of the bends. Test pieces of soft steel must bend cold 180° flat without any sign of fracture on the convex side of the bend. They must also stand the same bend after being heated to a light cherry red and quenched in water whose temperature is 82° Fahrenheit.

Soft steel must be used for rivets and medium steel for all other material. All steel must have a smooth surface and must be free from all faults or defects of any kind or of any indication of unsoundness. Each piece must be straight, free from wind and of proper section. A variation in weight either way of more than 2 per cent. from that specified shall be cause for rejection.

Eye bars used in trusses must comply with the following specifications in full size tests:

Ultimate strength in lbs. per sq. in. not less than................58,000
   "   "   " "   " " " " more " 66,000
Elastic limit in lbs. per square inch not less than 29,000
Elongation in 2 ft. length of bar nearest fracture..:........ ...... 15%
Reduction of area 40%

The fracture must take place in the body of the bar and must be generally silky. The mill requirements for material for eye bars must be as specified for by the manufacturers of the bars.

All wrought iron used in this building must have an ultimate strength of not less than 48,000 lbs. per square inch, an elastic limit of not less than 26,000 lbs. per square inch, and an elongation of 20 per cent. in 8 inches. The wrought iron required for bolts and rivets must be so ductile that test pieces will bend cold 180 degrees flat without any sign of fracture on the convex side of the bend. All the wrought iron must be perfectly welded in rolling, fibrous, uniform and free from all defects. · Each piece must be straight and of proper section.

All the cast steel used in this building must have an ultimate strength of not less than 60,000 lbs. per square inch, an elastic limit of not less than 32,000 lbs. per square inch, and an elongation in 8 inches of not less than 15 per cent. All castings must be annealed and all test pieces must be cast as coupons and detached after annealing.

All the cast iron used in this building must be tough gray iron, free from cold shuts, blow holes or other serious defects. Its quality must be such that sample bars 1 inch square cast in sand moulds must be capable of sustaining on a clear span of 4½ feet a central load of 500 pounds when tested in the rough bar.

Painting: All iron must receive a coat of pure raw linseed oil at the rolling mills just before being loaded on the cars.

The covered surfaces (surfaces in contact and surfaces enclosed) of all parts of riveted members must receive one good coat of graphite paint, after the pieces are punched and before they are assembled. All finished members must receive one complete coat of the graphite paint before they are taken from the shop or exposed to the weather. All surfaces that can be reached must have one coat of the graphite paint after erection. All truss members must have two coats of paint in the shop and the enclosed surfaces of these members must have the two coats before they are assembled.

Foundation beams and connections must have two coats of paint at the shop. All bolts used in erection and remaining permanently in the building must be dipped in graphite paint before being placed in position.

· All pins and bored pin holes or other planed surfaces in the trusses must be coated with white lead and tallow before leaving the shop.

All painting must be done on dry surfaces and preferably warm ones. All dirt and foreign matter of any kind must be removed from the iron before painting. All scale must be removed from finished members before painting the first coat in the shop. All scale must be removed from material required for the trusses before it is oiled at the rolling mill.

The paint used must be the superior graphite paint, prepared and mixed by the Detroit Graphite Manufacturing Company, of Detroit, Michigan.

Inspection: The inspection hereby provided will be made by inspectors employed by John Downey.

The contractor for the structural iron must furnish full and ample means for the inspection of all the materials called for by these specifications, and of all the work required in fitting such materials for erection; and to this end, he shall admit the architect, engineers, and inspectors to any part of the mills or shops where work under these specifications is being carried on.

To secure proper material, as herein specified, one pulling test must be made from every heat or blow of steel or rolling of iron, and one bending and one quenching test; when such requirements are specified, if these are satisfactory, the whole will be accepted. If they are not satisfactory, others may be made as the inspector may deem expedient. All test pieces must be prepared at the expense of the contractor for the structural iron. The test pieces of rolled steel and wrought iron must be cut out of finished material, and must not be less than ½ square inch in section. They must be at least 10 inches long between fillets when turned down. When possible they must be cut from the full thickness of the section, from which the tests are taken. The method of selecting test pieces for material for eye bars must be as required by the manufacturer of the bars.

The number of test pieces of cast steel must be fixed by the inspector.

Full sized tests of eye bars must be made as required by the architect or engineers. Test bars for such tests will be selected by the inspector from the lot after forging and before boring, the results of the test to determine the acceptance or rejection of the entire number which the test bar represents. Other full sized tests must be made if required.

The material used for all full sized tests will be paid for at cost, less the scrap value of the material to the contractor when the pieces are tested to destruction, and the test proved satisfactory; otherwise it must be solely at the cost of the contractor. The use of testing machines capable of testing both specimens of material and the full sized members, together with all necessary assistance in handling and operating the same, must be furnished by the contractor free of all expense.

All surfaces of all materials must be carefully examined by the inspectors, and all pieces that are of full section—free from flaws—straight and in every way satisfactory, must be accepted. This inspection will not, however, prevent the rejection of any piece at any later time, but before it is riveted in place in the building, if it is discovered that the piece is in any way unsuitable. Ample assistance must be given by this contractor to the inspector in making this examination.

All material manufactured under these specifications must be tested and examined as herein provided before the same is oiled or loaded on the cars for shipment from the mill, and as soon after rolling as may be convenient for the mill, and failure to comply with these specifications will be sufficient cause for the rejection of the material.

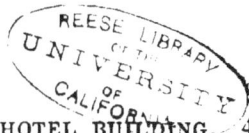

The inspection in the shop must, in general, cover the identification of material, the accuracy of work, and fulfillment of specifications and drawings in every respect, and reports of finished weights and progress of the work, in all of which the inspector must have ample opportunity to do his work. All rejected material must be made good to the satisfaction of the inspector.

All long measurements in the shop made by the inspector, must be made with a steel tape which must be compared with the shop's standard measure to assure their agreement.

In case of any disagreement between the inspectors and the contractors regarding the inspection, appeal may be ha l to Purdy & Henderson, Consulting Engineers, but their decision shall be final.

Beams: In general not more than $\frac{1}{8}$ of an inch will be allowed by the drawings for clearance at each end of beams connecting to beams and not more than $\frac{1}{4}$ of an inch at each end of beams connecting to columns. All beams supported by connection angles riveted to the webs when finished, must measure out to out of such connection angles, not more than the length given on the drawings, and not more than $\frac{1}{8}$ of an inch less than that length. All beams connecting to columns may be $\frac{1}{2}$ inch shorter than shown on the drawings, but must not be longer.

All open holes must be true to the drawings, and an error in the distance from end to end, between the open holes in the flanges at the ends of beams of more than 1-16 of an inch must not be approved by the inspector.

Where connections are marked standard, the standards adopted for this particular job must be used. Beams or other material used in floor construction, excepting bent plates used in connections, must not be heated for bending, cutting, or fitting, unless so marked on the drawings.

Beams split or permanently injured by work in the shop must not be used.

Beams which are required to be bolted together with separators in the building, must be assembled and bolted together in shop when practicable.

Columns: The distance from the center of the columns out to the open holes required for the connection of beams, must be verified by the inspector. If, on account of the material overrunning in weight or on any other account, these distances are wrong more than 1-16 of an inch, the error must be remedied, as the inspector may des re.

All columns must be milled or ground at each end to a smooth bearing surface at right angles to the axis of the column, and the inspector must verify from time to time, the adjustment of the machinery used in this work.

All columns must be exactly true to length, and any discrepancies in such lengths of more than 1-32 of an inch, must be reported promptly to the engineers. If more than 1-32 of an inch too long, they must be milled shorter.

Where columns coming over each other are designed to have the same exterior dimensions, a filler about 1-32 of an inch thick must be put under the splice plates where they are riveted to the columns. These fillers must cover the entire area covered by the splice plates. They will not be drawn on the drawings, but will be noted in the bill of material on each drawing where required.

Columns must all be straight and out of wind.

Riveted Girders: Web plates must be arranged so as not to project above or below the flange angles. The lines showing the edges of such plates will be omitted from the drawings.

In general, all stiffener angles must fit tight at both ends.

Open holes in flanges must have the same accuracy required for beams.

All riveted girders must be out of wind before leaving the shop.

Trusses: Eye bars must be entirely free from flaws and of full section. The heads must be so proportioned that the bars will break in the body of the original bar and the process of manufacture and the form of the head must be subject to the approval of the engineers. No welding will be allowed in the body of the bars. They must be perfectly straight before boring and the pin holes must be centered through the center line of the bar. The lengths back to back of pin holes must not vary more than $\frac{1}{64}$ of an inch from the figured lengths when the bars are 20 feet long or less; not more than $\frac{1}{32}$ of an inch when more than 20 feet long. Bars which go side by side in the trusses must be so perfectly bored that the pins will pass through the holes at both ends without driving when the bars are placed in a single pile. The holes must not be more than $\frac{1}{32}$ of an inch larger the pins. All eye bars must be annealed.

Compression members must have all butting ends planed smooth and exactly square to the center line of the member, and they must be assembled in the shop for the fitting of the splice plates and to assure perfect contact throughout. Such members must be entirely free from twists or bends and all work must be neatly finished and first-class in every respect. Pin holes must be bored $\frac{1}{32}$ of an inch larger than the pins, exactly perpendicular to a vertical plane passing through the center line of each member, when placed in a position similar to that which it should occupy in the finished structure.

Pins must be turned straight and smooth and to exact size.

Castings: The cast bases required in the column must be planed smooth on top and to exact dimensions. All holes for the bolts connecting to the columns must be drilled also to the exact measurements given, and the holes in the other castings, both steel and iron, must be drilled when so marked. All surfaces marked planed must be planed smooth and true for a perfect bearing as designed.

Rivets: Drifting that is liable to injure the material must not be allowed anywhere in erection.

Shop rivets must be machine driven as far as possible.

Rivet heads must be concentric with the necks of the rivets and all rivets when driven must completely fill the holes and be t ght.

Rivets will be used in erection wherever possible.

All rivets must be uniformly heated.

Holes that do not match sufficiently to admit the rivet without drifting, in assembling work in the shop, must be reamed.

All riveting must be done to the satisfaction of the engineers.

Erection: If beams are used in the construction of the foundations, the contractor for the structural iron must put them in position, both as to plan and as to height, using a surveyor's level for the purpose, but the grouting and covering of the beams will be done by the contractor for the masonry.

The outside building lines will be given, but the contractor for the structural iron must determine and fix the interior lines, and each cast base must be set in its exact position, both as to alignment and to height, supported on wooden wedges, before the bedding is run in. The center of each base must be true to the column center, as given on the plans, within $\frac{1}{16}$ of an inch, and its height must be adjusted exactly, using a surveyor's level and referring to a fixed bench mark. Each base must be bedded with a Portland cement grouting, by pouring the same through the center until all the spaces under and inside the base are filled. The cement must be of some imported brand which must be approved by the architect, and the sand must be clean and sharp and fine. The two must be mixed dry in equal quantities in a box—all that is required for one base at one mixing. Enough water must then be added to make the whole just flow under its own weight. The whole operation of mixing and setting must be done as rapidly as possible. After the bases are set their heights will be inspected by the engineers, and if they are found to vary more than $\frac{1}{8}$ of an inch from the correct height they must be taken up and reset.

The use of iron sledges in driving or hammering beams or columns or other structural iron will not be allowed where it

can be avoided.   Wooden mauls must be used wherever their
use is possible.   Care must also be exercised to prevent the
material from falling or from being in any way subjected to
heavy shocks.

Especial care must be used to keep the columns plumb and
in proper line during erection, and they must be plumbed to
the satisfaction of the architects and engineers as often as may
be desired.   In case the columns are not kept plumb the entire
work of erection shall stop at the written order of the architect
to that effect, and the measures to be employed to remedy the
defect must be approved by the architect before the erection
proceeds.*

The sections of columns, truss members, beams or girders
must nowhere be cut without first obtaining the approval of the
engineers.

Every failure of the material to come together properly
must be noted and reported daily to the engineers.   If any
serious difficulty occurs during erection, it must be reported to
the engineers before any unexpected measures are used to meet
the difficulty.

The plan or scheme for the erection of the trusses, and the
material connected to the trusses must be submitted to the
engineers, before the iron work is erected above the ground
floor, for their approval.

Pilot nuts must be used in entering all pins.

After the truss members are put in position, before they
are materially shadowed by temporary flooring or any other
construction, and after all surfaces are thoroughly dried by the
heat of the sun, they shall be protected by waterproof canvas,
tarred paper, or other materials from further exposure to the
weather.   Such protection to continue until those parts of the
building are under the cover of the other construction of the
building.   Such protection is desired to prevent water from
lodging and remaining in the concealed parts of the work.   Any
inaccuracy in the matching of the holes in the column splices
must be removed by reaming and not by drifting.

Temporary timber bracing must be put in the building
wherever required by the architect or the engineers.

The entire work of erection must be done to the satisfaction of both the architects and the engineers.

*Probably the worst practice in the erection of architectural iron work is the very
common use of shims in the joints between the successive column sections, thus con-
centrating the loads on the opposite sides of the cross-section.   The columns are
usually kept plumb in this manner, but the practice is extremely vicious and should not
be allowed.   If the faces of the ends are properly planed or milled off, and the base
plate is set exactly level, it will not be necessary to use shims.   The greatest difficulty
is in setting the bed plate in a truly horizontal plane.   The ordinary carpenter's level is
not sufficiently delicate for this purpose.   These specifications are not explicit on these
points.—AUTHOR.

CONTRACTOR'S BONDS.

175. Contract Bond or Surety. It is a very general custom in all important work to require the contractor to furnish a bond for the faithful and complete performance of his contract. Sometimes these bondsmen or sureties sign with the contractor, as in the case of the St. Louis contracts, exemplified in article 16S. It is more usual, however, to make this bond a separate document, following immediately the signatures of the contract itself.

Bonds are always executed under seal, and are therefore special contracts, since the bondsmen are not usually paid a consideration for the service rendered, and a sealed contract does not require a consideration to enforce it.

In case the original contract and specifications are deviated from in the execution of the work to any material extent, without the consent of the bondsmen, these latter are thereby released from their bond. Since such changes are nearly always made in the execution of engineering work after the contracts are signed, and since these are usually made without consulting the bondsmen, these latter are as a rule thereby released from all obligations, and the bond becomes of no effect. Even though the bondsmen be consulted in the matter of changes, they are not obliged to give their consent, and usually perhaps would not, in which case material changes could be made only by releasing the bondsmen. The practice, therefore, of securing the faithful performance of engineering contracts by means of bonds is a very unsatisfactory one. It would be better always to confine the contract strictly to the principals to the agreement, and to secure guaranties of faithful performance in some other manner than by the execution of a bond by outside parties, so far as engineering and building contracts are concerned. The form of bond given below is that used by the city of Boston, and may be taken as a general type of such a document.

CONTRACT BOND OR SURETY.

Know all Men by these Presents,

That we...

....

....

....

....

are held and firmly bound unto the City of Boston, in the sum of ....

....

dollars to be paid to the City of Boston, or its certain attorney, its successors and assigns, for which payment, well and truly to be made, we bind ourselves, our heirs, executors, and administrators, jointly and severally, firmly by these presents.

The Condition of this obligation is such that if the above-bounden...

....

....

shall well and truly keep and perform all the terms and conditions of the foregoing contract for excavation for stripping and shallow flowage and for building two roads, at Basin No. 5, in Southborough, on .... part to be kept and performed, and shall indemnify and save harmless the said City of Boston, as therein stipulated, then this obligation shall be of no effect; otherwise it shall remain in full force and virtue.

In Witness Whereof, we hereto set our hands and seals on this .... day of .... in the year eighteen hundred and ninety four.

.... [SEAL.]

.... [SEAL.]

.... [SEAL.]

.... [SEAL.]

.... [SEAL.]

Signed and Sealed in presence of

....

....

176. Indemnity Bond. The following is a common form of bond to cover all liens which may arise from a failure of the contractor to pay for his labor and materials.

Know all Men by these Presents: That ——— of ——— as principal, and ——— of ——— as surety, are held and firmly bound unto the ——— in the penal sum of ——— dollars, to the payment of which well and truly to be made we bind ourselves, our heirs, executors, administrators and assigns firmly by these presents.

Signed this ——— day of ——— 189—.

The Condition of the above Obligation is such that:

WHEREAS, the said ——— has this day entered into a contract in writing with the said ——— for the grading and construction of a certain ———'with ditches, roadways, and other works connected therewith, as more specifically set forth in said contract:

NOW, THEREFORE: If the said ——— shall well and truly perform his part of said contract, and each and every covenant and agreement therein contained, and shall indemnify and save harmless the said ——— from and against all damages which it may sustain by reason of liens for labor and materials furnished for said work, or by reason of the failure of said ——— to pay the wages and earnings of any of the——— laborers or mechanics employed by him as such contractor, in and about said work; or by reason of his failure to pay for any materials, provisions or goods of any kind furnished, or by reason of any just debts incurred in carrying on said work; and if the said ——— shall pay to the said ——— all sums of money, damages, or costs and expenses which it may be compelled to pay, or which it may sustain by reason of his failure as aforesaid, and if the said ——— shall pay all laborers, mechanics and material men, and persons who may have supplied provisions or goods of any kind, all just debts due to such persons, or to any person to whom any part of such work was given, then this obligation shall be void, otherwise of full force and effect.

——————————— [SEAL.]
——————————— [SEAL.]

APPENDIX A.

The following Instructions to Assistant Engineers are used by the engineer of bridges and buildings on the C., M. & St. P. R'y, and are inserted here as an illustration of the scope and character of the inquiries and investigations necessary for an intelligent solution of the problem in hand. It is only by means of such complete and detailed information that all future contingencies can be foreseen and provided for, so that there shall be no "unexpected" to happen. It is a common saying that "the unexpected always happens." In good engineering, "It is only the unexpected which can happen," since what was anticipated has been fully provided against. In the *best* engineering designs, however, every possible contingency has been foreseen and provided for, so there is no unexpected left which can happen, and hence security and permanence are assured in advance. The following instructions are a good illustration of this kind of preliminary survey of the problem which puts the engineer in a position to perfectly fit the design to all the conditions of the problem:

Instructions to Assistant Engineers in Regard to Surveys for the Renewal of Wooden Bridges with Perma-nent Structures.

(1) Gather information from the chief engineer's office and from the office of the engineer and superintendent of bridges and buildings relative to the grade, alignment, right-of-way for embankment and borrow pits, second track construction, contracts relating to crossings or cattle passes, recommendations already made by others as to style of reconstruction and any other matters that are liable to have a bearing on work in question.

(2) Determine the elevation of base of rail above an assumed datum across the bridge and for a distance of 1,000 feet on each side of it, at intervals of 100 feet, or less when the irregularities of the track make it necessary.

(3) Consider the question of changing grade and note the kind, condition and depth of ballast as well as other points

that will assist in determining the expense and practicability of making a change.

(4) Obtain particularly notes of the ground surface that will be covered by the proposed structure or embankment, by determining its elevation on the center line of bridge and when necessary on each side of same. These heights may be measured from the base of rail at each bent or panel point but should refer to the datum used in the survey, and additional notes should be made of intermediate irregularities that would concern the height of pedestals located between bents.

(5) Establish and note two bench marks on solid objects, conveniently located, one each way from the bridge, and which are unlikely to be disturbed during the construction of the permanent structure. For ordinary cases a track spike driven in a telegraph pole will be suitable.

(6) Note the alignment of the track at the structure and consider whether there is any evident reason for changing same.

(7) Consider the question of second track construction as concerning any change in alignment or in location of bridge. Conclude on which side of the present track the second track should be constructed and make note of the grounds for your conclusion.

(8) When track across the bridge or near the bridge is curved make full notes of elevation of outer rail. If the point of curve is so located that the elevation of outer rail on bridge is varying, determine by eye the location of point of curve and of the point where the elevation is commenced. On iron bridges the elevation should be constant when practicable.

(9) Take notes for a sketch of the water course for a sufficient distance on each side of the bridge, to determine whether a change in location of channel or an improvement in the channel is advisable, and indicate your recommendations in this regard, remembering that the most favorable condition for a bridge is usually a deep channel at right angles to the railway for some distance above and below the bridge. Contours in the immediate vicinity of the bridge should be sketched in. Ordinarily this can be done with sufficient accuracy by the eye, or by taking a few offsets.

(10) Ascertain the nature of foundations, whether soft, requiring pile foundations, or of sand, or of hard clay, or of rock. Reports should state the character, depth and dip of the strata.

(11) Ascertain present, ordinary and extreme high water marks. Inquire into cause of high water; whether by ordinary heavy rains, by water-spout, by damming from accumulations of drift or ice, or by overflow from other water courses, or from other causes which may be apparent.

(12) Note the probability of ice, drift-wood, hay, corn-stalks, fencing, etc., lodging against the proposed iron bridge.

(13) Take notes of the size of channel, area of waterway required, direction of current, etc. ·

(14) Ascertain if there is to be provided under the bridge a public or private roadway, wagon-pass or cattle-pass, with dimensions and conditions controlling the same.

(15) If any portion of the bridge is to be filled, make an examination of the ground and state where the material can be obtained, and whether inside of the boundaries of the right-of way, or on land which will have to be purchased.

(16) Ascertain whether any additional right-of-way is required for any purpose connected with the work, and if so note location and amount.

(17) Examine as to a suitable location for a stone yard, and for the storing of piles, timber and iron-work; also as to convenient locations for derricks and what provision will be required for suitable anchorage for derrick guys.

(18) If the proposed reconstruction involves any question of purchasing land or privileges, report the situation with advice, but avoid conversation with property owners which would in any way interfere with relations that may be established later between them and an agent authorized to make purchases or settle claims.

(19) Inquire as to the accommodations for boarding and lodging for workmen and how they can get to and from their work.

(20) Inquire into the condition of train service at the location with regard to the frequency of trains and the speed at which they ordinarily run over the bridge.

(21) If piles are to be driven, make your recommendation as to whether they should be driven with a land or track driver, and if with a track driver, state the nearest side-track to which it must retreat for passage of trains.

(22) Make preliminary estimates of the cost of the permanent structure, taking your prices from the tables of cost of iron bridges and abutments which are furnished you and from them make your recommendation for the permanent bridge.

(23) Make your recommendations as to the angles of piers and abutments, remembering that a square span is one in which its ends are at right angles to its longitudinal axis, and in a skew span the angle of skew is the enclosed angle between the end of the span and a line at right angles to its longitudinal axis.

(24) Make your recommendation as to what riprapping is required, with the amount and method of using it.

(25) Advise what is the best season of the year in which to do the work with reference to high water, ice, cold weather, interruption of traffic, facility for obtaining labor and material, etc.

(26) Report any information you can obtain with reference to using local material in the work, such as piles, timber, lumber, stone, sand, brick, etc.

(27) Avoid confusing terms in your notes. For instance the term "base of rail" is preferable to "grade." See B. & B., Rule 7 g.

FINALLY. After obtaining information on the points hereinbefore mentioned and all other data which you can find within your reach, consider the question of renewal just as if you had to make the full decision and were responsible for building the best bridge with the greatest economy and least risk; and make your report in such shape that the draughting office will have all the instruction which it requires for making the plans. This information may be furnished in writing and on a profile and map, and you are cautioned that your work will be judged by your giving the fullest accurate information with the fewest notes and the least amount of drawing.

O. B.